Proceedings of the International Symposium on

The Impact of Oxygen on the Productivity of Non-Ferrous Metallurgical Processes

Pergamon Titles of Related Interest

Battino	OXYGEN AND OZONE Solubility Data Series, Vol. 7
Macmillan	QUALITY AND PROCESS IN REDUCTION AND CASTING OF ALUMINUM AND OTHER LIGHT METALS
Rigaud	ADVANCES IN REFRACTORIES FOR THE METALLURGICAL INDUSTRIES
Ruddle	ACCELERATED COOLING OF ROLLED STEEL
Salter	GOLD METALLURGY
Tyson	FRACTURE MECHANICS

Related Journals

(Free sample copies available on request)

ACTA METALLURGICA

AEROSOL SCIENCE

CANADIAN METALLURGICAL QUARTERLY

Proceedings of the International Symposium on

The Impact of Oxygen on the Productivity of Non-Ferrous Metallurgical Processes

Winnipeg, Canada

August 23–26, 1987

Co-Sponsored by the Non-Ferrous Pyrometallurgy and Hydrometallurgy Sections of the Metallurgical Society of the CIM

Vol. 2. Proceedings of the Metallurgical Society of the Canadian Institute of Mining and Metallurgy

Edited by

GEORGE KACHANIWSKY

Noranda Minerals, Inc., Rouyn-Noranda, Quebec, Canada

and

CHRIS NEWMAN

Kidd Creek Mines Ltd., Timmins, Ontario, Canada

PERGAMON PRESS

NEW YORK · OXFORD · BEIJING · FRANKFURT
SÃO PAULO · SYDNEY · TOKYO · TORONTO

U.S.A.	Pergamon Press, Maxwell House, Fairview Park, Elmsford, New York 10523, U.S.A.
U.K.	Pergamon Press, Headington Hill Hall, Oxford OX3 0BW, England
PEOPLE'S REPUBLIC OF CHINA	Pergamon Press, Room 4037, Qianmen Hotel, Beijing, People's Republic of China
FEDERAL REPUBLIC OF GERMANY	Pergamon Press, Hammerweg 6, D-6242 Kronberg, Federal Republic of Germany
BRAZIL	Pergamon Editora, Rua Eça de Queiros, 346, CEP 04011, Paraiso, São Paulo, Brazil
AUSTRALIA	Pergamon Press Australia, P.O. Box 544, Potts Point, N.S.W. 2011, Australia
JAPAN	Pergamon Press, 8th Floor, Matsuoka Central Building, 1-7-1 Nishishinjuku, Shinjuku-ku, Tokyo 160, Japan
CANADA	Pergamon Press Canada, Suite No. 271, 253 College Street, Toronto, Ontario, Canada M5T 1R5

First edition 1987

Library of Congress Cataloging in Publication Data
International Symposium on the Impact of Oxygen on the Productivity of Non-Ferrous Metallurgical Processes (1987: Winnipeg, Man.)
Proceedings of the International Symposium on the Impact of Oxygen on the Productivity of Non-Ferrous Metallurgical Processes, Winnipeg, Canada, August 23–26, 1987
1. Nonferrous metals—Metallurgy—Oxygen processes
Congresses. I. Kachaniwsky, George. II. Newman, C. E. (Chris E.) III. Title.
TN758.1533 1987 669 87-7310

ISBN 0-08-035767-9

Printed in Great Britain by A. Wheaton & Co. Ltd, Exeter

INTRODUCTION

Whereas the use of oxygen in metallurgical processes is not new, its application has been expanding to meet industry's needs of improved productivity and reduced costs. With production costs rising relative to base metal prices, the trend has been away from heavy capital investment in new installations and towards increasing the productivity of existing plants. This symposium was put together to offer a forum for both oxygen consumers and suppliers to discuss their recent experiences and developments. Topics covered include increased process efficiencies, higher production rates, reduced energy consumption, and new and emerging processes. This is a truly international symposium with eleven countries represented, bringing together papers by industrial users of oxygen - both pyrometallurgical and hydrometallurgical - major oxygen producers, engineering firms, and leading experts in the field. The editors express their appreciation to the authors, session chairmen, institutes, universities, and companies who through their willingness to participate and contribute made this symposium possible.

George Kachaniwsky
Noranda Minerals Inc.

Chris Newman
Kidd Creek Mines Ltd.

August 1987

CONTENTS

Oxygen in Non-Ferrous Metallurgical Processes
Past, Present and Future

P.J. Mackey*

*Noranda Inc., Pointe Claire, Quebec, Canada

ABSTRACT

The use of tonnage oxygen in non-ferrous extraction metallurgy is now established practice and part of conventional technology, however it was not always so. While technically understood in the scientific literature decades before major developments were commercialized, widespread application, historically, is comparatively recent. In this paper, the progress of oxygen applications is traced and factors governing the rate of commercialization and acceptance within the industry are examined.

Canada has in fact played a leading role in developing oxygen-driven processes in non-ferrous metallurgy. As examples, Cominco, at Trail B.C., was the first to use oxygen-enrichment in zinc and lead pyrometallurgy in the nineteen thirties and forties; Inco pioneered oxygen smelting twenty-five years ago when its new flash furnace was commissioned in Sudbury with oxygen produced in the world's third largest plant. Testwork at Noranda's Horne smelter in the sixties and seventies confirmed production improvements with oxygen enrichment in the Noranda Process which now operates with over 400 tonnes of oxygen per day. The first commercial zinc oxygen-pressure leaching process was commissioned at Cominco's Trail plant in 1981 following initial laboratory work carried out nearly 30 years ago by Sherritt Gordon Mines Limited and subsequent joint piloting by Sherritt and Cominco.

An appreciation of these and many other developments is also useful in putting into perspective the state of innovation at the present time. Concepts which may lead to new future developments are also introduced.

KEYWORDS

Oxygen; oxygen smelting; non-ferrous metallurgy; pyrometallurgy; smelting; refining; oxy-fuel; copper and nickel smelting.

INTRODUCTION

Oxygen, so to speak, is the central pivot round which the whole of chemistry revolves - J.J. Berzelius.

At a 1986 North American metallurgical conference, following the presentation of a paper on new smelting technology, the author of the paper was asked whether the new process had been tested at high oxygen enrichment. The author replied that this was in the testing program underway. The present writer remembers thinking at the time that such a question would have been quite uncommon five years ago, that oxygen smelting processes now represent conventional technology, and that the selection of oxygen-enriched continuous smelting technology in any new non-ferrous smelter project today is commonplace and almost automatic; but this is only a recent phenomenon, a decade and a half ago oxygen usage in non-ferrous metallurgy was not practiced widely, even though testwork and evaluation of oxygen use in this field of metallurgy has been on-going since before 1940.

The Inco flash furnace (Inco Staff, 1955) which commenced operations in 1952 was one of the first new oxygen-driven processes, testwork on many other techniques for oxygen enrichment had commenced. It was only in 1971 that Outokumpu first introduced oxygen enrichment for commercial operation at the Harjavalta plant (Laurila, 1973) having been commenced the previous year at Japanese smelters (Kubota and Yasuda, 1973); four years later, oxygen enrichment was used on a commercial basis in the Noranda Process reactor at the Horne smelter, following pilot tests begun in 1970 (McKerrow, Themelis, Tarassoff and Hallett, 1972; Tarassoff, 1984).

At the present time, roughly over half the copper, nickel (sulphide) and lead smelters worldwide use oxygen enrichment, based on a recent survey (Taylor, 1987), Table 1. A little over ten years earlier, the average is estimated to be less than about 20%. In their 1979 survey of copper and nickel converter practice, Johnson, Themelis and Eltringham (1979) indicated that about 30% of the plants reporting data (47 plants) employed oxygen enrichment in the converter.

Table 1

Worldwide Non-Ferrous Smelters Using Oxygen Enrichment

Smelter	Proportion Using Oxygen Enrichment, %
Copper	57
Nickel	50
Lead	63
Average	57

Source: Taylor (1987)

The rather late widescale adoption of tonnage oxygen for non-ferrous metallurgical processes is at first surprising, considering that oxygen plant technology was well developed and that oxygen processes were beginning to be widely employed in the iron and steel industry, Figure 1 (U.S. Bureau of Mines, 1975). The Journal of Metals in 1961 (Starratt, 1961) heralded these changes in an editorial by announcing "this is the age of oxygen - an age of revolution in pyrometallurgy, perhaps as far reaching as anything since the dawn of the blast furnace."

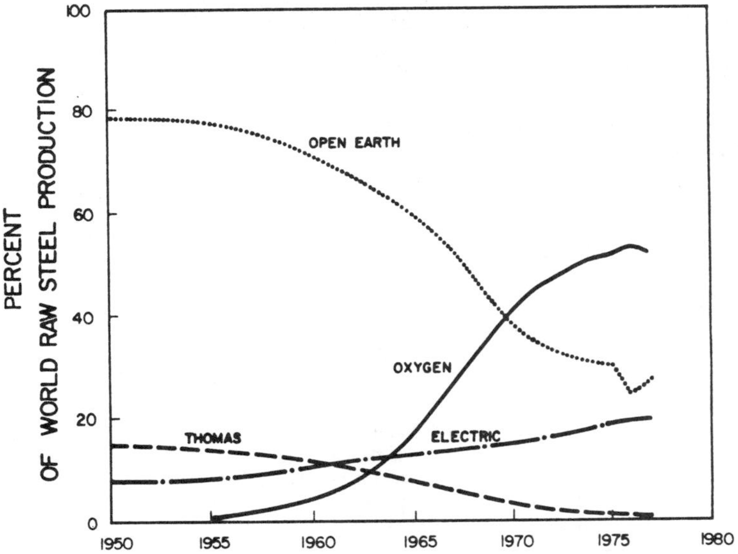

Figure 1. World Steel making production by process between 1950 and 1980. The application of oxygen steel making processes for increased worldwide ten-fold in the 1960's. Courtesy of U.S. Department of Interior, Bureau of Mines, Mineral Facts, and Problems (1975).

The editorial further stated,

> "Metallurgists have long since realized that any process requiring air could be greatly enhanced by enriching the oxygen content of that air to the point of using oxygen itself. But for use in metallurgical operations, large quantities of oxygen would be required, and progress was at a standstill until tonnage quantities of low-cost oxygen became available. And even then it took some time to adapt thinking toward oxygen usage. There were a few apostles of oxygen usage during the late 30's and early 40's but blossoming came about with the development of the LD steelmaking process in Austria. Since then, there has been a complete reorientation of thinking, and oxygen

has been squirted into blast furnaces, rotating furnaces, converters, and so forth. It has been used in lancing techniques for metal baths in ladles, open hearths, and electric furnaces."

But it was the requirement to reduce fuel costs in the face of higher fuel prices in the early seventies, improve productivity and meet environmental regulations, associated with the availability of newer technology having better adaptability to the use of oxygen (Noranda Process, oxygen-enriched Outokumpu furnace, QS Process, Kivcet Process, oxy-fuel burners, etc.) that provided the impetus leading to rapid growth of oxygen applications in non-ferrous metallurgy in the period 1970 - 1987.

Within the past decade, there have been a number of excellent reviews concerned with the contribution that tonnage oxygen has made to the improvement of metallurgical processes. The 1977 Extractive Metallurgy Lecture for The Metallurgical Society of AIME presented by Paul E. Queneau, was entitled "Oxygen Technology and Conservation" (Queneau, 1977), In this comprehensive and eloquent presentation, Queneau reviewed the role of tonnage oxygen in the development of metallurgical processes from a historical and operating perspective and discussed the economics of oxygen production and metallurgical processes including effects on sulphur dioxide level in off-gas and acid-making. More recently, the application of tonnage oxygen in copper smelting was discussed by Eacott (1983) in an exhaustive state-of-the-art review of the effects of tonnage oxygen on plant productivity, fuel requirements, off-gas volume and gas strength. Verney (1985) also reviewed oxygen usage in his 1985 Extractive Metallurgy Lecture, "Developments in Copper Extractive Metallurgy."

In this paper, past and present developments of oxygen usage in non-ferrous metallurgy in particular are reviewed and the chronology of major developments is presented; in order to provide some perspective to the excellent collection of papers in the present symposium volume, the progress of oxygen applications not completely covered in earlier reviews is traced, and factors governing the rate of commercialization and acceptance within the industry are examined using actual information available at the time developments were occurring. Comments on future directions are also made.

Consider the following situation that may have occurred in the 1960's when a North American plant was reviewing plans for expansion.

> The plant, which operates an oil-fired reverberatory furnace, is preparing a report for management on the recommended technology to adopt for the expansion.
>
> The feeling at the plant is that a second reverberatory furnace is probably the best choice of available technology, since cost estimates indicate it is the lowest cost option. However, there is also some interest in the use of oxygen as an alternative mode of

> operation in a reverberatory, converter or new flash furnace. Available published literature, for example, Schuhmann (1952) indicates a fuel saving of between 0.3 and 0.5 ton of coal equivalent per ton of oxygen added is possible. The technology for introducing this oxygen to the reverberatory furnace is not well known, although it is recognized there are several recent publications from the U.S.S.R. on the subject; if adopted, the oxygen would probably be added in with the primary air; the plant manager and the operators are worried about roof wear... Oxygen plant manufacturers have quoted (US) $2 million for the oxygen plant and total operating costs (including amortization of capital cost) of about (US) $6/ton of oxygen. Bunker C oil costs about (US) 3/US cents gallon. The plant estimates the savings with oxygen to be quite small, less than about 60c/ton of oxygen and it decides not to use oxygen enrichment for this application.

The fact is that the company could see no clear capital and operating cost advantage, there was no have proven technology readily available to enable oxygen to be seriously considered. The Inco flash furnace (Inco Staff, 1955) had been commercialized for eight years but evidently was not available to other parties. Oil and coal were so cheap that there was no real economic incentive to change. Even though there were developments occurring in oxygen usage in ferrous and non-ferrous metallurgy, that may have been applicable to the company, such developments probably did not represent "proven technology" and the company probably did not consider these applicable.

DEVELOPMENTS IN OXYGEN USAGE IN NON-FERROUS METALLURGY
HISTORICAL PERSPECTIVE

The practice of metallurgy, consisting of the traditional methods of smelting, refining, melting and working with metals and alloys can be traced back to the earliest times. The classic medieval text by Georgius Agricola, De Re Metallica (Hoover and Hoover, 1950), published in 1556 described the mining and metallurgical industries then carried out in Europe, more than 200 years before the discovery of oxygen. The four element theory of Aristotle (384-322 B.C.) stating that all matter consisted of ''prime matter'' which contained the ''four elements'': earth, air, fire and water, held the field for nearly 2000 years. All the best run smelters in Agricola's time evidently knew how to govern these ''four elements'':

> "They combine in right proportion the ores, which are part earth, placing no more than is suitable in the furnaces; they pour in the needful quantity of water; they moderate with skill the air from the bellows; they throw the ore into that part of the fire which burns fiercely. The master sprinkles water into each part of the furnace to dampen the charcoal slightly, so that the minute parts of ore may adhere to it, which otherwise the blast of the bellows and the force of the fire would agitate and blow away with the fumes. But as the nature of the ores to be smelted varies, the

> smelters have to arrange the hearth now high, now low, and to place the pipe in which the nozzles of the bellows are inserted sometimes on a great and sometimes at a slight angle, so that the blast of the bellows may blow into the furnace in either a mild or a vigorous manner."

The discovery of oxygen and related developments concerning the role of oxygen in combustion were important scientific advances. In the 1770's the nature of air was examined by several scientists, including Cavendish, Priestley, and Scheele, the latter two independently prepared the gas that is now called oxygen. To A.L. Lavoisier of France, however, goes the chief credit of isolating and demonstrating the two main components of air in 1774 (Figure 2). One of these gases was named ''vital air'', later called oxygen from the belief that it was an essential element in all acids (named from *oxus*, Greek, meaning ''sharp taste'' and *gen*, to produce growth); the other gas was originally referred to as ''azote'' meaning ''not fit to support life'', now called nitrogen.

Oxygen, eighth element in the periodic table, average atomic weight 16, the gaseous element that constitutes 20.94% of the earth's atmosphere, is a colourless gas without taste or smell. Oxygen can be liquified to a very pale blue liquid, boiling at -183°C at atmospheric pressure. The fact that liquid oxygen boils at -183°C and liquid nitrogen 13°C colder at -196°C affords a method for separation and production of these gases by a distillation process.

Figure 2. A.L. Lavoisier, who first demonstrated the real importance of oxygen, in his laboratory, an experiment on respiration is being conducted. From a drawing by Mme. Lavoisier, who is seen seated and taking notes (Sherwood-Taylor, 1946).

About eighty years after Lavoisier's experiments, Henry Bessemer in England invented the steelmaking converter which carries his name (Bessemer, 1905). In this process, compressed air was forced beneath a liquid iron bath to oxidize carbon; this principle of submerged air injection was later adopted for non-ferrous metals. The use of oxygen to speed up his process was recognized by Bessemer in patents dating to 1855 (Wilder, 1956). These and related developments can be considered in many ways to mark the beginning of modern extractive metallurgy.

INTRODUCTION OF OXYGEN PLANTS TO METALLURGY

The development of techniques for the low cost separation of oxygen from nitrogen in air and the construction of low cost oxygen plants to produce oxygen in sufficient quantities at a cheap enough cost for metallurgical use, competing with low cost fuels, represents an important next phase of metallurgical development. The first commercial uses of tonnage oxygen in metallurgy were in the iron and steel industry. The use of oxygen to enhance carbon elimination from pig iron was recognized by Bessemer but actual application was not possible until processes were develoed for the manufacture of low cost, tonnage oxygen.

Initially oxygen was made either chemically or by electrolysis of water. The first commercial plant for liquifying air was built by Karl von Linde in Munich, Germany in 1895 and was based on a cooling cycle produced by expansion of cold, compressed air; later development work by the French engineer, George Claude introduced a more efficient cooling cycle. In the nineteen twenties, cylinder oxygen could be supplied to steel plants and was used for burning out tapholes; uses were further developed for cutting, burning and dressing steel. At that time, oxygen (99.5% purity) cost about US $250/ton in bottles. Later liquid oxygen was produced at a central oxygen plant and trucked to the steel plant for piping around the plant in steel or copper tubes. At the end of World War II, the price of 99.5% purity oxygen had dropped to about US $80/ton as regenerative heat exchangers were adopted in oxygen plants to improve overall efficiency. The next stage was to build the oxygen plant at or near the point of use and with improvements in oxygen plant design, tonnage oxygen plants of the order of 200 tonnes of liquid oxygen per day and higher became economically possible. Oxygen plants were then able to supply gas at good prices, but not really low enough to demonstrate fuel cost savings; other benefits - higher throughput and productivity, more suitable off-gases (for acid making) - were not widely appreciated at the time.

THE MODERN OXYGEN PLANT - THE TONNAGE OXYGEN INDUSTRY

In the U.S., oxygen production in terms of tonnage is the third largest chemical produced after sulphuric acid and lime. There is no substitute for oxygen in its uses and it is not recycled or reclaimed; there is no problem of resource depletion of oxygen. Despite the rising cost of electrical energy, the chief operating expense in oxygen production, oxygen usage is expected to continue to increase. Total world oxygen output by country in 1980, about 80 million s. tons, is presented in Table 2; this is projected to increase to about 148 million tons by the year 2000. About 70% of the oxygen presently produced in the U.S. is consumed in the manufacture of iron and steel, less than one-tenth of this amount, or about 6% of the oxygen produced, is used in the non-ferrous industry; the chemical industry uses about 12% of oxygen produced, Table 3 (U.S. Bureau of Mines, 1975). Coal gasification and waste treatment represent the two areas for major growth in oxygen demand; a significant increase in oxygen usage in non-ferrous metals production is also forecast (Table 3).

Table 2 - Total World Oxygen Output (1980)

(thousand short tons)

Country		Output
North America		
Canada		1,200
United States		25,000
Other		250
	Total	26,450
South America		
Brazil		1,000
Other		1,000
	Total	2,000
Europe		
Belgium		2,400
France		3,400
Italy		2,000
United Kingdom		3,900
U.S.S.R.		8,400
West Germany		8,200
Other		4,500
	Total	32,800
Africa	Total	700
Asia		
Japan		15,800
Other		1,000
	Total	16,800
Oceania	Total	1,800
World	Total	80,550

Source: U.S. Bureau of Mines (1975).

In the U.S., most of the steel industry users purchase their oxygen from an industrial gas company, however, the largest industrial gas companies both manufacture plants and own and operate producing units. The trend in Canada, at least in the non-ferrous industry, is for the customer to purchase an oxygen plant on a turnkey basis. One factor probably influencing this approach is the general remoteness of Canadian non-ferrous plants from existing oxygen pipeline routes.

Table 3 - End Use of Oxygen in the U.S. 1973 and projection for year 2000

Industry	Year			
	1973		2000	
	10^6 st	%	10^6 st	%
Iron and steel	11.3	70	24	38
Chemicals	2.0	12	30	50
Non-ferrous metals	1.0	6	4	6
Fabricated metal products	1.0	6	2	3
Other	0.9	6	2	3
Total	16.2	100	62	100

Source: U.S. Bureau of Mines (1975).

In the modern plant, oxygen is produced from air in five major processing steps. A simplified diagram is presented in Figure 3; details of modern oxygen plant practice are given in several papers in the present volume.

1. Air Compression. Air is compressed to 75 to 200 psi in contrifugal compressors and then cooled.

2. Purification. Dust is filtered out and carbon dioxide and moisture are removed either by adsorption or by condensation.

3. Cooling in Heat Exchangers. The purified compressed air is further cooled in heat exchangers with outgoing cold products.

4. Refrigeration. The air is allowed to expand in an expansion turbine thus providing the Joule-Thompson cooling effect which partially liquifies the air.

5. Distillation and Separation. Nitrogen boils off first from liquified air in a two-stage distillation column, allowing separation of oxygen from nitrogen.

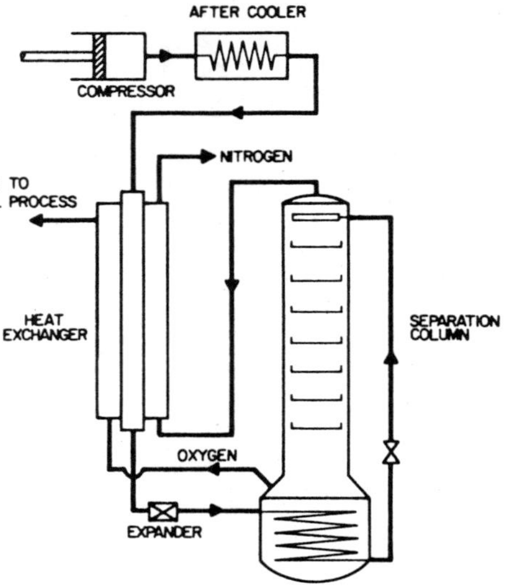

Figure 3. Simplified diagram of gaseous oxygen process after Kayan and Johnson (1963).

The average cost of production of oxygen depends on several factors including plant size and type of plant, oxygen purity, whether the product is gaseous or liquid, oxygen delivery pressure and most important of all power cost (power requirements represent 60 - 75% of direct cost at present day prices). The relationship between power cost and direct operating cost (US $/s.ton), is illustrated in Figure 4 for a typical 1987 plant used in the non-ferrous industry, producing commercial grade gaseous oxygen (95%O_2) at a delivery pressure of 20 psig (for plant size in range 300 - 600 stpd). The capital cost curve for a plant of this nature as a function of capacity is illustrated in Figure 5. These curves follow similar relationships presented previously by Slinn (1967) and Queneau (1977). The capital cost breakdown by plant area is illustrated in Table 4. (U.S. Bureau of Mines, 1975).

Table 4 - Components of Oxygen Plant Capital Cost

Plant Area	% of Cost
Air Compressors	40
Cryogenic Equipment	40
Oxygen Compressors	20
Total	100

A typical breakdown of direct operating costs for oxygen production in a 300 - 600 stpd unit (95% O_2, 20 psig) for a non-ferrous metallurgical plant is given in Table 5.

Table 5 - Direct Costs for Production of Oxygen

(300 - 600 stpd, 95% O_2, 20 psig)

Area	Direct Cost (US)$/s.ton	%
Power (260 kWh/s.ton, 3 c/kWh)	8	57
Labour (0.1 man-h/ton, $20/h)	2	14
Maintenance, chemicals, lubricants, etc. [1]	4	29
TOTAL	14	100

[1] Note: Range is 1 to 4 depending on type of plant, location, etc. A cost of $1 in this area gives a total cost of US $11/s.t.

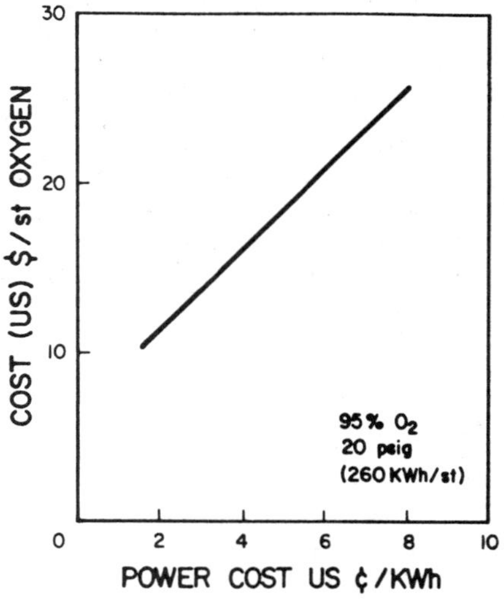

Figure 4. Effect on power cost on direct cost of oxygen, refer also to Table 5 for cost breakdown.

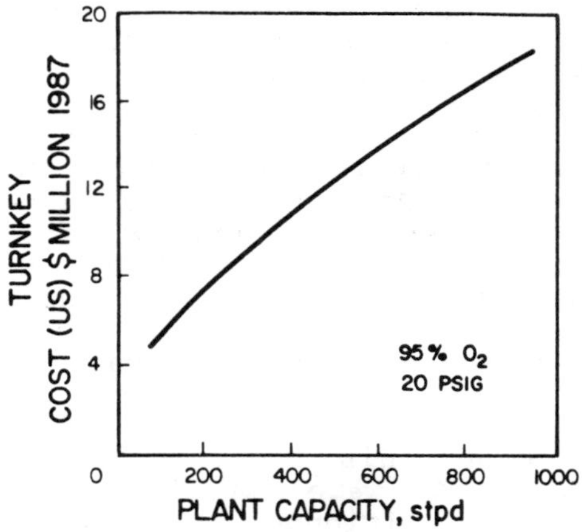

Figure 5. Turnkey oxygen plant cost as a function of oxygen plant capacity.

APPLICATIONS OF OXYGEN ENRICHMENT IN EXTRACTIVE METALLURGY

First Uses - Ferrous Metallurgy

The first commercial applications of oxygen in the metallurgical industries were for steelmaking processes. Testwork on these applications in the twenties and thirties, and the first use of oxygen in the Bessemer converter in 1931 followed a 1923 report issued by the U.s. Bureau of Mines' Committee for the Application of Oxygen or Oxygenated Air in Metallurgical Industries, (Davies 1923). This study, discussed by Queneau (1977) in his Extractive Metallurgy Lecture, was quite accurate in its forecast of the benefits of tonnage oxygen applications in the iron blast furnace, the open hearth and furnaces for copper and zinc processing. Davis stated, "the application of oxygen will revolutionize the art of smelting and it will probably change the whole operation and equipment." The report also evaluated the oxygen industry and projected that with the type of technology then available, oxygen at a cost of $3/s.ton would be possible and be competitive with fossil fuels; this forecast was evidently reliable and oxygen was available at about this price for several decades (Nagel, 1935; Inco Staff, 1955)[2].

A major drawback to the use of high oxygen enrichment in bottom-blown Bessemer converters was apparently rapid wear of the tuyeres then in use. A breakthrough in the furnace design came with the introduction of the top blown stationary converter initially tested by Durrer in Switzerland in 1947 and first operated commercially on a 15 tonne heat at the Donawitz steel plant in Austria in 1949, following earlier experimental trials at the Linz plant. The new technique was called the LD process. One of the driving forces behind the Linz-Donawitz work was evidently the need to correct a raw materials supply imbalance in post World War II Austria; steel scrap was in short supply and improvements to existing processes for the refining of the local high phosphorus pig iron were needed. Top blowing with oxygen was conceived as one approach.

The idea was successful and the rate of introduction of the LD Process was phenomenal (Figure 1). (The process is now referred to by various names in different parts of the world such as Basic Oxygen Furnace (BOF), LD-AC, OLP, etc.). A full size LD plant was commissioned at Linz in 1952 and at Donawitz in 1953; a year later, the first LD plant outside Austria came into operation at Dofasco's Hamilton plant in August 1954 (Wilder, 1956).

At about this time, oxygen roof lances and oxy-fuel burners (Figure 6) were employed in open hearth furnaces to improve process efficiency. This burner was no doubt the forerunner of the oxy-fuel burners first introduced to copper reverberatory furnaces at Almalik in the U.S.S.R. in 1972, later adopted and improved upon at the Caletones Smelter, Chile.
By the early nineteen sixties, tonnage oxygen (metallurgical

---

[2] Nagel (1935) reported an oxygen cost of $3.50/s.ton based on power at 0.5 cents/KWh; in 1955 a price of $4/s.ton based on 0.4 cents/KWh power was quoted (Inco Staff, 1955).

grade for ferrous industry, 99.5% O_2) was being produced for about $15/ton and so-called commercial purity oxygen (about 95% O_2) - suitable for non-ferrous plants - evidently cost of the order of $5/ton. A later development in the basic oxygen process for steelmaking is the bottom blown converter also referred to as the OBM or Q-BOP process, using a gas-sheathed tuyere invented in Canada by Savard and Lee (1966) and developed in West Germany.

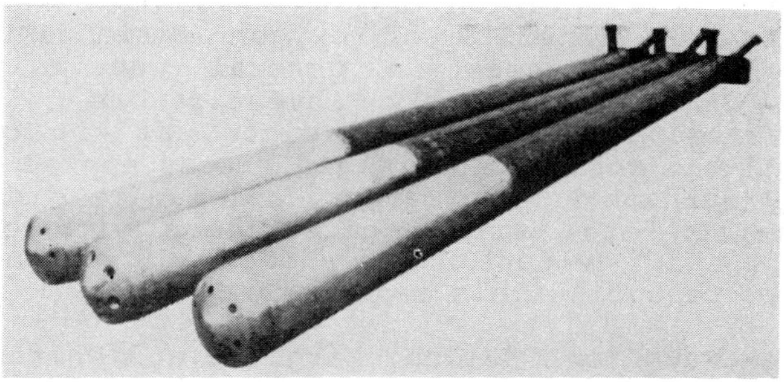

Figure 6. Typical oxy-fuel burners employed, developed by Air Products Inc., at the Ford Motor Co. open hearth furnaces in 1961 (Ferns, 1961). These burners had a capacity up to 2.5 mtpd of oxygen, similar to the capacity of present-day oxy-fuel burners on copper reverbatory furnaces.

It is perhaps of interest to briefly review the factors affecting the development of oxygen-driven processes in the steel industry particularly the LD Process, which was the first big oxygen development. While the advantages of using oxygen were evidently recognized for over a hundred years, the introduction of a successful commercial process depended on a combination of several factors, including:

- Plants for low cost production of large tonnages of high purity oxygen.
- Rapid techniques of chemical analysis.
- Techniques for measuring steel temperature (immersion pyrometers).
- Refractories to withstand furnace conditions
- Availability of suitable gas and fume cleaning technology

Different requirements, however, were emerging in the non-ferrous industry and the growth of oxygen-driven processes in non-ferrous metallurgy did not really commence until some 10-15 years after the expansion in oxygen usage in the iron and steel industry. Certainly the first factor - availability of low cost oxygen - existed: commercial grade oxygen (95%) was evidently being produced in a 300 tpd oxygen plant in Sudbury, Ontario in 1952 at $4/ton to feed Inco's newly developed flash furnace.

The pattern of developments concerning use of tonnage oxygen in non-ferrous metallurgy will now be reviewed. Commencing with the publication by Norman (1936), the application of oxygen at Cominco and later the Inco flash furnace and the Noranda Process, it will be seen that Canada has contributed significantly to these developments.

Developments in Oxygen Usage for Non-Ferrous Metallurgy

Industrial implementation of the technical knowledge and ideas concerning the benefits of oxygen enrichment in non-ferrous metallurgical processes in general was relatively slow to develop, Table 8. Early investigators recognized the attractiveness of oxygen was related to the cost of producing oxygen and the cost of fuel saved. Less appreciated was the need to modify furnace and/or burner design, originally intended for air operation, to perform effectively with oxygen-enrichment, however, it is evident that some early patents and technical papers did recognize this requirement.

Cominco (McNaughton, Weldon, Hargrave and Whiton, 1949; Landucci and Fuller, 1961), in Trail began testing the application of oxygen-enriched air in its zinc suspension roaster, this being used on a continuous basis commencing in 1937. A significant factor contributing to this development was the availability of by-product oxygen from the company's air liquefaction plant originally installed to provide nitrogen for the production of fertilizer. Oxygen addition rates were typically 18 stpd per roaster, handling about 200 stpd of zinc concentrate. This level of oxygen enrichment provided an increase in throughput corresponding to about 1.7 tons of concentrate/ton of oxygen. Within the next ten years, Cominco had tested and was ready to implement oxygen-enriched air for both the lead blast furnace and the slag fuming furnace.

At about this time, investigations commenced in the U.S.S.R. on oxygen enrichment in non-ferrous metallurgical plants (Sutulov, 1967). It seems that initial application of oxygen enrichment developed faster in the U.S.S.R. than elsewhere and by the early sixties, there were blast furnaces, reverberatory furnaces and converters operating with oxygen enrichment (as well as a flash furnace evidently adapted from western design).

In 1936, Norman (1936) proposed the use of oxygen or oxygen-enriched air for autogenous flash smelting of copper concentrates. Norman proposed a new process based on oxygen "flash" smelting of 1000 tons per day of concentrate in a vertical shaft-type smelting furnace (8.5 ft. in diameter by 25 ft. high for 1000 tpd), following initial work carried out by Anaconda, or in a horizontal reverberatory type furnace (Norman also considered a 3000 tpd unit). The economic attractiveness of this concept aginst the then current practice of a coal fired reverberatory smelting was examined by Norman; a summary of these results is given in Table 6. This suggests that the oxygen cost for autogenous smelting with 95% oxygen did not quite compete with the reverberatory furnace at the then prevailing fuel costs (the actual value of waste heat credit had some bearing on the conclusion), but that autogenous smelting with 43.3% O_2 and preheated air was in fact cheaper, Table 6.

Norman's concept essentially amounts to a forerunner of the Inco flash furnace and the Outokumpu process (with oxygen). But this idea essentially remained just that for twenty-six years until Inco commenced its 500 tpd oxygen flash furnace in Sudbury, Ontario in January 1952 following seven years of testing. The new furnace used 300 tpd oxygen produced at a cost of \$4/ton, evidently based on 0.4c/kWh power (Inco Staff, 1955). Prevailing coal costs were about \$7-9/ton, and assuming a net fuel saving compared to reverberatory smelting of about 0.4 - 0.6 coal/t of oxygen, this fuel saving would have been less than \$1/ton of oxygen. Additional benefits would have resulted from the sale of liquified sulphur dioxide and operating a smaller plant than a

Table 6 - Cost comparison developed by Norman (1936) for autogenous Smelting with oxygen enrichment

	CONCENTRATE TYPE			
	NORANDA (3.8% Cu, 33.6% Fe, 24.3% S)		ANACONDA (27% Cu, 30.8% Fe, 31.5% S)	
ITEM	95% O_2	43.3% O_2 +preheated air	95% O_2	43.3% O_2 +preheated air
Proposed Process				
Throughput, stpd	1000	1000	1000	1000
Matte grade, % Cu	12.8	12.8	38	38
Oxygen required, t/t conc.	0.2	0.14	0.17	0.12
Oxygen cost, cents/st conc. (\$3.50/t O_2, 0.5 c/KWh)	70	49	70	43
Waste heat credit	-	9	-	9
Net oxygen cost	70	40	70	34
Reverberatory furnace-comparison				
Fuel cost (coal, 11% of charge, \$7/ton of coal), c/st. conc.	78	78	78	78
Assumed waste heat credit, c/st conc.	31	31	31	31
Fuel cost, c/st conc.	47	47	47	47
Net saving with oxygen c/st conc.	(-23)	(-7)	(-23)	13

reverberatory furnace. The oxygen for the new process was produced in a plant supplied by Canadian Liquid Air, evidently the third largest oxygen plant then built in the world. The development and start-up of the Inco flash furnace process truly represents a milestone in extractive metallurgy in Canada, if not the world.

Post-1950 Developments

Developments in oxygen enrichment applications for metallurgical processes tended to focus on the following broad areas:

- Oxygen enriched air used for conducting primary pyrometallurgical reactions, such as roasting, converting, (eg, oxidation of sulphur), or reduction (eg reduction of metal oxides by coke and oxygen).

- Oxygen enriched air for the combustion of fossil fuel in pyrometallurgical furnaces, e.g. coal-fired reverberatory furnace, refining or melting furnace.

- Oxygen usage in newly designed oxy-fuel burners.

The first area represents an application that is analogous to the oxygen use in steelmaking (in this case carbon removal). Much early investigation from the 1950's onward focused on oxygen enrichment in converters, including oxygen smelting in converters in Japan, U.S.S.R. and elsewhere.

Norman's approach (Norman, 1936) to compare cost of oxygen with cost of fuel saved is still valid today, however, it is also necessary now to take into account other benefits of lesser importance in Norman's day - stronger sulphur dioxide gas improving acid economics, higher productivity and operating cost savings, and improved chemistry resulting from higher oxygen partial pressures. Because steelmaking temperatures are higher than in non-ferrous metallurgy, benefits due to fuel savings alone in steelmaking were more significant, hence oxygen applications in that industry advanced faster than in the non-ferrous field. Therefore the relatively low cost of fuel, power and labour that prevailed in the fifties and sixties compared to the cost of oxygen, plus the absence of any real incentive to produce high strength gas, significantly reduce fuel consumption and improve real productivity, probably retarded development of oxygen driven processes. Further, the need for suitable burners and/or furnace/boiler modifications to handle oxygen enriched air had not been well appreciated so that the full benefits of oxygen enrichment could not be widely demonstrated. As an example, Okazoe, Kato and Murao (1967) report the poor results obtained in 1956-62 when testing with oxygen as follows:

> "A series of tests were carried out for the purpose to prevent the tubes in the recuperator from corrosion by V_2O_5 in oil, and also to decrease the volume of the exhaust gas. In these tests, the air temperature was kept at 80°C to 300°C by adding oxygen to the hot air, and the oxygen content in the air was 32% to 36%. However, they resulted in no remarkable prospect for improvement with 80% of oxygen efficiency and little decrease in exhaust gas volume."

The Outokumpu flash furnace installed in 1956 at the Ashio smelter originally operated with preheated air. Oxygen enrichment at Ashio was operated commercially in 1978 when a shorter reaction shaft was introduced to better operate with highly oxygen-enriched air (Fujii and Shima, 1981).

In his textbook, "Metallurgical Engineering", Professor R. Schuhmann (1952) discussed the economic benefits of using oxygen enriched air in a pulverized coal burner. Schuhmann estimated that one ton of oxygen would save the equivalent of about 0.52

tons of coal, and that with 95% oxygen anticipated at \$5/ton, there would be an advantage in using oxygen with coal costing at about or in excess of \$10.[3] Schuhmann discussed the benefits as follows:

> "One ton of oxygen will eliminate 7500 lb of nitrogen from the combustion products, and at a critical temperature of about 1400°C (2550°F) this quantity of nitrogen would require about 6 x 10^6 Btu of sensible heat. Thus, the cost of securing available heat through enrichment with oxygen at \$5 per ton would be about \$1.00 per 10^6 Btu. Compared with using coal (12,000 Btu/lb) under conditions giving 40% available heat, the energy cost of using oxygen is equivalent to coal at \$10 per ton, which is not expensive energy."

Schuhmann pointed out that the initial addition of oxygen to burner air gives the most saving, subsequent additions of oxygen have a lesser effect, Figure 7 reproduced from Schuhmann's text illustrates the relative fuel consumption with increasing levels of oxygen enrichment for different flue gas temperatures. (The fuel was producer gas having the composition 25% CO, 5% CO_2, 15% H_2 and 55% N_2 with a calorific value of about 130 Btu/ft^3). This information and other theoretical and experimental data regarding benefits of oxygen remained essentialy unapplied insofar as active commercialization of oxygen-driven non-ferrous processes for more than fifteen years after publication of Schuhmann's text. During that period, the economic benefits of oxygen were mainly weighed against the fuel savings and thus the cost of fossil fuels had an important bearing on the progress of implementation oxygen enriched processes.

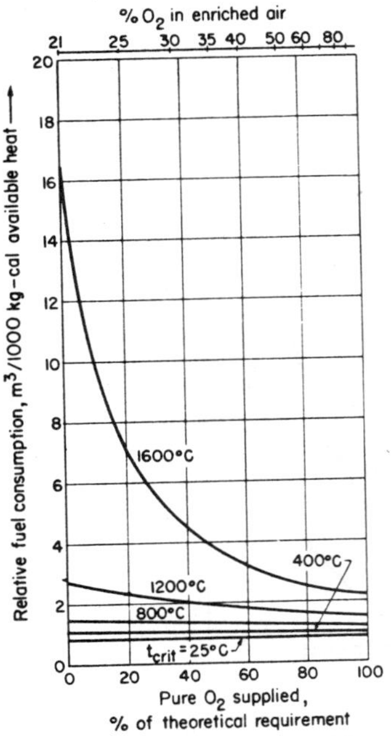

Figure 6.
Illustration of effect of oxygen enrichment and flue gas temperature on fuel consumption from 1952 publication. Curves apply to producer gas, 25% CO, 5% CO_2, 15% H_2 and 55% N_2. After Schuhmann (1952).

(3) For a flue gas temperature of 1400°C, value is smaller for lower flue gas temperature.

The historical trends in fuel and power costs over the period since about 1930 are illustrated in Figure 8. The now well-known dramatic price rise in the early seventies following the Middle East oil cutback is evident. It was about this time that oxygen usage became economically attractive in almost all applications. The historical trend in the cost of oxygen is presented in Figure 9. The more recent increases in oxygen cost reflect both increases in electrical power and cost of capital (the direct costs (1987) are taken from Table 5).

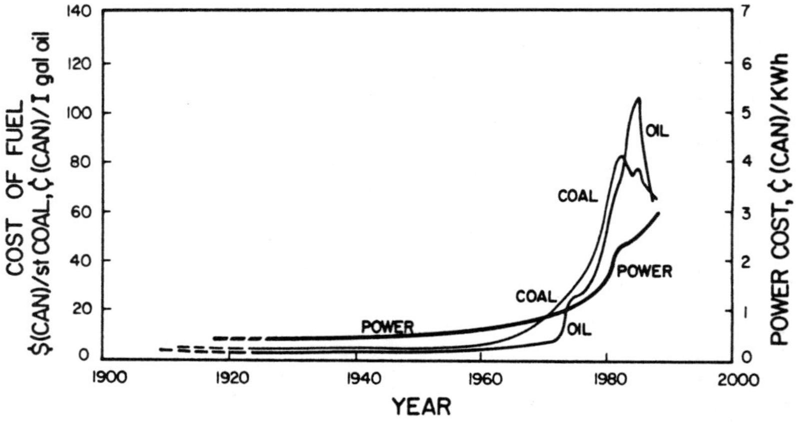

Figure 8. Historical trends in cost of fuel and power (typical values, Eastern Canada).

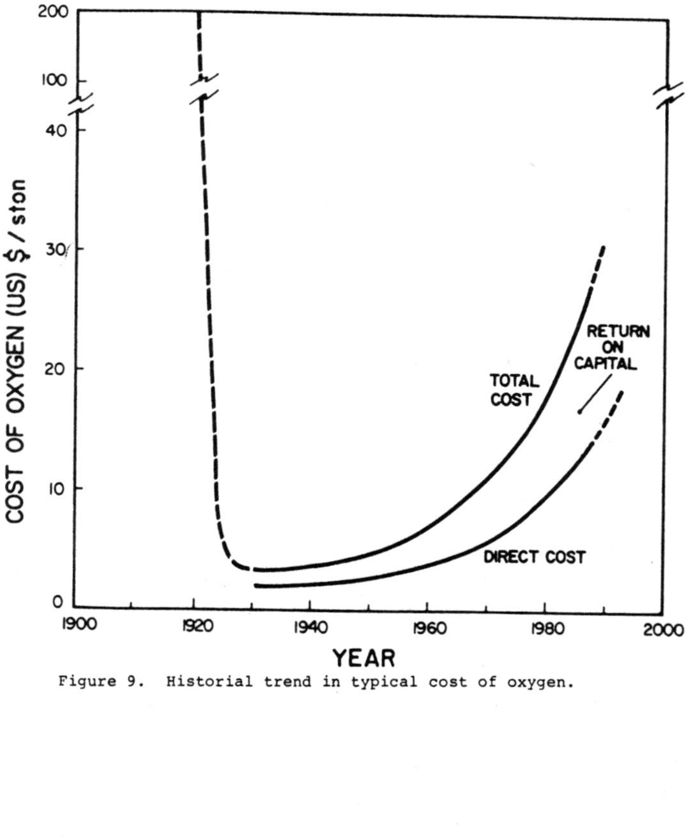

Figure 9. Historial trend in typical cost of oxygen.

The general type of trade-off relationship between cost of oxygen and fossil fuel cost (coal is used in example) is shown in Figure 10. The shaded area represents net fuel benefits due to oxygen of between 0.4 and 0.7 t of coal/t of oxygen. Above this area, oxygen would be cheaper, below the area fuel would be cheaper for a specified level of saving. The cross-hatched areas represent typical data for indicated periods. It is evident that the value of the fuel saving alone due to oxygen was quite small (Table 7) until the 1973 oil price increase.

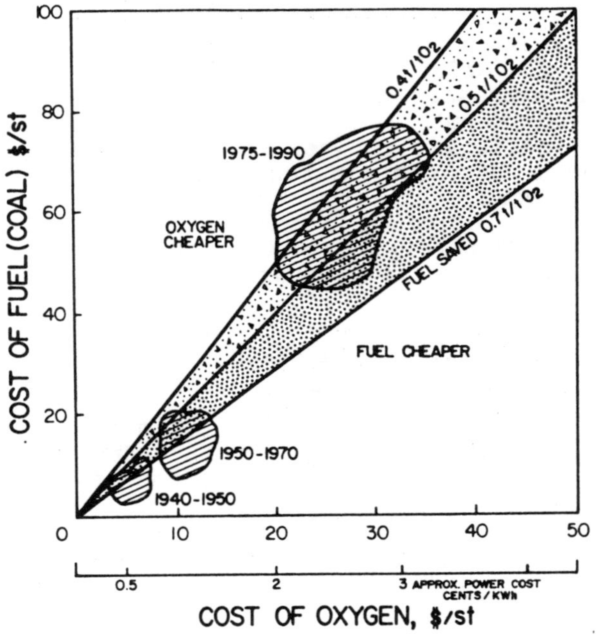

Figure 10. Approximate relationship between cost of oxygen and cost of fuel (coal equiv.); above shaded area oxygen is cheaper on basis of line for specified amount of fuel saved due to oxygen.

Subsequent testwork on new processes has shown that improved fuel savings due to oxygen are possible, secondly metallurgical and productivity gains and amenability of off-gas for acid production soon became benefits far exceeding simple fuel savings. Thus, gradually as testwork and technical knowledge advanced, it became evident that a second benefit after fuel savings, namely significant productivity gains, were possible using oxygen enrichment and with increasing labor and capital costs, this benefit needed to be included in the cost/benefit analysis. It should be noted that while it became widely recognized at this time that oxygen addition permitted higher tonnages in the same furnace, this was demonstrated by Cominco (1949) in the nineteen-thirties on a zinc suspension roaster, however this oxygen advantage was not widely exploited until much later.

A third factor, environmental control, subsequently also became important as mandatory controls required sulphuric acid plants to collect all or some of the emitted sulphur dioxide gas; oxygen enrichment produced a stronger gas which required less capital for acid-making facilities. The improvement in gas strength with oxygen enrichment was demonstrated by Cominco (1949) as far back as the nineteen-thirties when oxygen enrichment for zinc roasting commenced.

The individual developments concerning the introduction of oxygen processes following the fifties and sixties are too numerous to discuss individually. Therefore, a general chronology of the key oxygen developments in non-ferrous processes is presented in Table 8. Much pioneering work was done in this period and many new and enduring processes employing oxygen enrichment techniques were created, tested and commercialized. The most active period for these developments was between about 1960 and 1980. As noted previously, Canada has contributed significantly to new oxygen processes and techniques for using oxygen in non-ferrous metallurgy. These include the pioneering work by Cominco and Inco prior to the fifties, subsequent developments include the Noranda Process and the zinc pressure leaching process (Table 8).

Table 7 - Estimated Fuel Savings with Oxygen Over Four Decades

Decade	Estimated Fuel Savings $/ton of oxygen
1930-1940	-0.7
1950-1960	-1
1960-1970	-4
Post 1973	5-10+

Based on simple fuel saving of 0.52 st coal equivalent/s ton oxygen as projected by Schuhmann (1952). Present day operations offer additional benefits other than fuel savings (in increased throughput and stronger off-gas) and are not included in above.

It is of interest to note that much early work on oxygen enrichment was carried out in the U.S.S.R. In a 1973 publication "Bibliography on Copper Smelting", Malhotra (1973) lists about 100 papers dealing with the subject, "Oxygen in copper smelting and refining" covering the period 1940-1970. About 70% of the papers listed are of Soviet origin, the balance from the U.S., Canada, Japan and Europe. The listing was not necessarily intended to be an exhaustive one, however it does suggest an early emphasis on oxygen usage in the U.S.S.R.. The Soviet work covers oxygen addition to the blast furnace, roaster, converter and in flash smelting. Evidently by the early sixties (and probably before), three large smelters were operating with oxygen enrichment on the blast furnace, reverberatory furnace and converter (Sutulov, 1967). Early U.S.S.R. testwork on oxygen enrichment in the converter showed much promise - in terms of improved productivity - allowing additional concentrates to be smelted and giving an improved gas strength for acid making. Evidence had been developed that oxygen enrichment of conventional burner air in reverberatory furnaces was not particularly advantageous and was very dependent on the fuel cost. Sutulov (1967) in his review of Russian copper metallurgy commented as follows: [(4)]

> "It (oxygen enrichment) is justified in Russia if the oxygen cost is less than 0.7 cents/m^3, which could happen if electrical energy is available at 0.3 cent/KWH. It was calculated (33) that, if the price of oxygen rises to 1.3 cents/m^3 its cost is higher than the advantages obtained. In this case, pre-heating of combustion air to 300°C or 500°C is more favorable than the use of 25 percent and 29 percent oxygen respectively."

Sutulov (1967) projected that by 1970, "fully 22% of the total copper produced in Russia will use technical oxygen for smelting purposes". Later efforts at Almalyk developed the oxy-fuel burner, subsequently installed and improved upon at the Caletones Smelter in Chile and now used elsewhere in Chile, Canada and Japan, refer Table 8.

OXYGEN APPLICATIONS - PRESENT AND FUTURE TECHNOLOGY

As noted earlier (Table 1), oxygen enriched air is now used in about half the copper, nickel and lead smelters worldwide. At these plants, oxygen is usually employed in the primary smelting furnace; about one-third of copper smelters also use oxygen enrichment in converters, Pannell (1988). A discussion of present trends in oxygen usage at copper smelters worldwide is to be given by Pannell (1988) in a review titled "Survey of World Copper Smelters." It is shown that present oxygen enrichment levels in copper smelters (smelting unit) range from 27% to 98% O_2.

The effects of oxygen enrichment on copper smelter productivity, fuel savings, off-gas volume and sulphur dioxide level for different processes presently in use have been reviewed in detail by Eacott (1983) and will not be duplicated here. More recent applications and operating data on specific processes and plants are available in the individual papers within this symposium

(4) Sutulov's ref. (33) is: A.K. Shakhnazarov: Technical progress and economy of the copper industry, Tsvetnyie Metally 1966, 3, p. 26.

volume to which the reader is referred. However, to illustrate the dramatic effects of oxygen enrichment on production increases and fuel reduction, reference is made to Figures 11 and 12 which illustrate the effects of tonnage oxygen on the performance of two different types of copper smelting furnaces.

Figure 11 shows the almost two-fold increase in throughput with oxy-fuel burner technology installed on a copper reverberatory furnace, together with the resulting reduction in fuel ratio, while Figure 12 illustrates in the form of a nomograph developed by Bailey and Storey (1979), the effect of tonnage oxygen on the concentrate smelting rate and fuel ratio in the Noranda Process. It can be deduced from data such as those presented in Figures 12 and 13 that the general level of fuel savings estimated by Schuhmann twenty-five years ago (Schuhmann, 1952) is still valid today, as demonstrated in Table 9. Far greater benefits and savings are now recognized in the form of increased smelting capacity of the same furnace, also illustrated in Table 9,(for typical comparison purposes only). While Table 9 refers only to two oxygen processes, in general, similar parameters also apply to other types of smelting furnaces and systems now commonly used in non-ferrous metallurgy (with due correction to concentrate type, level of oxygen, etc.).

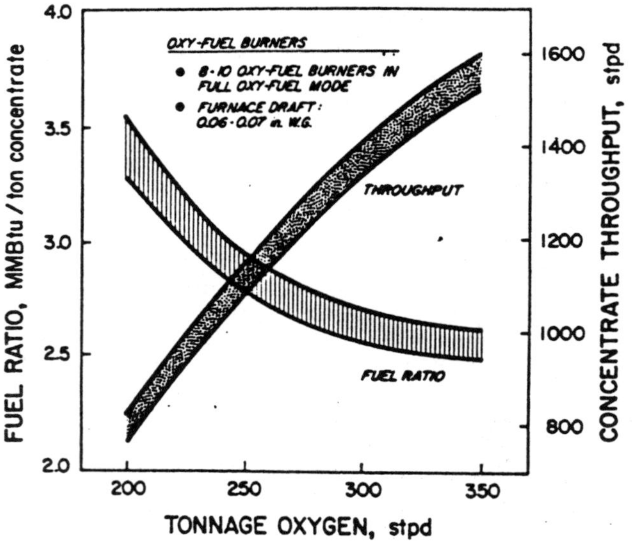

Figure 11.

Effect of tonnage oxygen (oxy-fuel burners) on copper reverberatory performance.

The effects of oxygen-enriched air on sulphur dioxide levels in smelter off-gas and the economics of off-gas treatment have been included in the reviews by Queneau (1977) and Eacott (1983). The reader is referred to these reviews and also to individual papers in this volume for further details. It is evident, however, that smelter gas in the range of 7 -10% SO_2 with a suitable O_2/SO_2 ratio for acid making is now routine (sulphide concentrates only, lower SO_2 levels for non-sulphide components in charge), a higher gas strength in the final gas at the acid plant is also possible. Perhaps sulphuric acid technology requires a fresh review concerning the economic feasibility of handling much stronger gas than in current industry practice.

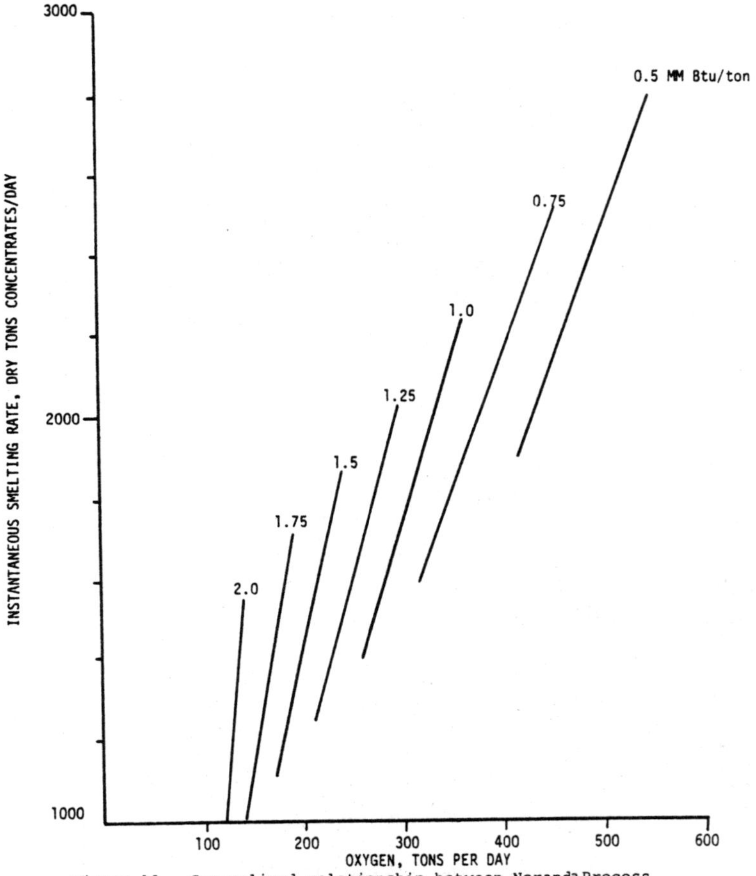

Figure 12. Generalized relationship between Noranda Process smelting rate and tonnage oxygen at different fuel ratios (operating conditions: Chalcopyrite copper concentrate, 73% Cu matte grade, blowing rate 76,500 Nm^3/h fuel supplied by coal). After Bailey and Storey (1979).

Table 8 - Chronology of Major Developments
Oxygen usage in Non-Ferrous Metallurgy

1895 First commercial air separation plant built in Munich Germany.

1920 - 1930

Various test work and investigations undertaken on use of oxygen enriched air for roasting and smelting pyrite and low grade concentrates.

U.S. Bureau of Mines issues report on potential applications of oxygen in metallurgical processes (1923).

NOTE - 1931 First commercial use of oxygen in Bessemer converter.

1930 - 1940

1933 Cominco at Trail, B.C. commences testwork on oxygen enrichment in zinc suspension roasting.

1936 Publications by Norman proposing autogenous smelting of copper concentrates with high levels of oxygen enrichment.

1937 Cominco commences commercial operations with oxygen in zinc suspension roasting.

1940 - 1950

1940 Ideas develop for "oxygen pressure leaching" of Au, Cu, Ni, Co, Zn, etc.

1945 Inco commences testwork on oxygen smelting of copper concentrates, Sudbury, Ontario.

1949 Cominco commences use of oxygen in lead blast furnace, and slag fuming, Trail, B.C.

1950 - 1960

1950 Investigations on "oxygen pressure leaching" underway for Au, Cu, Zn, Co, and Ni.

1951 First converter smelting tests with oxygen at Hitachi smelter, Japan (using bottled oxygen).

Note - 1952 First commercial LD plant, Austria (oxygen steelmaking); 1954 - First LD plant outside Austria operated in Canada.

1952 Inco flash furnace commences operation, Sudbury, Ontario, Canada.

Semi-commercial testing of converter smelting at Hitachi using 4 mtpd oxygen plant; smelting rate and gas strength for acid-making investigated.

1958 Commercial operation of Hitachi oxygen smelting process in converter with 60 mtpd oxygen plant, tuyere enrichment up to 38% oxygen.

Inco uses oxygen in converters, Sudbury, Ontario, Canada.

1959 Research on high temperature, oxygen pressure leaching of zinc concentrates originated by Foward and Veltman, Sherritt Gordon Mines Limited, Canada.

1960 - 1970

1960	Use of oxygen in copper blast furnace in U.S.S.R. (Irtyshski).
1961	Testwork on oxygen enrichment in Outokumpu flash furnace at Ashio smelter, Japan.
	Fire refining of secondary copper by oxygen top-jetting tested in U.K.
1962	Asarco commences testwork with oxygen enriched air on lead blast furnace.
	Applications of oxygen enrichment for non-ferrous melting furnaces develop.
1963	Oxygen usage at Balkhash, U.S.S.R.
1965	Active R&D activities related to oxidative pressure leaching of metal sulphides.
1966	Design of 100 tpd Noranda Process pilot plant initiated, including option for oxygen enrichment.
	Commence operation of blast furnaces and Peirce-Smith converters using oxygen-enriched air at Saganoseki smelter, Japan.
1967	Testwork at Rhokana(now Nkana) on oxygen addition to conventional burners in copper reverberatory furnace.
	Experiments with oxygen addition to conventional burners in copper reverberatory furnace at Almalik, U.S.S.R.
	Hoboken testwork commences on oxygen enrichment of blast furnace and Hoboken converter.
	Inco uses oxygen on reverberatory furnaces, TBRC and converters.
1967	Demonstration Kivcet plant in operation, 25 mtpd of lead concentrates/residues at Vniitsvetmet UST Kamenogorsk, U.S.S.R.
1968	Outokumpu tests oxygen enrichment in Outokumpu flash smelting process, Hajarvalta, Finland.

1970 - 1980

1970	Oxygen usage at Hoboken commercialized.
	Oxygen enrichment tested in Noranda Process pilot reactor at Horne smelter, Noranda, Québec, Canada.
	Start-up of Saganoseki flash furnace in Japan, first Outokumpu furnace operating commercially with oxygen.
	Oxygen enrichment tested in WORCRA pilot plant.
	Commercial scale Kivcet plant in operation, 500 mtpd of complex Cu-Zn-Pb concentrates, Glubokoye, U.S.S.R.
1971	Start-up of 72 mtpd Mitsubishi pilot plant with oxygen enrichment at Onahama smelter, Japan.
	400 mtpd oxygen plant installed at Caletones smelter, Chile.
	230 mtpd oxygen plant built at Harjavalta for copper and nickel flash furnaces.
	Application of oxy-fuel burners at Onahama smelter, Japan.

1974 Use of oxy-fuel burners in reverberatory furnace at Caletones smelter, Chile (later also adopted at Chuquicamata).

Start-up of Mitsubishi continuous process with oxygen enrichment, Naoshima smelter, Japan.

1975 85 mtpd oxygen plant on-line at Horne smelter, Noranda, Quebec, Canada for Noranda Process additional oxygen capacity added 1982.

1976 Start-up of Boliden's TBRC using oxy-fuel burner for treatment of lead bearing dusts at Ronnskar smelter, Sweden.

QSL pilot tests commenced.

1977 Cominco and Sherritt Gordon pilot zinc pressure leaching (with high pressure oxygen) on 3 mtpd pilot plant.

Start-up of first CMT bath smelting vessel with oxygen enrichment at Caletones smelter in Chile.

1978 Start-up of Boliden's TBRC for smelting copper concentrates, at Ronnskar smelter, Sweden.

<u>1980 - 1990</u>

1980 Commissioning of 120 kg/h lead Isasmelt pilot plant at Mt Isa, Australia.

1980 Application of oxygen enrichment for precious metals processing (slimes leaching, Doré furnace, etc.).

1981 Contop Process test on 24 tpd pilot plant in Germany.

Start-up of world's first commercial zinc pressure leaching plant (1300 kPa) by Cominco, Trail, B.C. (170 mtpd zinc concentrate, 0.25 t O_2/t conc.).

Pilot testing of flame cyclone process at Norddeutsche Affinerie underway in 10 mtph unit.

Start-up of 10 tph QSL demonstration plant at Berzelius, Germany.

Operation of 1500 mtpd foam smelting furnace at Norilsk smelter, U.S.S.R., plant designs for other units underway.

1982 Oxygen sprinkle smelting tests at Morenci smelter, Arizona, U.S.

1983 Start-up of St. Joe's Flame Reactor for treatment of metal oxide feed.

1984 Outokumpu reportedly reach 90% oxygen enrichment.

1986 Commissioning of Isasmelt demonstration plant on copper concentrates.

Commercial scale lead Kivcet plant commences operation, 350 mtpd lead concentrates, UST Kamenogorsk, U.S.S.R.

1987 First lead Kivcet plant outside U.S.S.R., 600 mtpd of lead concentrates, Portovesme, Italy.

<u>1990 - 2000</u>

Increases in intensity of tonnage oxygen usage for all processing steps in non-ferrous metallurgy.

Development of improved instruments for monitoring oxygen-driven processes.

All major non-ferrous smelters and plants will utilize tonnage oxygen technology.

Table 9 - Benefits of Oxygen Enrichment for Different Copper Processes

Process	Fuel Savings t coal equiv./t oxygen	Productivity Gain t conc/t oxygen
Oxy-fuel Burners(5)	0.4	6
Noranda Process(6)	0.45	8

(5) Applicable for case for addition of first 250 tonnes of oxygen; subsequent oxygen addition give lower fuel savings, of the order of 0.15 t/t.

(6) Typical data only; case-by-case values depend on level of oxygen enrichment, throughput and other operating conditions.

All other areas of non-ferrous metallurgy in which oxygen is required for the process, are expected to find increased application of tonnage oxygen technology in future. A recent new development for enhancing copper melting and refining processing is the Submerged Melting and Refined Technology ("SMART") process of the Linde Division of Union Carbide Corporation (Masterson, 1987). This process employs an oxy-fuel shrouded tuyere somewhat akin to the Savard-Lee tuyere, to inject tonnage oxygen and fuel in required proportions for heating, oxidation or reduction. Typical tonnage oxygen consumption data are given in Table 10. Another recent development is the metal oxide flash smelting reactor piloted by St. Joe (Pusateri, 1986). This process uses oxygen enriched air and fuel under reducing conditions to flash smelt metal oxide feed.

The present levels of tonnage oxygen usage in non-ferrous metallurgical plants can be compared in several ways, such as referring to % O_2 in enriched air or by reducing the tonnage oxygen used to a ratio expressing intensity of use, as given by tonnes of bulk oxygen employed per tonne of metal produced. For the purposes of comparison, the latter ratio applicable to non-ferrous metallurgical processes is illustrated in Table 10. This shows values ranging from 0.1 to about 0.8 for various copper, zinc and lead processes; a typical value for basic oxygen steelmaking is also indicated.

Table 10 - Typical Values - Intensity of Tonnage Oxygen Usage In Non-ferrous Metallurgical Processes

Process	Tonnage Oxygen Intensity t oxygen/t metal
Zinc-pressure leaching	0.50
Zinc suspension roaster[7]	0.20
QSL Process (lead)	0.51
Kivcet Process (lead)	0.73
Noranda Process	0.75
Outokumpu flash furnace	0.30-0.70
Inco Process	0.75
CMT Converter	0.56
Oxy-fuel burners	0.7-0.8
Mitsubishi Process	0.55-0.65
Linde SMART Process	0.1
Basic Oxygen Steelmaking	0.1

[7] No longer used.

The limit on oxygen enrichment has not yet been reached in most cases, it is expected that with new research and development, ways will be found to increase the intensity of oxygen usage in the future; this will require new R & D efforts by all interested parties - metals producers, gas companies, engineering companies and so on. Changes in equipment to utilize higher oxygen levels are likely. As an example, new tuyere technology is one approach as discussed by Noranda authors El-Barnachawy, Kachaniwsky, Persson and Poggi (1987) in a paper in the present volume.

As higher and higher levels of oxygen enrichment are used, metallurgical reactions occur at a faster rate. In this regard, improved instrumentation techniques will be required to monitor and record process variables. One such instrument for measuring temperature, the Noranda Tuyere Pyrometer (Noranda Inc., 1987), is finding application in the Noranda Process and converters. Other applications where quick and reliable melt temperatures are required are possible.

In Agricola's time, the best plants properly controlled the "four elements" then believed to constitute matter - today, it is necessary for a plant to replace process air by tonnage oxygen and maximize its use before the plant can be considered efficient. Quoting Queneau (1977), the "dead hands of nitrogen" can thus be removed, allowing the many benefits of tonnage oxygen to be realised as reviewed and discussed in this paper.

Further tonnage oxygen applications and increases in oxygen enrichment levels in all processes will in future require the most sophisticated techniques of metallurgical engineering, advanced smelting, refining and extraction technology along with improved understanding of metallurgical processes, to attain the full value offered by tonnage oxygen technology.

REFERENCES

Bailey, J. B., and A.G. Storey (1979). The Noranda Process after six years' operation, paper presented at the 18th Annual CIM Conference of Metallurgists. The Canadian Institute of Mining and Metallurgy. 19-23 August, Sudbury, Ontario.

Bessemer, H. (1905). Sir Henry Bessemer, F.R.S. - An Autobiography. Offices of "Engineering". London.

Davis, F. W. (1923). The use of oxygen or oxygenated air in metallurgical and allied processes. U.S. Bureau of Mines, R.I. 2502, U.S. Department of Interior. Washington, D.C.

Eacott, J. G. (1983). In H. Y. John, D. B. George and A. D. Zunkel (Ed.). Advances in Sulfide Smelting. Vol. 2, Technology and Practice. The Metallurgical Society of AIME, New York. pp. 583-634.

El-Barnachawy, S., and G. Kachaniwsky, H. Persson and D. Poggi (1987). Oxygen use at Noranda's Horne smelter, paper included in present volume.

Ferris, G. A. (1961). Oxy-fuel increases open hearth potential. J. Metals, 13, 4, pp. 298-299.

Fujii, O., and M. Shima (1981). Reaction shaft of Ashio Smelter's flash furnace utilizing high oxygen enriched air. In D. B. George and J. C. Taylor (Ed.), Copper Smelting - an Update. The Metallurgical Society of AIME. New York. pp. 165-171.

Hoover, H. C., and L. H. Hoover (1950). Georguis Agricola - De Re Metallica. Dover Publications Inc. New York. pp. 379-380.

Inco Staff (1955). The oxygen flash smelting process of The International Nickel Company. Trans. C.I.M., LVIII, pp. 158-166.

Johnson, R. E., N. J. Themelis, and G. A. Eltringham (1979). A survey of worldwide copper converter practices. In R. E. Johnson (Ed.). Copper and Nickel Converters. The Metallurgical Society of AIME. New York. pp. 1-32.

Kayan, C. F., and V. J. Johnson (1963). Refrigeration. In J. H. Perry (Ed.). Chemical Engineers' Handbook. McGraw Hill Book Company, New York. pp. 12-1 - 12-41.

Kubota, Y., and M. Yasuda (1973). Recent developments in copper smelting at the Hitachi smelter and refinery. In C. Diaz (Ed.), The Future of Copper Pyrometallurgy. The Chilean Institute of Mining Engineers. Santiago, Chile. pp. 147-154.

Landucci, L., and F. T. Fuller (1961). Oxygen-enriched air in lead and zinc smelting. J. Metals, 13, 10, pp. 759-763.

Laurila, A. (1973). Some aspects of Outokumpu flash smelting. In C. Diaz (Ed.). The Future of Copper Pyrometallurgy. The Chilean Institute of Mining Engineers. Santiago, Chile. pp. 147-154.

Mackey, P. J., and P. Tarassoff (1983). New and emerging technologies in sulfide smelting. Advances in Sulfide Smelting Vol.2, Technology and Practice. pp. 399-426.

Malhotra, S. C. (1973). Bibliography on Copper Smelting. Insight Printing and Graphics. Salt Lake City, Utah, U.S.A. pp. 136-141.

Masterton, I., D.G.George and F.A. Rudloff. 1987. The Linde-Kennecott Submerged Melting and Refining Technology (SMART) Process for Copper Fire-refining, Paper presented at the 26th Annual Conference of Metallurgists, Winnipeg, Manitoba, 23-26 August, 1987.

McKerrow, G. G., N. J. Themelis, P. Tarassoff, and G. D. Hallett (1972). The Noranda Process. J. Metals, 24, 4, pp. 25-32.
McNaughton, R. R., T. H. Weldon, J. H. Hargrave, and L. V. Whiton (1949). The use of oxygen enriched air in the metallurgical operations of Cominco at Trail, B.C. Trans. AIME, 185, pp. 446-450.
Nagel, T. (1935). Low-cost oxygen for metallurgical operations, Mining and Metallurgy, 16, May, 34.
Noranda Inc., (1987). Noranda tuyere pyrometer. Technical Brochure. Noranda Inc. Pointe-Claire, Québec.
Norman, T. E. (1936). Autogenous smelting of copper concentrates with oxygen-enriched air. Eng. Min. J., 137, 10, pp. 499-503; Eng. Min. J., 137, 11, pp. 562-567.
Okazoe, T., T. Kato and K. Murao (1967). The development of flash smelting process at Ashio Copper Smelter, Furukawa Mining Co., Ltd. In J. N. Anderson and P. E. Queneau (Eds.), Pyrometallurgical Processes in Non-ferrous Metallurgy. Metallurgical Society Conferences, Vol. 39, Gordon and Breach Science Publisher. New York. pp. 175-195.
Pannell, D. G. (1988). A Survey of World Copper Smelters. The Metallurgical Society of AIME to be published.
Pusateri, J.F. (1986). Development of a Metal Oxide Flash smelting Reactor. In D.R. Gaskell, J.P. Hager, J.E. Hoffman, and P.J. Mackey (Eds.). The Reinhardt Schuhmann International Symposium on Innovative Technology and Reactor Design in Extraction Metallurgy, The Metallurgical Society of AIME, New York, N.U. pp 131-148.
Queneau, P. E. (1977). Oxygen technology and conservation. Met. Trans. 8B. pp. 357-369.
Savard, G. and R. Lee (1966) French Patent 1,450,718.
Schuhmann, R. (1952). Metallurgical Engineering. Vol. I - Engineering Principles. Addison Wesley Press Inc. Cambridge, Mass. pp. 126-128.
Sherwood-Taylor, F. (1946). Inorganic and Theoretical Chemistry, Eighth Edition, William Heinemann Ltd., London.
Starratt, F. W. (1961). The Age of oxygen (editorial). J. Metals, 13, 4, pp. 257.
Slinn, P. C. (1967). Economics of oxygen production for non-ferrous metallurgy. In J. N. Anderson and P. E. Queneau (Eds.). Pyrometallurgical Processes in Non-Ferrous Metallurgy. Metallurgical Society Conferences, Vol. 39, Gordon and Breach Science Publisher. New York. pp. 333-347.
Sutulov, A. (1967). Copper Production in Russia, University of Conception. Chile.
Tarassoff, P. (1984). Process research and development - The Noranda Process. Met. Trans. B, 15B. pp. 411-431.
Taylor, J. G. (1987). Non-ferrous pyrometallurgy - a changing industry. Paper presented at 89th CIM Annual Meeting, Toronto, Ontario. 3-7 May 1987. Paper No. 148.
U.S. Bureau of Mines (1975). Mineral Facts and Problems,U.S. Bureau of Mines Bulletin No. 667. U.S. Department of Interior. Washington, D.C. pp. 761-768.
Verney, L.R. (1985). Developments in copper extractive metallurgy. The 1985 Extractive Metallurgy Lecture, The Metallurgical Society of AIME, New York, February, 1985.
Wilder, A. B. (1956). One hundred years of Bessemer steelmaking. J. Metals, 8, 6, pp. 742-753.

OXYGEN UTILIZATION FOR THE MITSUBISHI CONTINUOUS SMELTING PROCESS AT THE NAOSHIMA SMELTER

Moto Goto*, Mineo Hayashi** and Susumu Okabe*

* Process Technology Department, ** Metallurgy Department,
Mitsubishi Metal Corporation,
5-2, Ohte-machi, 1-chome, Chiyoda-ku, Tokyo, Japan

ABSTRACT

The first use of oxygen for copper pyro metallurgy in Japan was reported in the 1950's. Since then, the oxygen utilization has become one of the most common techniques to increase the operation efficiency. Today, copper smelters in Japan have approximately 50 t/hr of oxygen production capacity in total.

The total off-gas treatment system of the Naoshima Smelter was also rationalized with the oxygen utilization in 1983. The lead sintering off-gas was mixed with oxygen and introduced into the reverberatory furnace as a secondary burner air. The oxygen concentration in the blast air of the Mitsubishi Continuous Process was increased to 40 % for the smelting furnace and 30 % for the converting furnace. Since then, the continuous process has had more than 7,800 t/month of blister copper production capacity which is approximately twice the original design, and considerable amount of energy consumption of the smelter has been saved. These modificationshave been realized without remarkable reinforcements of acid plants.

INTRODUCTION

The oxygen utilization has been conducted as one of the most effective techniques to improve the production efficiency of modern copper pyro metallurgy. The principal effect of the oxygen enrichment of combustion air or blast air is to decrease nitrogen content in exhaust gas. This results in a decreased volume of exhaust gas and hence :

(1) decreased carry out of sensible heat in the exahust gas lessens fuel consumption;

(2) increased flame temperature improves melting rate of the charge;

(3) increased SO_2 concentration in exhaust gas makes the sulfuric acid plant operation easier and more economical;

(4) a small amount of gas does not need large size of gas handling facilities and energy.

Other than these advantageous features there are some effects on material balances. That is :

(5) decreased volume of exhaust gas reduces distribution of volatile matters to gas phase;

(6) increased reaction temperature shifts the equilibrium conditions.

Recently, most of the copper smelters in Japan are using tonnage oxygen at various stages of their operations, i.e. reverberatory furnaces, flash furnaces, converters and Mitsubishi Continuous Process. The production capacity of tonnage oxygen is estimated to be approximately 50 t/hr in total as shown in Table 1.

In this paper, history of the oxygen utilization for copper pyro metallurgy in Japan is reviewed, then expansion program of the Mitsubishi Continuous Process and the Naoshima Smelter is described.

TABLE 1 Production Capacities of the Oxygen Plants of Copper Smelters in Japan

Smelter	Oxygen Concentration (%)	Oxygen Production Capacity (Nm^3/hr)	(t/hr)	Smelting Process
Naoshima	80.0	12,000	14	Reverb. & Continuous
Toyo	90.0	7,000	9	Flash
Tamano	95.0	6,700	9	Flash
Saganoseki	95.0	6,200	8	Flash
Kosaka	90.0	4,500	6	Flash
Ashio	95.0	3,000	4	Flash
Onahama	-	-	-	Reverb.
Hibi	-	-	-	Blast
Miyako	-	-	-	Blast
Total	-	-	50	

OXYGEN UTILIZATION FOR COPPER PYRO METALLURGY IN JAPAN

In the 1950's, many innovations in copper extractive metallurgy were conducted in Japan. The reasons for this were:

(1) facilities were wornout through hard operations during wartime, and they needed to be renewed;

(2) rapid growth of copper demand due to economic recovery after world war II required expansions of smelters;

(3) recovery of sulfur oxide in exhaust gas was becoming much more necessary.

The first oxygen utilization for copper pyro metallurgy in Japan was "The Oxygen Smelting Process" of Nippon Mining Company, the basic research of which was started at Saganoseki Smelter in 1951 (Tsurumoto, 1961).

In 1952, fluosolid roasting with hydro metallurgy process for copper-zinc concentrates was put into operation at Kosaka Smelter. In 1955, direct smelting of copper concentrate with a blast furnace started at Shisakajima Smelter. The first Outokump-type Flash Furnace in Japan was introduced to Ashio Smelter in 1956 (Okazoe, 1965).

In 1958, abovementioned "Oxygen Smelting Process" was put into operation at Hitachi Smelter (Tsurumoto, 1965), with which blended and pelletized copper concentrates were directly smelted in P.S. converters. Insufficient heat to melt concentrates was compensated by oxygen enrichment of blast air, and for this purpose the oxygen plant with a capacity of 2,000 m^3/hr (as 95 % O_2) was installed. The Oxygen Smelting Process at Hitachi Smelter had been operated until a new flash furnace took its place in 1972. The operation of the process is summarized and its flow sheet is shown in Fig. 1. Five small P.S. converters with a diameter of 2.8 m and a length of 6.15 m were employed to this operation, and three of them were kept hot and the other two were for repair. Approximately 15 tons of matte at 40-50 % matte grade produced by a blast furnace was charged into a converter as starting matte, then pellets of blended concentrates were fed into the furnace through a retractable chute. The amount of pellets for the 1st and the 2nd matte blowing stages was 35-40 tons in total. Blast air with 35-38 % oxygen was blown at the rate of 175 m^3/min for approximately 70 minutes for each stage. The amount of slag produced through these stages was around 30 tons, and more than 90 % of copper contained in the slag was recovered by flotation method. These matte blowing stages were succeeded by a copper blowing stage, at which blast air with 25-29 % oxygen was blown at 190 m^3/min for 70 minutes. Blister copper produced in each cycle was approximately 23 tons, and monthly production was 4,500 tons.

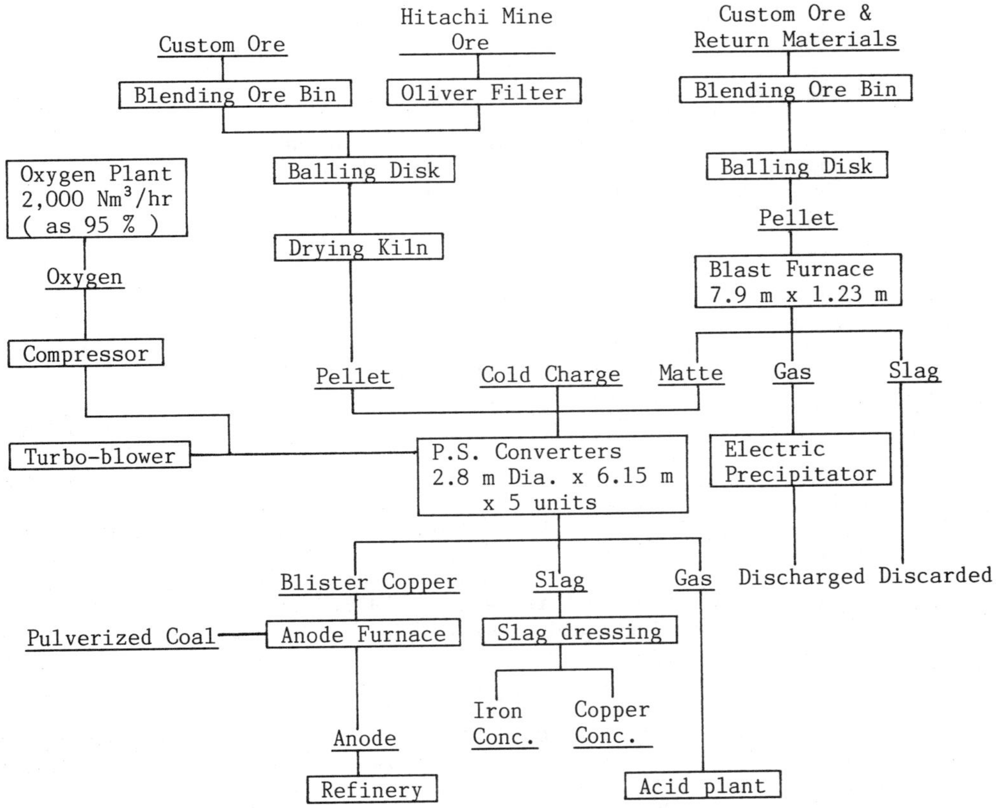

Fig. 1. Flowsheet of the Oxygen Smelting Process at Hitachi Smelter (Tsurumoto, 1965).

In 1961, Mitsubishi Metal Corporation started the fundamental research on direct and continuous smelting and converting process for copper concentrates at their Central Laboratory (Suzuki and Nagano, 1972).

In the 1960's, however, the most common process for copper smelting was still blast furnace, and some of them applied hot air blasting techniques to increase their efficiencies. In the late 1960's, constructions of large scale smelters were promoted. In 1965, a green charge type reverberatory furnace was put into operation at Onahama Smelter, of which treatment capacity was 16,700 tons per month and production capacity of anode copper from concentrates was 4,000 tons per month (Morisaki, 1966). In 1967, the oxygen plant with production capacity of 2,500 m^3/hr (as 95 % O_2) was installed at Saganoseki Smelter supposing the construction of large size flash furnace in the near future, and its oxygen was used for the "Oxygen Smelting Process" same as at Hitachi Smelter. This process had been operated until the new flash furnace was constructed.

In 1969, the Naoshima Smelter opened the operation of a calcine charge reverberatory furnace with 7,000 tons of anode copper production per month.

In 1971, the oxygen fuel burner was applied to a reverberatory furnace at Onahama Smelter. Two oxygen fuel burners with 22.5 % O_2 combustion air which were installed in addition to the existing conventional oil burners increased the smelting capacity by 6,000 t/month (Goto, 1976).

In those days, several Outokump-type flash furnaces were constructed and old blast furnaces were replaced at Kosaka Smelter in 1967, Saganoseki Smelter in 1970 (Fujii and others, 1972), Toyo Smelter in 1971 (Ogura and others, 1974), Tamano Smelter and Hitachi Smelter in 1972 (Kitamura and Shibata, 1973; Yasuda, 1974). The flash furnaces constructed at Saganoseki Smelter and Hitachi Smelter were applying pre-heated and oxygen enriched blast air. The pre-heated temperature was around 1000 °C and oxygen concentration was around 30 %.

In 1974, Mitsubishi Metal Corporation completed the development of the new direct continuous smelting and converting process for copper concentrates named "Mitsubishi Continuous Copper Smelting and Converting Process", and started the commercial operation at the Naoshima Smelter. As the Continuous Process was originally designed to use tonnage oxygen, the oxygen plant with production capacity of 4,000 m^3/hr (as 80 % O_2) was put into operation at the same time (Nagano and Suzuki, 1976).

In 1974, after the 1st oil crisis which raised the price of fuel oil three times as much as before, cheaper fuel i.e. coal, coke and scrapped tire etc. were used in place of oil (Kohno and others, 1982) and the oxygen utilization became of major interest again. Ashio Smelter, Tamano Smelter and Toyo Smelter commenced the oxygen utilization for the flash furnaces in 1978, 1981 and 1982 respectively (Fujii and Shima, 1982; Hashiuchi and others, 1982; Tomono and others, 1984).

Tonnage oxygen is used not only for smelting furnaces but also for converting furnaces. At Saganoseki Smelter, heat deficit in converter operation due to high grade matte is made up by oxygen enriched blast air. Oxygen concentration is approximately 25 % for the matte grade of 60 % (Fujiwara and others, 1979). At Toyo Smelter, blast airs for matte blowing stage and copper blowing stage of converter are oxygen enriched to 26.5 % and 24 % respectively, and coolant treatment capacity is expanded to twice as much as before (Tomono and others, 1984). Tamano Smelter also uses tonnage oxygen for matte blowing stage of converter, and its oxygen content is about 24 % (Okada and others, 1983). Although the oxygen enriched blast air causes local excess heat and it may damage the refractory around tuyeres, abovementioned operations with under 30 % enrichment are reported to have no problem . Tsurumoto (1965) concluded through his experiences in the oxygen smelting process that upper limit of oxygen concentration in blast air of P.S. converter was 40 % . Recently, however, Kimura and others (1985) succeeded in increasing the oxygen concentration in blast air for a P.S. converter-like furnace to approximately 50 % on average without refractory damages by complex blasting technique.

In 1982, the oxygen plant of Naoshima Smelter was reinforced from 4,000 to 12,000 m^3/hr production. The purposes of it were the expansion of treatment capacity of the Mitsubishi Process and the rationalization of exhaust gas treatment system. This expansion program is described in detail later. Recently, a part of the oxygen is used for converters of conventional process too, and it enhances secondary resources treatment.

At Onahama Smelter, the test operation of the smelting with the P.S. converter has been conducted since 1983. This is the application of the injection smelting technology of the Mitsubishi Continuous Process, and dried concentrate is injected into the furnace with oxygen enriched air through submerged coaxial double pipe structured tuyeres.

In 1985, Kosaka Smelter installed an oxygen plant. Hence, all of the smelters other than two blast furnaces in Japan are using tonnage oxygen now.

THE OPERATION OF THE NAOSHIMA SMELTER

The Naoshima Smelter has two copper smelting processes. One is the Mitsubishi Continuous Process and the other is the conventional process with a fluosolid roaster, a reverberatory furnace and P.S. converters. Adjacently to the smelter, the lead smelter of Mitsubishi Cominco Smelting Company Limited is operating. Exhaust gases from these proceses are treated together by three acid plants installed in the Naoshima Smelter.

The Mitsubishi Continuous Process

In 1974, the Mitsubishi Continuous Process started its operation at Naoshima Smelter at the originally designed feed rate of 25 t/hr and with anode copper production capacity of 4,000 t/month. Since then, several improvements and modifications such as pulverized coal utilization with blended concentrates (Goto and others, 1980), refractory protection by water cooled copper blocks (Goto and Echigoya, 1980) and the more amount of oxygen utilization (Goto and Kikumoto, 1984) have been conducted in order to establish more stable and intensive operation. The Mitsubishi Continuous Process is distinguished from other processes by its unique features such as gas-solid mixture injection through top blowing lances, transportations of molten products between furnaces with launders and Cu_2O-CaO-Fe_3O_4 ternary system slag for the converting furnace .

The process consists mainly of three furnaces, namely the smelting furnace, the slag cleaning furnace and the converting furnace. The general layout of the process is shown in Fig. 2. Dried copper concentrates mixed with fluxes, pulverized coal and process recycled materials are pneumatically conveyed and injected through top blowing lances which have coaxial double tubed structure. Raw materials fed through the inner tube and oxygen enriched blast air introduced through the outer tube are mixed at the lower part of the lance and injected into molten bath at high velocity. Feeding system for the smelting furnace is schematically shown in Fig. 3. Injection of gas-solid mixture causes deep penetration into bath and hence the instantaneous absorption of raw materials, the high oxygen utilization efficiency, small amount of mechanical dust and vigorous agitation of molten bath are realized without submerged lances. Blast air is controled to maintain the matte grade at 65 % copper. Produced matte and slag are taken out by overflow together and transported to the slag cleaning furnace through a launder. The slag cleaning furnace is three electroded electric furnace, in which the slag is separated from matte and cleaned to 0.5–0.6 % copper concentration during its retension time of about one hour. The overflown slag is water-granulated and discarded. The matte is siphoned out and transferred to the converting furnace through a launder.

At the converting furnace, limestone is fed with oxygen enriched blast air through lances in a same way as the smelting furnace, and matte is continuously converted to blister copper with making Cu_2O–CaO–Fe_3O_4 ternary system slag. In order to keep CaO and Cu concentrations in slag at 15–17 % and 12–15 % respectively, additional limestone and blast air are adjusted. Blister copper is siphoned out and stored temporarily in a holding furnace, then delivered to anode furnaces by a ladle car. Slag is tapped through a slag hole, water-granulated, dried and returned to the smelting furnace continuously.

The exhaust gases from the smelting furnace and the converting furnace are cooled to approximately 350 °C through respective boilers, cleaned by electrostatic precipitators and then sent to the acid plants.

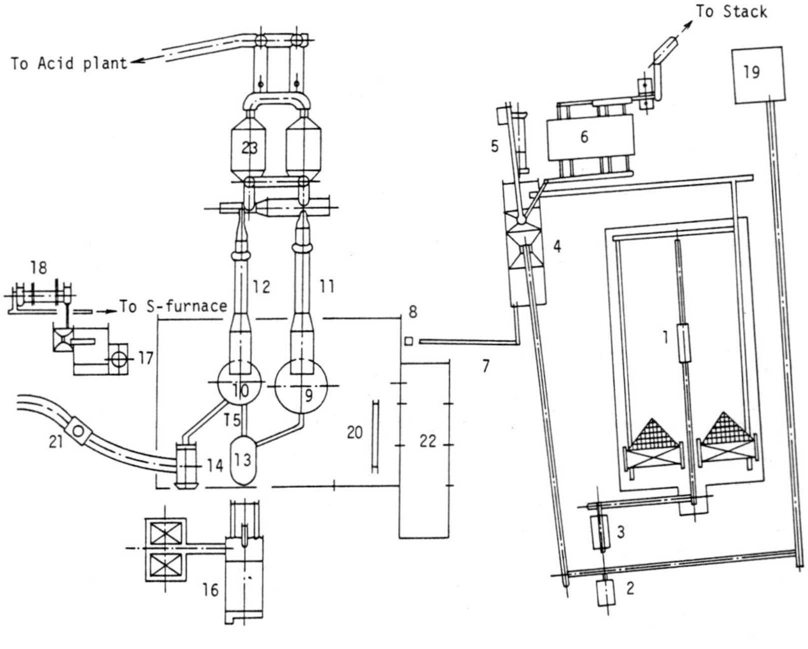

1-Blending yard 2-Flux receive 3-Concentrate receive 4-Daybins 5-Concentrate dryer 6-Bag houses 7-Chain conveyors 8-Bucket conveyors 9-Smelting furnace 10-Converting furnace 11,12-Boilers 13-Slag cleaning furnace 14-Blister copper holding furnace 15-Launders 16-CL-slag granulation 17-C-slag granulation 18-C-slag dryer 19-Flux crushing 20-Hoist 21-Ladle car to anode furnace 22-Control room 23-Electrostatic precipitators

Fig. 2. General layout of the Mitsubishi Continuous Process at the Naoshima Smelter

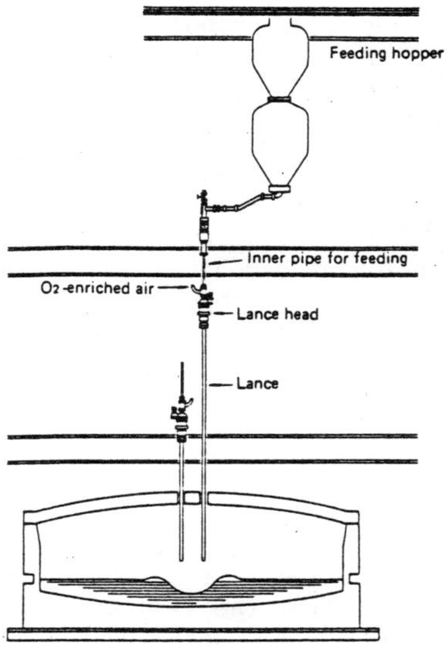

Fig. 3. Schematic drawing of feeding and lancing system.

The Conventional Process

The conventional copper smelting process has been operating since 1969.

The fluosolid roaster is a Dorrco-type. Copper concentrates blended with fluxes and process recycled materials are charged at the feed rate of 55 wt/hr, and then roasted at around 600 ° C. Approximately 500 m^3/min of exhaust gas containing 11.4 % of SO_2 is constantly generated by desulfurization at 35-40 %. Produced hot calcine is transported by drag conveyors and charged into the reverberatory furnace.

The reverberatory furnace is hot calcine-charge and deep bath type with smelting capacity of 1200 tons of calcine per day. Since fuel consumption ratio is very high as 100-110 l/t conc. on bunker C-oil basis, a large amount of exhaust gas with thin SO_2 is generated. This results in an uneconomical operation of acid plants. The volume and SO_2 concentration of exhaust gas are 1080 m^3/min and 2.1 %.

Three P.S. converters are installed and two of them are kept hot and blown alternately, and the other one is for repair. One cycle of the operation consists of three slag making stages and a copper making stage. Since the volume and the SO_2 concentration of the exhaust gas change periodically, off-gases from the converters are mixed with off-gas from the roaster and then processed by acid plants. Average volume and SO_2 concentration of the exhaust gas of converters are 850 m^3/min and 12.6 %.

The Lead Smelter

The lead smelter of Mitsubishi Cominco Smelting Company Limited, which is jointly owned by Cominco Limited in Canada and Mitsubishi Metal Corporation, has been operated since 1966. The smelter treats solely high purity Pine Point lead concentrate by a conventional updraft sintering machine and a blast furnace with the production capacity of 3,500 t/month of refined lead. Since the concentrate originally contains little impurities, refined lead with the purity over 99.999 % is obtained with only simple refining method such as copper removal with Al-Zn alloy and succeeding clean up with sodium hydride.

Approximately 90 % of sulfur in concentrate is oxidized and removed through the sintering, and generated exhaust gas with 4.3 % SO_2 at the rate of 250 m^3/min is processed by an acid plant in the Naoshima Smelter. When the sintering off-gas was sent to an acid plant directly, however, a serious problem of black colored acid was inevitable. The reason for this was the sintering temperature was so low as 450 °C that considerable portion of pyrobitumens contained in the Pine Point lead concentrate at 0.3-1.0 % was not decomposed and contaminated the sulfuric acid. The black colored acid had to be bleached with hydrogen peroxide.

The exhaust gas from the blast furnace is desulfurized by scrubbing with calcium carbonate and then discharged.

The Acid Plants

The original off-gas treatment system of the Naoshima Smelter is shown in Fig. 4. In this system, the exhaust gas from the lead sintering machine and a part of the exhaust gas from the continuous process were treated together by the No.1 acid plant. The No.2 acid plant was operated for the roaster and the converters. The exhaust gas from the reverberatory furnace and the rest part of the exhaust gas from the continuous process were treated by the No.3 acid plant. This off-gas handling system, however, had several problems as follows.

(1) No.1 plant had a serious problem of black colored acid due to the lead sintering off-gas which required 40 t/month of hydrogen peroxide as bleaching material.

(2) The operation efficiency of No.3 plant was reduced by low SO_2 concentration of the reverberatory furnace off-gas.

(3) Total off-gas handling capacity limited the feed rate of the continuous process within 30 t/hr.

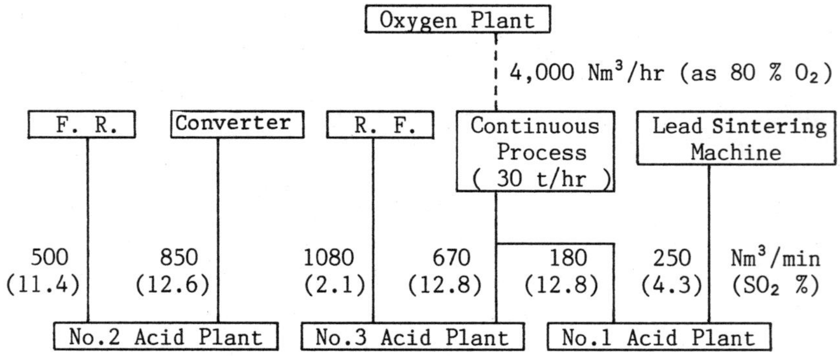

Fig. 4. Original Off-gas Treatment System at the Naoshima Smelter

EXPANSION PROGRAM OF THE NAOSHIMA SMELTER

In 1980, the expansion program of the Naoshima Smelter aimed to increase the anode copper production capacity of the continuous process to more than 7,000 t/month was commenced. The basic concept of this program was to keep the total volume of exhaust gases as little as possible with the oxygen utilization in order to minimize extension of off-gas handling facilities. The low SO_2 concentration of the exhaust gas from the reverberatory furnace and black colored acid due to the lead sintering off-gas were also intended to be improved at the same time. For these purposes blast air of the continuous process was planned to be oxygen enriched more and the lead sintering off-gas was planned to be mixed with oxygen and introduced into the reverberatory furnace as a secondary burner air. Therefore an oxygen plant with the production capacity of 8,000 Nm^3/hr as 80 % O_2 was installed additionally. The expansion program was divided into two stages as shown in Fig. 5.

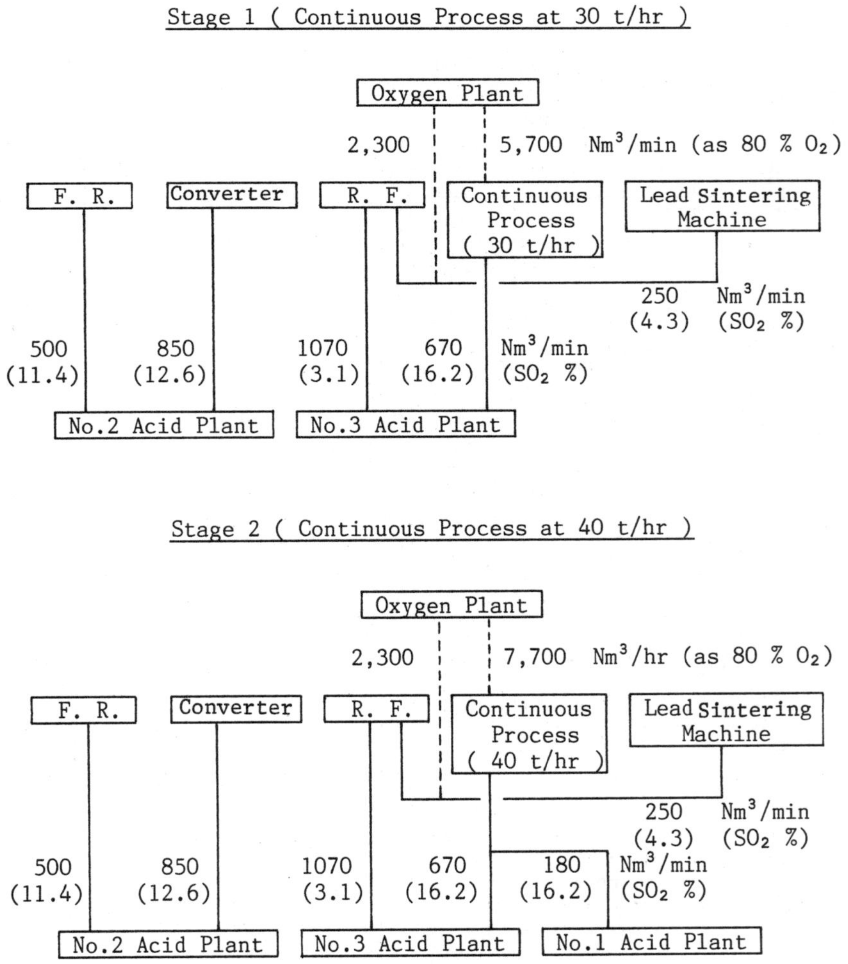

Fig. 5. Change of the Off-gas Treatment System at the Naoshima Smelter.

At the first stage, feed rate of the continuous process was maintained at 30 t/hr as before, but oxygen concentration of blast air was increased in order to reduce the volume of exhaust gas by 180 m^3/min, which corresonded to the amount of the continuous process off-gas treated by No.1 acid plant. The lead sintering off-gas which had been treated directly by No.1 acid plant was mixed with 2,300 Nm^3/hr of oxygen, introduced into the reverberatory furnace as the secondary burner air and then sent to No.3 acid plant as a part of the reverberatory off-gas. As a result, total consumption of oxygen was increased from 4,000 to 8,000 Nm^3/hr but inlet gas of No.1 acid plant was eliminated and No.1 acid plant was shut down.

At the second stage, feed rate of the continuous process was increased from 30 to 40 t/hr. To cope with the increased volume of the exhaust gas, No.1 acid plant was operated again. Since originally designed smelting and converting furnaces of the continuous process had been confirmed to have enough capability of treating 40 t/hr of concentrate through test works, only the feeding facilities and the slag cleaning furnace were reinforced. Total consumption of oxygen was increased from 8,000 to 10,000 Nm^3/hr. The typical operational figures at each stage are summarized in Table 2 and Table 3.

TABLE 2 Typical Operational Figures of the Continuous Process at Each Stage of the Expansion Program

		Original	Stage 1	Stage 2
SMELTING FURNACE				
Fed Concentrate	(t/hr)	30.0	30.0	40.0
Silica Sand	(t/hr)	5.3	5.3	7.1
Limestone	(t/hr)	0.2	0.2	0.3
Recycled C-Slag	(t/hr)	3.3	3.3	4.4
Lance Blast				
Compressed Air	(Nm^3/hr)	22,400	12,000	16,500
80 % Oxygen	(Nm^3/hr)	3,000	4,600	6,500
O_2 Concentration	(%)	28.1	37.6	37.9
Fuel				
Coal with Conc.	(DKg/hr)	1,000	500	900
Oil for Burner	(1/hr)	800	720	550
Matte Production	(t/hr)	13.6	13.6	18.1
Slag Production	(t/hr)	21.1	21.1	28.2
Off-gas Generation	(Nm^3/min)	570	410	480
SO_2 Concentration	(%)	12.2	17.1	19.2
CONVERTING FURNACE				
Limestone	(t/hr)	1.1	1.1	1.4
Lance Blast				
Compressed Air	(Nm^3/hr)	8,600	8,200	12,000
80 % Oxygen	(Nm^3/hr)	1,000	1,100	1,200
O_2 Concentration	(%)	27.3	28.1	26.5
Fuel Oil for Burner	(1/hr)	140	110	40
Blister Production	(t/hr)	8.3	8.3	11.0
Slag Production	(t/hr)	3.3	3.3	4.4
Off-gas Generation	(Nm^3/min)	180	170	230
SO_2 Concentration	(%)	18.3	19.3	19.8

TABLE 3 Composition of each materials

	Cu	Fe	S	SiO_2	CaO
Concentrate	27.5	25.0	30.0	6.5	0.5
Matte	65.0	11.0	22.0	-	-
Discard Slag	0.6	36.0	-	32.0	4.0
Converting Slag	13.0	44.5	-	-	17.0
					(%)

In 1983, the expansion program at the Naoshima Smelter was completed. Since then the total energy consumption at the smelter has remarkably decreased and the continuous process has had the smelting capacity of 40 t/hr. A record production of anode copper, 8,575 t/month, was achived through normal commercial operation in December, 1986. This figure was more than twice of original design. Furthermore, the capability of 50 t/hr feeding and 9,000 t/month of anode copper production was confirmed through short term test operation in June, 1985 (Goto and others, 1986).

EFFECT OF THE OXYGEN UTILIZATION

As a part of this expansion program, a number of test works were carried out (Goto and others, 1980; Goto and Kikumoto, 1981; Suzuki and others, 1983). Through these test works and actual commercial operation after the program, effects of the oxygen utilization on each process have been confirmed.

The Continuous Process

The oxygen utilization efficiencies at the smelting furnace and the converting furnace of the Continuous Process are approximately 98 % and 94 % respectively, and they are not affected by the increase of the oxygen concentration in blast air. On the other hand, the oxygen utilization efficiency is strongly dependent on the lance height and the blowing velocity. Hence, the decreased volume of blast air due to oxygen enrichment requires number of lances or diameter of lance to be reduced in order to keep appropriate blowing velocity.

Changes of fuel requirement and exhaust gas volume due to the oxygen enrichment agree well with theoretical calculations because the oxygen utilization efficiency is not affected by the increase of the oxygen concentration of the blast air. If the compressed air for blast is replaced by equivalent amount of tonnage oxygen, off-gas volume changes accordingly to the difference of their nitrogen content. For example, 1 Nm^3/hr of tonnage oxygen (as 80 % O_2) decreases 3.8 Nm^3/hr of compressed air and 2.8 Nm^3/hr of exhaust gas. At that time, decreased carry out of sensible heat in the exhaust gas reduces fuel oil requirement and combustion gas volume by 0.29 l/hr and 3.2 Nm^3/hr respectively. Therefore total volume of the exhaust gas decreases by 6 Nm^3/hr. As a result, electricity consumption at the oxygen plant, the air compressors and the exhaust gas blowers change by + 0.4 KWH, - 0.3 KWH and - 0.12 KWH respectively, and decreased calorific value of fuel oil is 2.9 Mcal. However, decreased volume of the exhaust gas lessens steam generation at the waste heat boiler, and electric power recovery at the turbine generator also reduces by approximately 0.35 KWH.

Since profit and loss of the oxygen utilization depends on local price of each energy, PFE proposed by Kellogg (1974) is useful for the universal evaluation. A conversion factor of electric power to PFE proposed by Kellogg is 11,075 KJ/KWH. As it represents the thermal power generation efficiency, in the case of cheaper power such as hydroelectricity the converting factor decreases toward physical equivalent value of 3,600 KJ/KWH. The changes of energy consumption of the continuous process due to additional 1 Nm^3 of oxygen utilization is summarized in Table 4. In this table, both of physical equivalent value and PFE are shown. After all, 10,950 KJ (8,480 PFE-KJ) is saved at the continuous process by additional 1 Nm^3 of oxygen utilization.

The relations among fuel requirements, off-gas volumes, feed rates and oxygen blowing rates of the smelting and the converting furnaces are calculated and shown in Fig. 6. For these calculations, all the intermediates such as converting furnace slag and dusts are assumed to be recycled to the smelting furnace. Other calculating conditions are as shown in the figures.

The temperature adjustment with the oxygen enrichment is easier and realizes quicker response than with fuel burning.

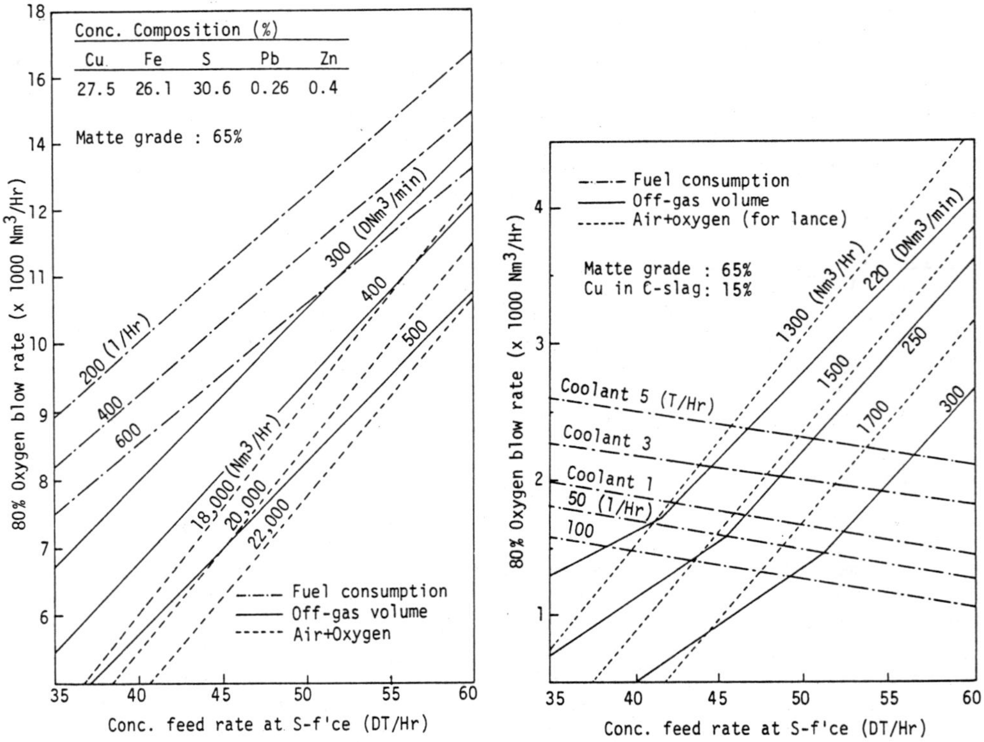

(a) the Smelting Furnace (b) the Converting Furnace

Fig. 6. Relations among various operation parameters.

TABLE 4 Effect of additional 1 Nm^3 of Oxygen Utilization on Energy Consumption of the Continuous Process

	Changes of Energy Consumption or Recovery	Physical Equivalent Value (KJ)	Process Fuel Equivalent (PFE-KJ)
Oxygen Production	+ 0.4 KWH	+ 1,440	+ 4,430
Air Compression	- 0.3 KWH	- 1,080	- 3,320
Fuel Oil Consumption	- 2.9 Mcal	- 12,140	- 12,140
Off-gas Transportation	- 0.12 KWH	- 430	- 1.330
Sub-total		- 12,210	- 12,360
Electricity Recovery	- 0.35 KWH	- 1,260	- 3,880
Total Consumption		- 10,950	- 8,480

Lance consumption and refractory wear. Local excess heat generated around oxygen enriched blast air has a danger of extra consumptions of tuyeres, lances and refractories. As a matter of fact, the oxygen concentration of blast air for the P.S. converter is reported to be limited within 40 % (Tsurumoto, 1965). In the case of the Mitsubishi Continuous Process, however, lances are not submerged and cooled enough by fed materials, hence consumption of lances is less affected by the oxygen enrichment of the blast air. No undesireable phenomena is observed in usual operation with blast air of 40-45 % O_2 for the smelting furnace and 30-35 % O_2 for the converting furnace. The highest oxygen concentration of blast air was 60 % which was confirmed at the test work of the smelting furnace in 1979. However, the long term effect on the whole furnace at such a high oxygen concentration has not been confirmed yet.

The distribution of minor elements such as Pb, Zn, As and Sb which transfer to gas phase very easily as shown in Table 5 are strongly dependent on volume of exhaust gas (Ohshima and Hayashi, 1986). In the case of arsenic, As_2, AsS and AsO are assumed as gas species, and then transfer amount into gas phase (X) is expressed as follows.

$$X = SM_{As}(2P_{As_2} + P_{AsS} + P_{AsO})V/RT \quad \ldots\ldots\ldots\ldots\ldots\ldots\ldots \quad (1)$$

here S ; saturation degree
M_{As} ; atomic weight of As
P_{As}, P_{AsS}, P_{AsO} ; partial pressure of each species
V ; volume of exhaust gas
R ; a gas constant
T ; temperature

As the equilibrium of these species between liquid and gas phases are considered in following reactions, and the equlibrium constants for them are represented by K1, K2 and K3, equation (1) is converted to equation (2).

$$2As(l) = As_2(g) \quad ; K1$$
$$As(l) + 1/2S_2 = AsS(g) \quad ; K2$$
$$As(l) + 1/2O_2 = AsO(g) \quad ; K3$$

TABLE 5 The Distributions of Minor Elements into Gas Phases (Observed Figures)

	S-furnace (%)	C-furnace (%)
Pb	54.0	44.0
Zn	14.5	19.0
As	74.0	25.0
Sb	13.0	3.0
Operating Conditions	Concentrate Feed Rate 38.0 t/hr Off-gas Volume 500 Nm^3/min	Matte Feed Rate 17.5 t/hr Off-gas Volume 240 Nm^3/min

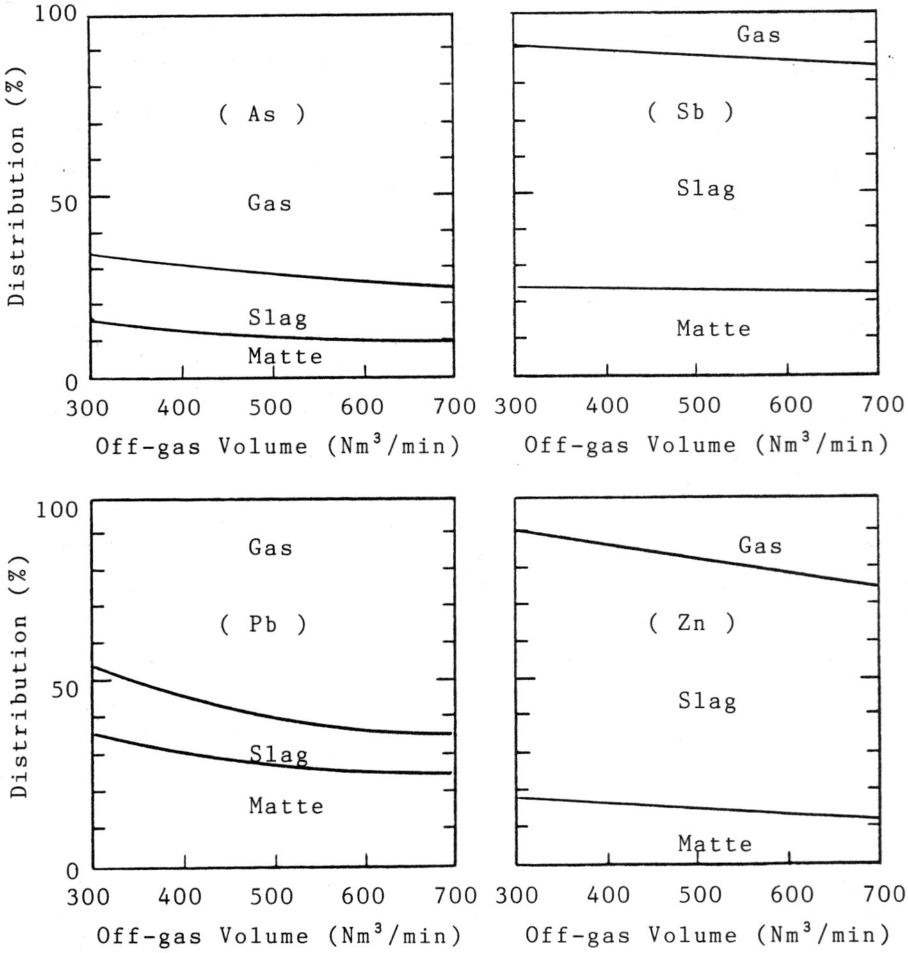

Fig. 7. Relation between distribution and off-gas volume of the smelting furnace. (at 1230 °C)

$$X = SM_{As}(2K1a_{As}{}^{2} + K2a_{As}P_{S_2}{}^{1/2} + K3a_{As}P_{O_2}{}^{1/2})V/RT \quad ....(2)$$

here a_{As} ; activity of As in liquid phase
P_{S_2}, P_{O_2} ; partial pressure of each species

Therefore transfer amounts of volatile matters into gas phase are dependent on volume of exhaust gas as well as oxygen potential and temperature. Since the activities of As in slag and matte is determined by actual operation data, the effect of the off-gas volume on the distribution of As into these phases can be calculated with equation (2). Other volatile matters can be considered in same way (Yanagida and others, 1983), and the calculation results are shown in Fig. 7.

The Reverberatory Furnace

Approximately 280 Nm^3/min of the lead sintering off-gas which originally contained only 12.4 % of oxygen is mixed with 38 m^3/min of oxygen in order to have same oxygen concentration as air and introduced into the reverberatory furnace as a secondary burner air. Detail on this oxygen enrichment is summarized in Table 6. Sensible heat of the lead sintering off-gas at 350 °C compensates a part of calorific value of burner fuel and coal consumption at burners of the reverberatory furnace is reduced by 90 t/month. As a result, the volume of the reverberatory furnace off-gas is reduced by 10 Nm^3/min. On the other hand, SO_2 concentration of the reverberatory furnace off-gas is increased from 2.1 to 3.1 % because of sulfur dioxide contained in the lead sintering off-gas at 4.3 %.

TABLE 6 Details of the Secondary Air for the Reverberatory Furnace

	Lead Sintering Off-gas (at E.P. Outlet)		Tonnage Oxygen (80 % O_2)		Balance Air		Total	
	Nm^3/min	(%)	Nm^3/min	(%)	Nm^3/min	(%)	Nm^3/min	(%)
O_2	30.5	(12.4)	30.6	(80.0)	6.5	(21.0)	67.6	(21.4)
N_2	199.5	(80.8)	7.7	(20.0)	24.6	(79.0)	231.8	(73.3)
SO_2	10.8	(4.4)					10.8	(3.4)
CO_2	6.0	(2.4)					6.0	(1.9)
Total Dry Gas	246.8	(100.0)	38.3	(100.0)	31.1	(100.0)	316.2	(100.0)
H_2O	33.2				0.6		33.8	
Total Wet Gas	280.0		38.3		31.7		350.0	
Temperature	350 °C						280 °C	

The Acid Plants

At the first stage of the expansion program, decreased total volume of exhaust gases enabled No.1 acid plant to be shut down. Therefore, whole power consumption of the No.1 acid plant, 800 MWH/month, was entirely saved. Furthermore, the capacity of No.1 acid plant became surplus which could be utilized for the expansion of the continuous process at the second stage of the expansion program.

On the other hand, increased SO_2 concentration in inlet gas of No.3 acid plant required the reinforcement of the heat exchanger for the converter, increase of the cooling capacity for the circulating acid in the absorption tower and an additional desulfurizing tower for tail gas. Therefore, the electricity consumption of No.3 acid plant has been increased by approximately 30 MWH/month since the expansion program. However, improved heat balance of the acid plant due to increased SO_2 concentration has reduced fuel oil consumption for deficit heat by 60 Kl/month.

The lead sintering off-gas introduced into reverberatory furnace is re-heated above 1200 °C and most of the hydro-carbon contained in the sintering off-gas is decomposed. This results in saving approximately 40 t/month of hydrogen peroxide as bleaching material.

Total Energy Consumption at the Smelter

At the first stage of the expansion program, decreased off-gas volume with more intensive oxygen utilization realized remarkable reduction of total energy consumption at the Naoshima Smelter. Since No.1 acid plant was shut down, its electric power consumption was entirely eliminated. Furthermore, consumptions of fuel oil and coal at the continuous process were reduced too. This modification, however, required 4,000 Nm^3/hr of additional oxygen, which consumed approximately 950 MWH/month more electricity. On the other hand, decreased steam generation at the waste heat boilers due to decreased exhaust gas volume of the continuous process reduced electric power recovery. After all, total energy consumption at the Naoshima Smelter was saved by 17,800 GJ/month, which corresponded to 16,140 PFE-GJ/month. These figures are summarized in table 7.

TABLE 7 Changes of Total Energy Consumption of the Naoshima Smelter at the First Stage of the Expansion Program

(GJ/month)

Process	Electric. Consum.	Electric. Recovery	Oil Consum.	Coal Consum.	Total
Oxygen Plant	+3,400				+ 3,400
Reverb. Furnace				-2,300	- 2,300
Continuous Process	-1,200	- 1,400	-13,800		-13,600
No.1 Acid Plant	-2,900				- 2,900
No.3 Acid Plant	+ 100		- 2,500		- 2,400
Total (GJ/month)	- 600	- 1,400	-16,300	-2,300	-17,800
(PFE-GJ/month)	-1,850	- 4,310	-16,300	-2,300	-16,140

CONCLUSION

1. The first use of oxygen for copper pyro metallurgy in Japan was reported in the 1950's. Since then, the oxygen utilization has become one of the most common techniques to increase the operation efficiency. Today, copper smelters in Japan have approximately 50 t/hr of oxygen production capacity in total.

2. The blister copper production capacity of the Mitsubishi Continuous Process in the Naoshima Smelter has been increased from 4,000 to more than 7,800 t/month without major reinforcements on acid plants by more intensive oxygen utilization.

3. Since the oxygen utilization efficiency of the continuous process is not affected by the oxygen concentration of blast air, changes of fuel requirement and exhaust gas volume due to the oxygen enrichment agree well with the theoretical calculations. Total energy consumption of the continuous process is saved by approximately 10 MJ accordingly to additional 1 Nm^3 of oxygen utilization.

4. Introduction of the lead sintering off-gas into the reverberatory furnace as a secondary burner air and more intensive oxygen enrichment of the continuous process reduces total energy consumption of the Naoshima Smelter by approximately 18,000 GJ/month.

ACKNOWLEDGEMENT

The authors wish to express their thanks to the management of Mitsubishi Metal Corporation for permission to publish this paper.

REFERENCES

Fujii, O. and Shima, M. (1982). Reaction shaft of Ashio Smelter's flash furnace utilizing high oxygen enriched air. In D. B. George and J. C. Taylor (Ed.), Copper Smelting - An Update -. The Metallurgical Society of AIME, New York. pp. 165-171.

Fujii, T., Ando, M. and Fujiwara, Y. (1972). Copper smelting by flash furnace at Saganoseki Smelter and Refinery. Proc. MMIJ-AIME Joint Meeting (Tokyo), T4d5.

Fujiwara, Y., Okura, T. and Maeda, M. (1979). Converter operation progress at Saganoseki Smelter. In R. E. Johnson (Ed.), Copper and Nickel Converters. The Metallurgical Society of AIME, New York. pp. 144-166.

Goto, M. (1976). Green-charge reverberatory furnace practice at Onahama Smelter. In J. C. Yannopoulos and J. C. Agarwal (Ed.), Extractive Metallurgy of Copper, Vol. 1. The Metallurgical Society of AIME, New York. pp. 154-167.

Goto, M. and Echigoya, T. (1980). Refractory practice and application of water jackets in Mitsubishi Process. Proc. AIME (Las Vegas), A80-19.

Goto, M., Kikumoto, N. and Igarashi, T. (1980). Utilization of coal and oxygen in Mitsubishi Continuous Process. Proc. MMIJ-AIME Joint Meeting (Tokyo),D-2-2.

Goto, M. and Kikumoto, N. (1981). Process analysis of Mitsubishi Continuous Copper Smelting and Converting Process. Proc. AIME (Chicago), A81-60.

Goto, M. and Kikumoto, N. (1984). Intensive operation of the Mitsubishi Process. In M. J. Jones and P. Gill (Ed.),Mineral Processing and Extractive Metallurgy. The Institution of Mining and Metallurgy, London. pp. 325-334.

Goto, M., Kawakita, S., Kikumoto, N. and Iida, O. (1986). Improvements in high intensity operation of the Mitsubishi Process at Naoshima Smelter. Proc. AIME (New Orleans), A86-1.

Hashiuchi, M., Okada, S. and Watanabe, T. (1982). Review of current operation at Tamano Smelter. In D. B. George and J. C. Taylor (Ed.), Copper Smelting - An Update -. The Metallurgical Society of AIME, New York. pp. 237-250.

Kawakita, S., Saito, M., Ohshima, E., Kamio, S and Hirano, M. (1985). Utilization of the lead sintering off-gas as the burner air for the copper reverberatory furnace. Proc. AIME (New York), A85-25.

Kellogg, H. H.(1974). Energy efficiency in the age of scarcity. J. Met., 26,25-29.

Kimura, T., Tsuyuguchi, S. and Ishii, Y. (1985). Developement of top blown Peirce-Smith converter. Proc. AIME (New York), A85-31.

Kitamura, T. and Shibata, T. (1973). Flash Smelting at Tamano Smelter, Hibi Kyodo Smelting Co. Proc. AIME (Chicago), A73-49.

Kohno, H., Asao, H. and Amano, T. (1982). Use of alternative fuel at Onahama. Proc. AIME (Dallas), A82-40.

Morisaki, T. (1966). Construction of Onahama copper plant. J. MMIJ, 82, 588-601.

Nagano, T. and Suzuki, T. (1976). Commercial operation of Mitsubishi Continuous Copper Smelting and Converting Process. In J. C. Yannopoulos and J. C. Agarwal (Ed.), Extractive Metallurgy of Copper, Vol. 1. The Metallurgical Society of AIME, New York. pp. 439-457.

Ogura, T., Fukushima, K. and Kimura, S. (1974). Process control with computer in Toyo Copper Smelter. Proc. AIME (Dallas), A74-3.

Ohshima, E. and Hayashi, M. (1986). Impurity behavior in the Mitsubishi Continuous Process. Metallurgical Review of MMIJ, 3, pp. 113-129.

Okada, S., Miyake, M., Hara, A. and Uekawa, M. (1983). Recent improvement at Tamano Smelter. In H. Y. Sohn, D. B. George and A. D. Zunkel (Ed.), Advances in Sulfide Smelting, Vol. 2. The Metallurgical Society of AIME, New York. pp. 855-874.

Okazoe, T., Kato, T. and Murao, K. (1967). The development of flash smelting process at Ashio Copper Smelter, Furukawa Mining Co., Ltd. In J. N. Anderson and P. E. Queneau (Ed.), Pyrometallurgical Process in Nonferrous Metallurgy. The Metallurgical Society of AIME, New York. pp. 175-195.

Suzuki, T. and Nagano, T. (1972). Development of new continuous copper smelting process. Proc. MMIJ-AIME Joint Meeting (Tokyo), T4e4.

Suzuki, T., Yanagida, T., Goto, M., Kawakita, S., Echigoya, T. and Kikumoto, N. (1983). Test operation for smelting more tonnages of copper concentrates at the Mitsubishi Continuous Copper Smelting and Converting Process. Proc. AIME (Atlanta), A83-34.

Tomono, M., Shibata, Y., Moriyama, K. and Kitamura, O. (1984). J.MMIJ,100,364-367.

Tsurumoto, T. (1961). Copper smelting in the converter. J.Met., 13, 820-824.

Tsurumoto, T. (1967). Improvements on the Oxygen Smelting Process at Hitachi Smelter. In J. N. Anderson and P. E. Queneau (Ed.), Pyrometallurgical Process in Nonferrous Metallurgy. The Metallurgical Society of AIME, New York. pp. 291 -305.

Yanagida, T., Kawakita, S., Kikumoto, N. and Hayashi, M. (1983). Lead and zinc behavior at S-furnace of the Mitsubishi Process and P.S. converter - Thermodynamic predications and practice. Proc. MMIJ-Aus.IMM Joint Symposium (Sendai), pp. 83-104.

Yasuda, M. (1974). Recent Developments of Copper Smelting at Hitachi Smelter. Proc. AIME (Dallas), A74-8.

OXYGEN SMELTING AND THE OLYMPIC DAM PROJECT

T.J.A. Smith*, I. Posener** and C.J. Williams***

* Seltrust Engineering Ltd., BP Minerals International Ltd., London
** BP Minerals International Ltd., London
*** Western Mining Corporation (UK), London

ABSTRACT

The metallurgical plant for the treatment of the massive copper-uranium-gold deposit at Olympic Dam in South Australia is currently under construction. This joint venture operation between Western Mining Corporation and the British Petroleum Group, incorporates a number of novel features which relate to the unusual nature of the mineral deposit. This paper discusses the copper smelting technology selected.

The need to cope with particularly high grade concentrates of low intrinsic energy value provides a unique opportunity for the application of high level oxygen enrichment in one-step flash smelting direct to blister.

The projected change, over the life of the mine, to higher tonnages of more conventional chalcopyritic feedstock presents a challenge for the primary smelting technology selected for initial operations. The paper discusses steps taken to develop a smelting strategy to meet these requirements. The relevance of recent developments in oxygen flash-converting to this strategy is also discussed.

KEYWORDS

Copper, flash smelting, direct blister, oxygen, one step smelting, flash converting.

INTRODUCTION

The Olympic Dam copper-uranium-gold deposit was discovered by Western Mining Corporation Ltd. (WMC) in 1975. Following initial exploration by WMC, a Joint Venture was formed by WMC and the British Petroleum Group (BP) in 1979, with the project being managed on behalf of the joint venturers by Roxby Management Services Pty Ltd. (RMS), a wholly owned subsidiary of WMC.

Extensive surface drilling, followed by more detailed drilling from underground development, has identified a resource of 2000 million tonnes containing 1.6% copper, 0.6 kg/t uranium oxide and 0.6 g/t gold. Within that resource a probable ore reserve of 450 million tonnes containing 2.5% Cu, 0.8 kg/t uranium oxide and 0.6 g/t gold, has been outlined.

Two main types of sulphide mineralisation exist. The ore to be mined in the initial stages of the operation is composed primarily of bornite with some chalcocite and chalcopyrite. In the later stages of the project the ore is primarily chalcopyrite with some bornite and minor chalcocite and pyrite.

At the time of writing, the metallurgical plant is in the early stages of construction. Detailed design is not yet complete and therefore some of the data referred to in this paper is preliminary.

The flowsheet for the initial phase of the Olympic Dam project is shown in Fig. 1 and involves:-

- Milling and flotation of a sulphide copper concentrate; direct smelting to a high sulphur blister copper; refining to anode copper and electrorefining to copper cathode in on on-site refinery.

- Recovery of uranium from the flotation tailings in a conventional acid leach, solvent extraction plant

COPPER CONCENTRATE

In considering treatment routes for the copper circuit, both the analysis of the copper concentrate and required scale of operation are relevant.

Concentrate Analysis

The ore to be treated in the initial years is predominantly bornitic whilst in later years the ore will be predominantly chalcopyritic.

In addition to differences in the primary copper minerals, the close association of copper and uranium mineralisation influences mineral separation requirements and hence final concentrate composition. In primary flotation, whilst the majority of the uranium reports to the tailings, some 10-15% reports to the copper concentrate. This uranium is mainly associated with entrained haematite which is the major gangue mineral.

Since revenue is derived from both copper and uranium (and to a lesser extent gold) early flowsheet development needed to consider recovery of this additional uranium from the copper concentrate.

Following testwork a further concentrate treatment step was developed enabling improved separation of uranium for subsequent recovery. This step involves cleaning of the rougher concentrate, regrinding to 80% passing 25 to 30 microns, recleaning, acid leaching and reflotation. Besides the leaching of uranium from the copper concentrate the majority of the haematite gangue is also rejected. This results in a final concentrate which is significantly upgraded, being a relatively pure copper sulphide mineral with low gangue content. This effect and the differences in mineralisation for initial and later years are illustrated by the indicative analyses for final concentrates given in Table 1 below.

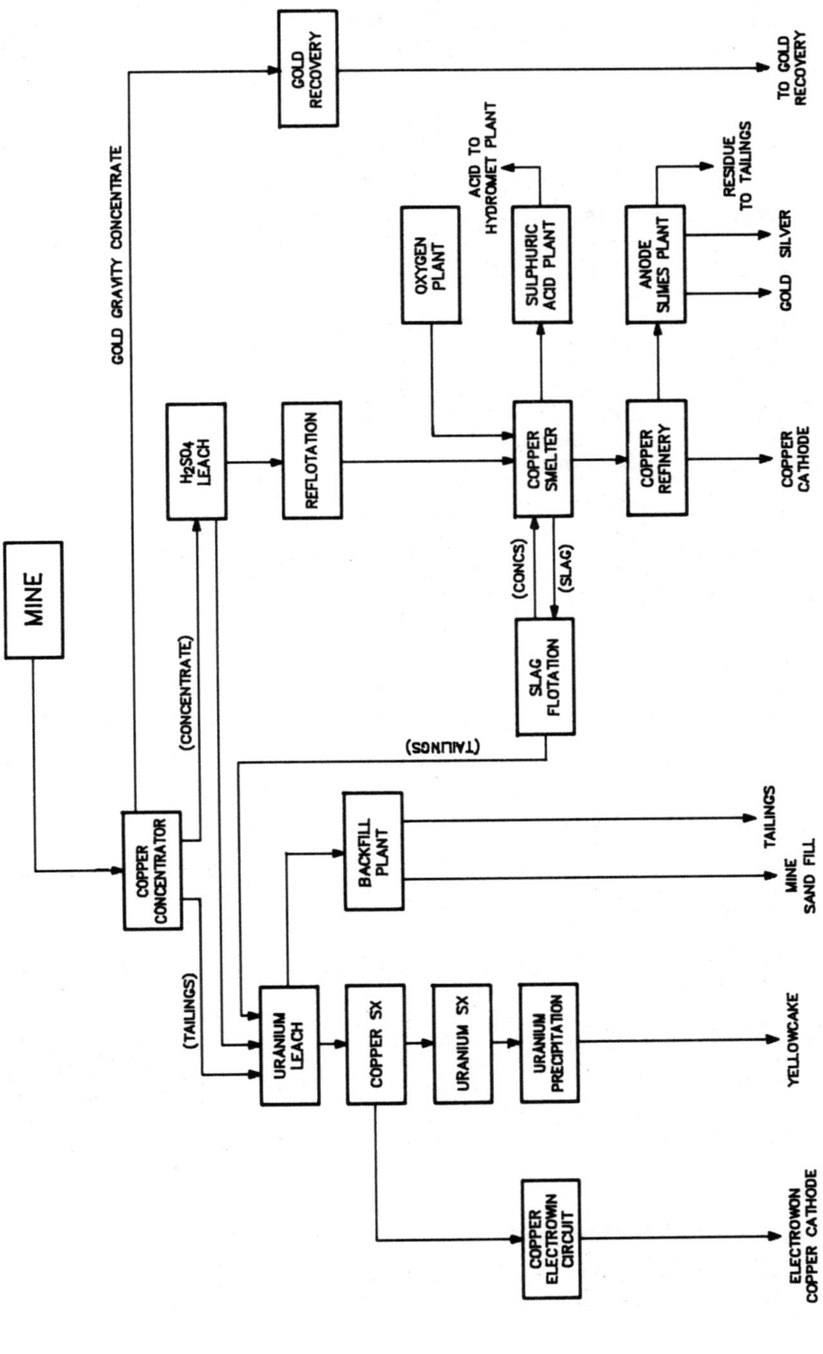

Fig. 1 OLYMPIC DAM GENERAL METALLURGICAL FLOWSHEET

TABLE 1 Indicative Concentrate Analyses - Weight %

	Initial Years	Later Years
Cu	56	28-45
Total Fe	14	25-35
S	24	25-30
Fe_2O_3	3	10-20
Other	3	

Of relevance to the smelting process selection is the high initial copper grade and the increase in iron, sulphur and gangue content in later years.

Scale of Operation

The Joint Venture Agreement and the Indenture Agreement with the State of South Australia envisaged an ultimate production rate of 150 000 tonnes/year of copper. Initially a smaller operation, in the region of 30 000 to 55 000 tonnes/year of copper, is being built, based on the high grade material available in the areas to be mined first.

Both changing mineralogy and scale of operation were key factors in the selection of processing routes.

SELECTION OF SMELTING TECHNOLOGY

Early studies had assessed hydrometallurgical routes for copper recovery but, as further information was obtained, these were rejected in favour of smelting.

A number of alternative established smelting practices were studied, both for the initial and ultimate production rates. Smelting rates in the range of 30 000 to 150 000 tonnes/year were evaluated, taking account of the higher grade of the initial concentrate.

The smelting technologies studied included:

Top Blown Rotary Converting
Electric Smelting
Noranda Continuous Smelting
Mitsubishi Continuous Smelting
Inco Flash Smelting
Outokumpu Flash Smelting

During the earlier studies, the additional concentrate treatment step for uranium recovery, had not been developed. Consequently uranium in copper concentrate would report to smelter slags and was to be recovered by pressure leaching of these slags. The ability to recover uranium from slags was a factor in comparison of smelting routes.

A further major consideration was the ability of the primary smelting unit to accommodate turndown for the initial operation with the added complication of significantly higher grade concentrate. This presented difficulties for most of the technologies considered and it was necessary in certain cases to contemplate a change in smelting technology between initial operating and ultimate large-scale operation. It was, however, becoming evident that direct blister smelting of initial concentrate was one potential option.

At the completion of these studies at the Pre-Feasibility level, Outokumpu Flash Smelting was identified as the favoured route. The nature of the Olympic Dam orebody with complex and changing mineralisation in a haematite matrix gives rise not only to a series of process steps but to a highly integrated overall process route. It was considered prudent to verify and optimise as much of the process route as possible by pilot plant testwork. A pilot plant was constructed at the Olympic Dam mine site to confirm the proposed uranium circuit and to produce copper concentrate for smelting trials. The ability to test the proposed smelting route at an appropriate scale was considered an important factor in the selection of smelting technology.

Smelting trials were carried out at the Outokumpu Metallurgical Research Centre in Pori, Finland. Both smelting to matte and direct to blister were evaluated with different concentrates produced at the Olympic Dam Pilot Plant. During this period the additional concentrate treatment step was developed at the Olympic Dam pilot plant.

The smelting pilot plant work is discussed in detail in a paper to be presented in London, UK during September 1987 (Asteljoki and Muller).

Based on the results of this testwork it was established that direct blister smelting of the initial concentrate in the Outokumpu Flash Furnace was feasible. Furthermore, smelting to matte resulted in production of very high matte grades (close to white metal) with the likely co-production of some blister copper under normal furnace operating conditions.

A further period of study followed in which the testwork results were fully evaluated together with additional information on the new concentrate treatment step. Other smelting strategies and the requirements of high grade matte converting were also investigated.

It was concluded that direct blister flash smelting offered the best technical and economic solution to the particular requirements of the initial processing of copper concentrates at Olympic Dam. This process also offered the best potential for a logical, staged expansion into the higher tonnage, chalcopyritic feed envisaged in later years.

INITIAL OPERATION - DIRECT BLISTER OXYGEN FLASH SMELTING

In selecting direct blister oxygen flash smelting for initial operations at Olympic Dam three key requirements for such an operation are satisfied by the initial concentrate. These requirements are:

Concentrate With a Relatively Low Intrinsic Energy Value.

Such concentrate permits use of high level oxygen enrichment and is a function both of the high copper/sulphur ratio and low sulphide iron content of the initial bornitic concentrate. A high level of oxygen enrichment (i.e. high oxygen partial pressure in the Reaction Shaft) is necessary to ensure a satisfactory operation in the blister smelting mode. Typical chalcopyrite copper concentrates, when oxidised directly to blister, generate reaction heat in excess of that required for autogenous operation, even at low levels of oxygen enrichment. The inability to dissipate this additional heat prevents or complicates the use of higher levels of oxygen enrichment.

Low Slag Production per Unit of Concentrate Smelted.

The low iron and low gangue content of the initial concentrate yield a lesser slag fall when compared with more conventional concentrate. The high copper content slag produced is treated by slow cooling, crushing, milling and flotation to produce a slag concentrate for recycle. A lower slag fall, relative to higher iron/gangue content concentrates, thus generates less slag concentrate for recycle. Since this slag concentrate displaces new feed, any reduction in the quantity of recycled slag concentrate produced increases the furnaces' primary smelting capacity. A more efficient smelting unit is therefore feasible with the initial Olympic Dam concentrate.

Clean Concentrate.

In the case of Olympic Dam, the initial concentrates are clean resulting in low levels of impurity elements in final blister. High levels of certain impurity elements (e.g. Arsenic, Antimony and Bismuth) tend to be undesirable for one step direct blister smelting, because their elimination is generally lower than by conventional smelting routes and consequently, unacceptably high levels may report to the product blister.

Testwork confirmed the feasibility of a direct blister operation at high levels of oxygen enrichment. Further confidence in the operation of a flash furnace at such high levels of oxygen enrichment was derived from the operation of the existing Outokumpu Harjavalta smelter copper flash furnace (in matte mode) at similar oxygen enrichment levels.

Subsequent to the Pori smelting trials for Olympic Dam concentrate, major pilot smelting work was undertaken by Outokumpu for Kennecott on the process today known as Kennecott-Outokumpu Flash Converting. This process was initially developed by Kennecott (K.J. Richards, D.B. George and L.K. Bailey, 1983) and later pilot tested by Outokumpu. (Asteljoki and co-workers 1985). The Pori trials into the application of this oxygen flash converting process to high grade copper matte gave encouraging results.

In chemical terms, particularly with the lower haematite levels, the initial Olympic Dam concentrate resembles a high grade matte, differing primarily in its physical characteristics. It has fine particle size and an active surface due to leaching. The successful Kennecott-Outokumpu flash converting trials on both granulated and milled matte, therefore gave added confidence in the technical feasibility of direct flash smelting of Olympic Dam initial concentrates.

The adoption of direct flash smelting technology with high level oxygen enrichment permitted a significant reduction in furnace and gas train dimensions with consequent savings in capital cost. Overall this process route showed the best economics, particularly in terms of reduced capital investment when compared to other alternatives.

LATER OPERATION - OXYGEN FLASH CONVERTING

As indicated it was considered that by selection of direct blister oxygen flash smelting for the initial operation, potential now exists to provide a logical transition to the later, higher tonnage operation with chalcopyritic material.

Continued operation of the initial furnace in direct blister mode would not be feasible with the chalcopyritic feedstock due to:

- Higher intrinsic energy value in concentrate resulting in excessive energy release in the Reaction Shaft (both due to more Fe/S oxidation and higher throughput). A lower oxygen enrichment and hence larger furnace and gas train would be necessary with increase in cost.

- Substantially increased slag fall due to the higher iron content. This results in increased slag for retreatment and increased furnace size for Settler residence time requirements, both of which increase cost.

For the above reasons the initial compact furnace designed for nominally 55 000 tonne/year copper from bornitic feed is unable to satisfy later primary smelting requirements. A new primary smelting unit will therefore be required at that stage.

The proposed strategy for the later stages of the operation at Olympic Dam developed logically from these observations. By the installation of a second, larger flash furnace, oxygen flash smelting of the chalcopyritic concentrate to a high grade matte could be carried out. This solidified matte could then be treated in the initial blister furnace which would act as an oxygen flash converter.

Besides the significant economic advantages of adopting this staged approach to expansion, a number of additional benefits are apparent, the most important of which include:

- the ability to continue utilisation of the initially installed plant

- no major change in the principal smelting technology - easier operator acceptance and minimal retraining requirements

- similar design requirements for both plants

- steady gas streams suitable for acid production from both smelting units

FURNACE DESIGN CONSIDERATIONS

A number of factors required consideration when sizing the initial furnace unit in order to satisfy both the initial direct blister smelting and the later flash converting requirements.

The most important of these are outlined below:

Molten Product Holding and Residence Time

Adequate Settler capacity is required to hold molten copper prior to the subsequent fire-refining step. In addition, sufficient residence time is necessary to permit completion of slag reactions and separation of phases. This latter requirement is critical. When considering this factor alone, the low slag fall with the bornitic, low gangue concentrates, indicates that an extremely compact furnace Settler is appropriate.

In the case of the Olympic Dam furnace the diameter of the Reaction Shaft dictates the width of the Settler. Aspect Ratio requirements (related to dust carryover) dictate the distance between Reaction shaft and Uptake relative to Settler freeboard. Thus, whilst for the initial operation the slag residence time requirements would permit design of a smaller Settler, this is not practical since the gas-related Reaction Shaft diameter and Aspect Ratio are the determining factors.

To give an indication of some preliminary data, the initial concentrate treatment rate (approx. 12.6 tonnes/hour) gives rise to an average slag production in the region of 7 tonnes/hour (or 10 tonnes/hour for the maximum duty design case). This relates to a 14% Fe content in concentrate. Copper production averages 7 tonnes/hour.

In the projected future flash converting operation, a matte grade of above 70% copper will be achieved in the new primary smelting unit (bearing in mind that this grade can be selected). Flexibility of the operation permits this choice.

Iron content in such matte will be in the region of 5% or approximately one-third of that for the initial concentrate. This is illustrated in Table 2 below by considering the primary feeds (excluding slag concentrate). Even though the copper production will increase by a factor of three, the average total slag production remains broadly similar. As a consequence, as far as molten products are concerned, copper tapping to the batch fire-refining operation will be three times as frequent; slag residence time, however, remains similar to that of the initial operation.

TABLE 2 Comparison of Primary Feeds to the Flash Furnace

		Initial Concentrate	High Grade Matte
Copper Feed Rate (t/y)		55 000	150 000
Concentrates (Matte) Feed Rate (t/y)		100 000	208 000
Analysis Wt %	Cu	56	72
	Total Fe	14	5
	S	24	21
	Other	6	2
Fe Content (t/y)		14 000	10 400
Sulphide Fe (t/y)		11 900	10 400
Sulphur Content (t/y)		24 000	43 700

Gas volume considerations. The initial direct blister smelting operation incorporates design which envisages potential operation with Reaction Shaft oxygen enrichment up to commercial oxygen purity levels (approx. 95%). In order to accommodate the necessary extra degree of freedom in operational control, the furnace and gas train are sized on a design basis of 75% oxygen enrichment.

Furthermore, the design is such that even at the highest enrichment level of 95% oxygen, some fuel combustion is necessary to satisfy heat balance requirements. Consequently, with the small initial throughput, a significant proportion of the offgas volume is derived from combustion products. The initial average gas volume to the waste heat boiler is in the range 5 to 6 000 Nm3/h of which, at 95% oxygen enrichment, approximately one half is process gas (primarily sulphur dioxide).

Under future flash converting conditions, the objective will be to produce a high grade matte, such that the flash converting furnace can operate autogenously.

Referring again to Table 2, it can be seen that when throughput and analysis are considered, sulphide iron and sulphur to be oxidised decrease by 13% and increase by 82% respectively. The net effect of this on the heat balance is that, even with the threefold increase in copper throughput, autogenous operation in the flash converting of matte can be achieved at an oxygen enrichment level of the order of 95% (i.e. commercial purity). Process gas (primarily sulphur dioxide) forms the offgas, and the volume to the waste heat boiler is again in the region of 5-6000 Nm3/h (i.e. approximately twice the mass of sulphur oxidised compared with the initial concentrate).

From consideration of the molten product and gas volume parameters discussed above, it can be seen that the same furnace can be designed to satisfy both the initial flash smelting and the later flash converting duties.

Furnace Construction

The furnace for Olympic Dam is an extremely compact unit, primarily because of the use of high level oxygen enrichment. The projected flash converting duty will require the installation of a new furnace to produce high grade matte as a feed. It is nevertheless of interest to compare the size of the Olympic Dam furnace with existing Outokumpu matte smelting Flash Furnaces in the same capacity range of 55 000 to 150 000 tonnes/year capacity.

Rio Tinto Minera operate a matte flash smelting furnace at Huelva in Spain which, with a shortened Reaction Shaft and oxygen enrichment, has a nominal capacity of 150 000 tonne/year of copper. This capacity compares with the previous capacity of 100 000 tonnes/year of copper without oxygen enrichment and a 9.8 x 6.5m Reaction Shaft. (De la Villa, Cevallos and Barrios 1986).

The Onsan matte flash smelting furnace in Korea also operates at a nominal capacity of 100 000 tonnes/year of copper with oxygen enrichment.

Figure 2 provides a size comparison of the Olympic Dam furnace with these two furnaces. The benefits of design for high oxygen enrichment are apparent even when allowing for the lower initial throughput, some of which are discussed in a recent paper. (Rodolff, Anjala and Hanniala 1986).

The furnace engineering was carried out by Outokumpu Oy and incorporates a number of features representative of current design concepts.

Some of the main features of the furnace and associated plant are briefly listed below:-

- Single concentrate burner
- Auxiliary heating by diesel burners; roof-mounted, short-flame, oxy-fuel Settler burners
- Water-cooled, copper element Reaction Shaft based on Kalgoorlie nickel smelter (WMC) design
- Copper cooling elements throughout Settler and critical Uptake areas
- Uptake offset relative to furnace centreline for reduced dust carryover
- Gas cooling by Ahlstrom waste heat boiler
- Blister copper laundered directly to one of two rotary anode furnaces

CONCLUSIONS

A smelting process, a furnace design and a strategy for the future for the Olympic Dam Project have been established.

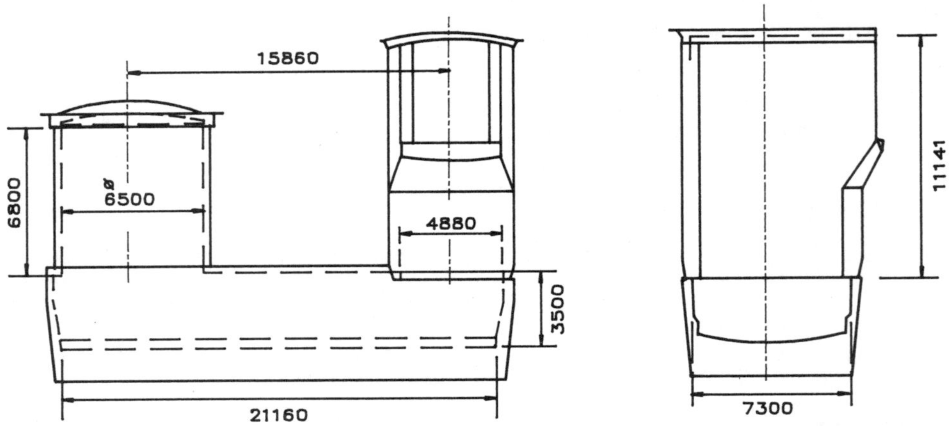

Rio Tinto Minera — 100 000 tonnes/year copper without oxygen
Matte Flash Smelting 150 000 tonnes/year copper with oxygen

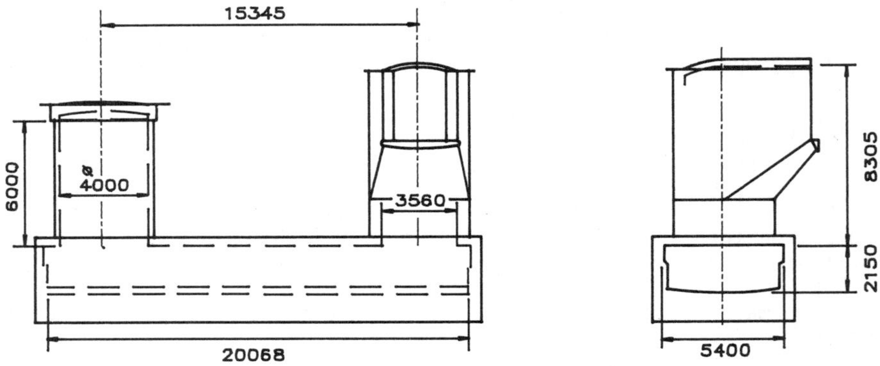

Onsan — 100 000 tonnes/year copper with oxygen
Matte Flash Smelting

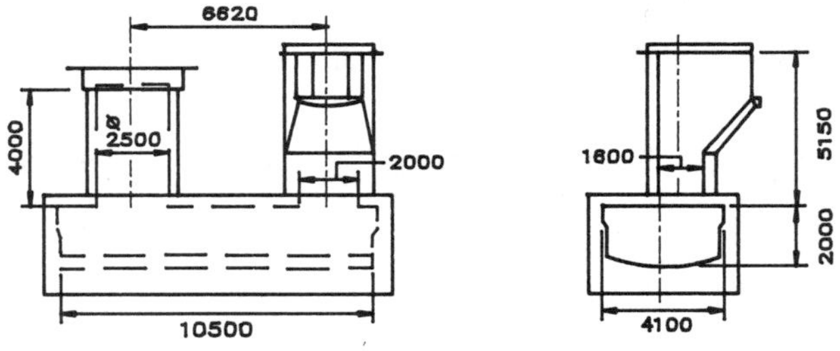

Olympic Dam
Direct Blister Flash Smelting — 55 000 tonnes/year copper with oxygen
Flash Converting — 150 000 tonnes/year copper with oxygen

Fig. 2 Comparison of Flash Smelting Furnaces

(all dimensions in mm.)

The Olympic Dam smelter, as with the other O.D. metallurgical process plants, is the result of the combined efforts of many metallurgists, engineers, operators and consultants from the Joint Venture partners and outside organisations. This combined input has been co-ordinated and added to by the RMS metallurgical and engineering personnel.

The outcome is felt to be a practically orientated plant and process design, which use "state of the art" technology to cater for an unusual smelting problem.

The smelter presently being constructed will:

- provide an effective technical solution to both initial and future smelting needs

- minimise both initial and future capital and operating costs

- take full advantage of the opportunities presented by the unique nature of smelter feed

- utilise direct blister smelting and then the option of solid matte flash converting.

Of particular relevance to this symposium is the realisation that the smelter design being utilised is only workable with the use of high levels of oxygen enrichment

ACKNOWLEDGEMENTS

The authors would like to express their gratitude to the management of Western Mining Corporation, The British Petroleum Group and Outokumpu Oy for permission to publish this paper.

REFERENCES

Asteljoki, J.A. and S.M.I. Kyto (1985). Alternatives for direct blister copper production. 1985 AIME Annual Meeting, New York February 1985.

Asteljoki, J.A., L.K. Bailey, D.B. George and D.W. Rodolff (1985). Flash converting - continuous converting of copper mattes. J. Metals, May 1985, 20-23.

Asteljoki, J.A., and H.B. Muller. Direct smelting of blister copper - pilot flash smelting tests of Olympic Dam concentrate. To be presented at the Institution of Mining and Metallurgy meeting Pyrometallurgy '87, London, September 1987.

De la Villa, D., F. Cevallos and P. Barrios (1986). Recent innovations at the Huelva Smelter. 1986 Annual meeting of AIME, New Orleans, March 2-6 1986.

Richards, K.J., D.B. George and L.K. Bailey (1983). A new continuous copper converting process. In H.Y. Sohn, D.B. George and A.D. Zunkel (eds.) Advances in Sulfide Smelting, proceedings of 1983 Fall Meeting of TMS-AIME, San Francisco, November 1983, 489-498.

Rodolff, D.W., Y.J. Anjala and P.T. Hanniala (1986). Review of flash smelting and flash converting technology. 1986 AIME Annual meeting, New Orleans, March 1986.

OXYGEN ENRICHMENT FOR ACCELERATED SMELT-REDUCTION OF NICKEL OXIDE

G.F.Garrido, L.Lauzon and R.G.H.Lee
Research and Technology Department
Canadian Liquid Air Ltd
Montreal, Quebec. Canada

ABSTRACT

Smelt-reduction of nickel oxide sinter by petroleum coke was carried out in a 185 tonne reverberatory furnace. Fuel oil burning rate was limited by potential damage to the refractory roof above 1650°C. The actual heat cycle using standard air/oil burners averaged 77 hours.

A demonstration program was carried out to evaluate the benefits of increasing heat transfer from the flame to the carbon-saturated molten nickel bath. Using selective oxygen flame enrichment, with molecular oxygen being lanced between the flame and the molten bath while maintaining an overall reducing flame, shorter heat cycles averaging 56 hours were attained without increasing roof temperature. Results are compared with pre-mixing molecular oxygen with the combustion air. Oxygen supply interlocks were designed with special attention to safety.

Techniques involving oxygen-enriched reduction processes in non-ferrous metallurgy offer significant savings in operating cost by decreasing unit fuel consumption and offer the operators an extra degree of freedom to optimize the smelt-reduction cycle.

KEYWORDS

Nickel oxide; nickel refining; non-ferrous pyrometallurgy; reverberatory furnace; oxygen enrichment; flame geometry; reduction kinetics.

INTRODUCTION

Nickel oxide sinter resulting from fluidized-bed roasting of sulphide concentrates at Inco's Copper Cliff refinery was charged to a reverberatory-type furnace to be smelted to blister metal using petroleum coke as a reductant. The metallic content of the charge was approximately 94% Ni, 3-4% Cu, 1-2% Co. Fe, Pb, As and other metals each occurred in concentrations of less than 1%.

Reduction takes place at temperatures between 1430°C and 1650°C. Heat was supplied by seven air/oil burners installed through the front wall of the furnace, to establish a lengthwise flow of combustion gases. The gas mixture then passes through a waste heat boiler and subsequently to an air preheater before being directed to the stack. A schematic of the furnace is shown in Fig. 1.

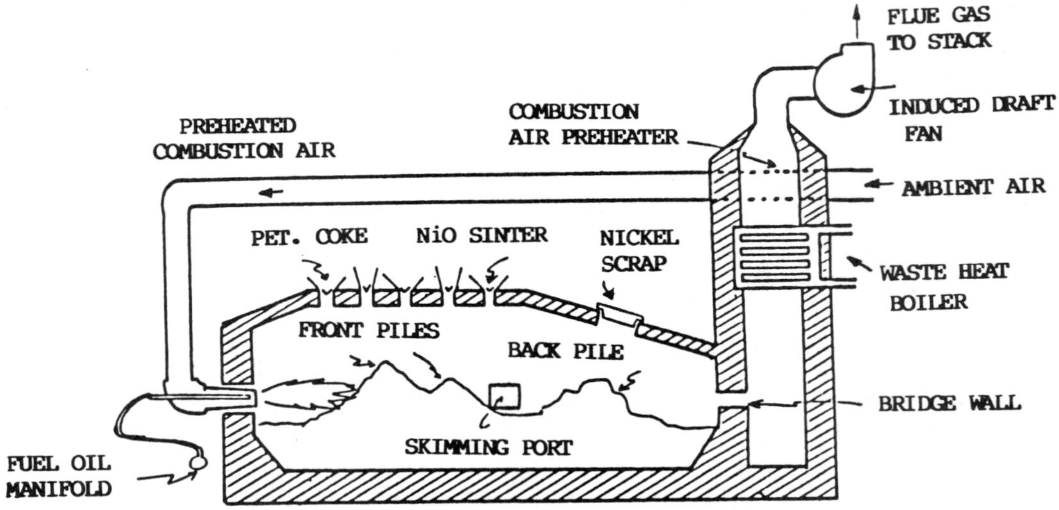

FIGURE 1. SCHEMATIC DIAGRAM OF INCO's #5 ANODE FURNACE

The Port Colborne Nickel Refinery of INCO Metals Ltd required a 10% increase in the smelting rate of Anode Furnace #5. To achieve this, it was decided to use oxygen enrichment.

Two methods of introducing oxygen into the furnace were planned:

(a) Phase I - Enrichment of the air/oil flame by selectively injecting molecular oxygen into the flame quadrant directly over the molten metal bath.

(b) Phase II - Pre-mixing molecular oxygen in the air blast at the main header supplying pre-heated primary air to the seven oil burners.

This paper emphasizes Phase I, the selective oxygen flame enrichment part of the program on Anode Furnace #5.

GROUNDS FOR SELECTIVE OXYGEN FLAME ENRICHMENT

Flame temperature increase by oxygen enrichment is well established in combustion technology [1-3]. Introduction of molecular oxygen can be done either by diffusing it into the air duct upstream the burner (pre-mixing) or selectively by means of a lance discharging molecular oxygen preferentially towards a predetermined zone of the flame.

Furnace #5 was limited by three main factors:

1 - Roof Temperature - The silica refractory must not exceed 1650°C.
2 - The Induced-Draft Fan was operating at maximum capacity.
3 - High Dust Entrainment Rate into the waste heat boiler and the air preheater tubes necessitated frequent shut-downs for cleaning.

The simplest procedure, pre-mixing the oxygen with air, creates a hotter and shorter flame which would result in roof temperatures higher than 1650°C unless the oil flowrate was reduced. Oxygen lancing, on the contrary, allows "tailoring" the flame geometry, so that the hotter region of the flame is closer to the bath, thus prevents overheating of the furnace roof.

Molecular oxygen is equivalent to a fivefold volume of combustion air. Hence, it alleviates the loaded induced-draft fan while decreasing dust carryover by the lower gas velocity through the furnace at equivalent smelting rates.

FURNACE CYCLE OPERATION

Measured weights of nickel oxide sinter and coke are pre-mixed and fed to the furnace from roof-mounted hoppers in 3 to 5 charges or "drops"per cycle.
A new charge was dropped only when the previous portion had been melted to a point where solid charge (or pile) did not obstruct the flow of molten nickel from the burners area (highest temperature) to the back of the furnace (lowest temperature). Spent anodes plus variable amounts of nickel scrap were also recycled into the furnace along with the first drop. When all the material inside the furnace was in a liquid state, the "melt-down" period had ended. A viscous slag layer covered the molten nickel bath. After the slag had been skimmed, the target carbon content (0.02%) must be reached by either adding coke when the bath was carbon deficient or charging extra nickel oxide when the bath contained excess carbon. This phase is referred to as the "refining" period. When the temperature of the bath was 1540°C, the furnace was tapped and the metal was cast into nickel anodes for the electrolytic refinery. A normal heat cycle is illustrated in Fig. 2, where the instantaneous oil flowrate (firing rate) is plotted against time.

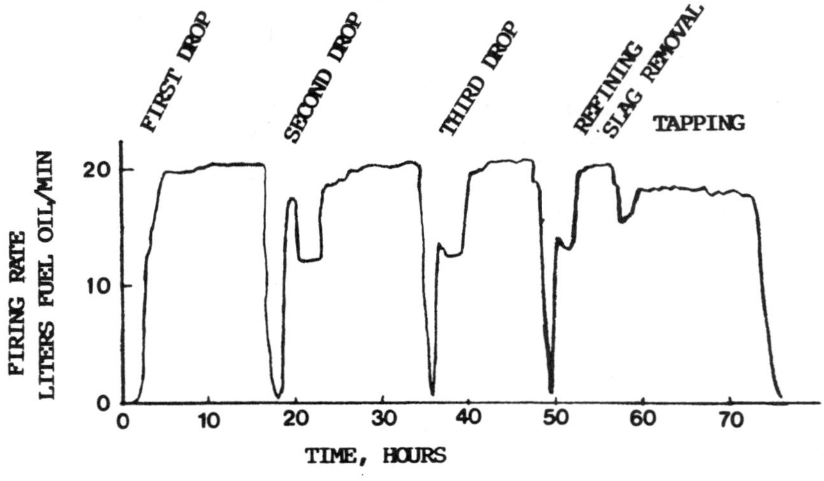

FIGURE 2. NORMAL FURNACE CYCLE BASED ON FUEL OIL FLOWRATE

MONITORING OF THE OXYGEN PROGRAM

To properly evaluate the effect of oxygen enrichment on the furnace cycle, it was necessary to establish a baseline for the normal operation . Existing instruments were recalibrated and control interlocks were installed to integrate the instrumentation needed to monitor the safe supply of oxygen.

Fuel Oil (Bunker #5)

Total oil flowrate was measured continuously and was shown in a circular chart/integrator recorder. Oil at about 80°C, was distributed to the seven burners via a 4.5 m long insulated manifold. Calibrated valves to each of the oil nozzles allowed independent adjustment.

Combustion Air Flow

Air preheated to 250°C was distributed to each of the seven burners from a main header. A water column manometer was used to monitor the total air flow. For the test period a Pitot tube was installed to improve the accuracy of the reading.

Roof Temperature

An infrared pyrometer, aimed at the roof approximately 3 m from the burners, was used to measure the refractory temperature. An alarm was set off when the roof temperature reached 1660°C.

Back-End Gas Composition

Excess O_2 and carbon monoxide were periodically analysed. A water-cooled tube inserted through the refractory near the bridgewall allowed aspiration of gas samples. CO_2 was measured on a number of occasions by means of an ORSAT analyser.

Molecular Oxygen Injection

A liquid oxygen storage tank was installed. Gasification was carried out with a steam vapourizer. A 3" dia. pipe carried the oxygen to the furnace area where it was distributed to seven lances, each of which was independently controlled and interlocked to the air/oil system. Oxygen flowrate through each lance was measured with a calibrated nozzle.

Figure 3 is a sketch of a lance positioned in the front wall of the furnace. Swivelling supports permitted the lance position to be altered during the test.

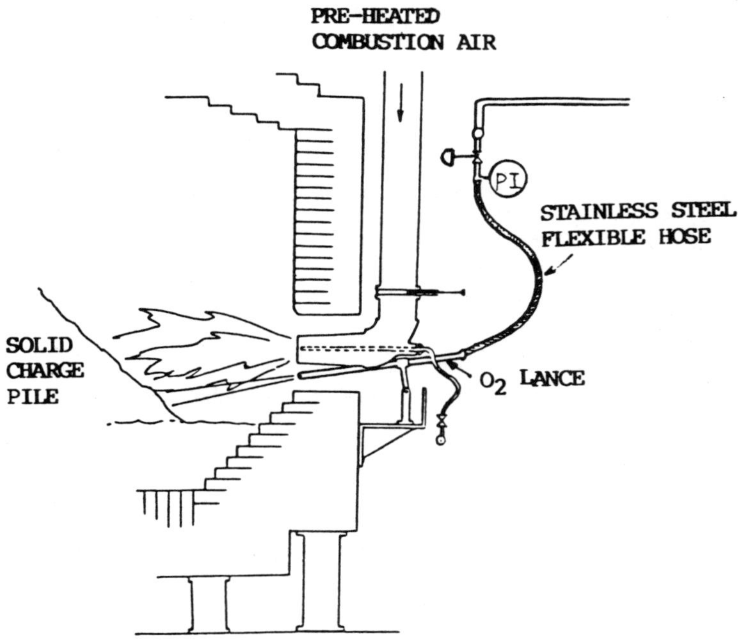

FIGURE 3. CROSS SECTIONAL VIEW OF THE FRONT OF #5 ANODE FURNACE SHOWING THE POSITION OF AN OXYGEN LANCE.

RESULTS

Oxygen enrichment was not used continuously throughout each heat. To ensure efficient oil and oxygen utilization, selective O_2 flame enrichment was only practiced during the "high-firing" regime. Therefore, whenever extra heat was unnecessary, the non-water-cooled lances were retrieved.

During each heat, the firing rate was decreased for a number of operations : charging, slag skimming, boiler tube cleaning, tapping, etc. On the average, these activities added up to about one third of the total furnace cycle. However, these varied from 25 to 42% of the total cycle. Since these activities could not be influenced by oxygen enrichment, care has been taken to first isolate the effect of oxygen on the "high-firing"regime, ignoring the rest of the cycle. Subsequently, in order to evaluate the impact of oxygen enrichment on the overall cycle, a normalized procedure was followed, whereby "low-firing" operations were assumed to last equal times for each of the heats evaluated.

Classification of heats

Twelve heats were monitored, namely #34 to #38 and #40 to #46. Of these, heats #34, #35 and #43 consisted of charges with high in nickel scrap content. This resulted in heat times shorter than the conventionally charged heats. Consequently, these will not be considered in the overall discussion on the effect of oxygen enrichment.

The other nine heats have been classified based on the oxygen enrichment level:

- Group C is the "Control" heats #41 and #46, without any molecular oxygen enrichment.
- Group A, consists of heats #36, #37, #38 and #40, where low oxygen enrichment level was employed.
- Group B, consists of heats #42, #44 and #45, where high oxygen enrichment level was employed.

Table 1 is a list of the "high-firing" periods and the normalized heat times. A third column shows as a reference, the actual total Heat times recorded in INCO's Furnace Log Book.

TABLE 1 Summary of Heat Times

HEAT #	Time of ACTUAL HIGH-FIRING regime	Time of NORMALIZED Heat	ACTUAL HEAT TIMES (from INCO's Furnace Book)
	hr	hr	hr
34	31.6	48.4	50.25
35	31.3	47.7	47.75
36(A)	42.0	59.3	59.58
37(A)	41.3	58.1	60.25
38(A)	37.6	54.4	55.25
40(A)	42.1	59.4	58.83
41(C)	49.6	66.8	82.02
42(B)	36.7	53.9	53.25
43	30.3	46.7	52.25
44(B)	34.0	52.2	53.17
45(B)	38.0	55.5	60.75
46(C)	50.7	67.9	71.93

GROUP C : CONTROL HEATS

"Melt-down" and "refining" times were determined directly from the oil circular charts. During Heat #41 a fifth charge was added to purposely extend heat cycle by about 8 hours. Twelve tonnes of NiO sinter premixed with coke were charged. Therefore eight hours have been removed from the actual melt-down time. This accounts for most of the difference between the actual Total Heat time from the Furnace Log Book and the normalized heat time.

The average normalized heat time of the Control heats was 67.4 hours. The average logged heat times was 77 hours.

By dividing the cast weight of the anodes nickel per heat by the heat time, the daily production was calculated. For control heats it was 66.7 tonnes Ni tapped per 24 hours. The production from these heats, as reported in the Furnace Log Book, averaged 59.5 tonnes/day.

GROUP A: LOW OXYGEN ENRICHMENT LEVEL

The average instantaneous oxygen flowrate during this 4-heat group was 4.2 Nm^3/min. This figure was calculated by dividing the volume of oxygen consumed during the heat by the period in which oxygen was effectively ON, that is, during the actual melt-down and refining regimes. The average normalized heat time for group A was found to be 57.8 hours. This representeda 9.6 hour decrease in heat time.

Considering an average air flowrate of 213.5 Nm^3/min, the oxygen injection at 4.2 Nm^3/min represented a 2.0% enrichment. The Group A production averaged was found to be 75.8 tonnes Nickel tapped per 24 hours, which represents a 13.6% increase over the "control" heats. The corresponding result from the Furnace Log Book is 73.2 tonnes Ni per day, equivalent to a 22.7% increase. Therefore, the normalized figure proves to be a fairly conservative evaluation of the effect of oxygen enrichment.

GROUP B: HIGH OXYGEN ENRICHMENT LEVEL

During this phase of the program two changes in operating parameters were made:

a)- The lances were inclined towards the bath at approximately 15 to 20° from the horizontal.

b)- A target of 3% carbon monoxide in the exit gases was maintained throughout most of the process time. This was attained by varying the oil and the primary air flowrates.

Pre-heated air flowrate average 173.7 Nm^3/min. The oxygen flowrate averaged 5.7 Nm^3/min or equivalent to a 3.2 % O_2 enrichment. The average normalized heat time for Group B was 53.9 hours. This represents 13.5 hours less than the control heat, equivalent to 20.0% improvement. The average daily production calculated from the normalized heat times is 83.4 tonnes Ni tapped per 24 hours, a 25% increase in productivity. The corresponding Furnace Log Book average production is 79.4 tonnes Ni per day, equivalent to a 33.7% improvement over the average "control heats". Table 2 is a summary of the effect of oxygen flowrate on heat time and daily production.

TABLE 2. Average Heat Times and Daily Nickel Anode Production

Group	Molecular O_2 Flowrate [Nm^3/min]	Normalized Heat Time [hours]	Daily Production Nickel Anodes [tonnes/day]
C: Control Heats	0	67.4	66.7
A: Low O_2 Enrichment	4.2	57.8	75.8
B: High O_2 Enrichment	5.7	53.9	83.4

DISCUSSION - REACTION MECHANISM

There are several mechanisms by which carbon may react with nickel oxide to produce metal. These are represented by the following equations:

i) Solid-solid reaction

$$C_{(s)} + NiO_{(s)} \longrightarrow Ni_{(l)} + CO_{(g)} \qquad [1]$$

ii) Gas-solid reaction

$$CO_{(g)} + NiO_{(s)} \longrightarrow Ni_{(l)} + CO_{2(g)} \qquad [2]$$

iii) Liquid-solid reaction, where the carbon, dissolved in the molten nickel, reacts with the solid sinter particle.

$$\underline{C}_{(l)} + NiO_{(s)} \longrightarrow Ni_{(l)} + CO_{(g)} \qquad [3]$$

The melting point of NiO is 1990°C, whereas the melting point of nickel is 1450°C. Since the reaction takes place between 1540°C and 1650°C it can be stated that molten nickel oxide is not involved in the reduction reaction.

Reaction mechanism shown in equations [1] and [2] are, however, unlikely to play a significant role.

Reaction [3] is the main mechanism for the reduction. The equilibrium solubility of carbon in molten nickel, as reported in the ASM Metals Handbook 1948, is shown in Fig. 4. At 1540°C, up to 2.5% carbon can be dissolved. The The measured carbon content of the molten bath was always less than 2%. Nevertheless, the opportunity for this reaction to proceed is greater than for mechanisms [1] and [2]. Once carbon has dissolved in molten nickel, it will readily reduce NiO in contact with it. The more molten nickel in contact with NiO, the large the number of sites for the reduction to occur.

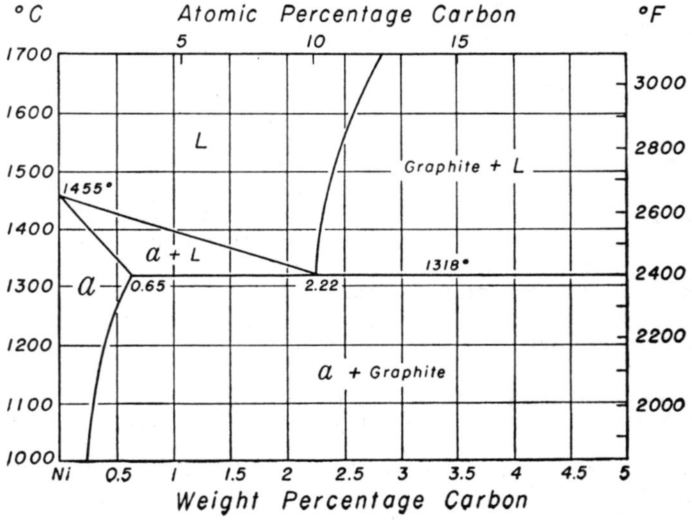

FIGURE 4. CARBON-NICKEL PHASE DIAGRAM
From Metals Handbook 1948

Because a reducing atmosphere is maintained at the front end, it would be beneficial to add oxygen towards the back of the furnace, to burn the CO to CO_2, thus transferring the generated heat to the back piles.

PREMIXED OXYGEN-ENRICHMENT

A series of tests were run to examine the effect of pre-mixing oxygen with the pre-heated air before distribution to each of the burners.

The main advantages of pre-mixing compared to selective oxygen flame enrichment were:

1. Less maintenance and simpler operation
2. No lance setting and retrieval from the part of the operators throughout the various stages of the cycle.
3. Less noise at the burner area.

The disadvantages were:

1. Higher roof temperature for equivalent oxygen enrichment levels.
2. Incapability of preferentially directing heat towards the molten bath, to increase its temperature and favour the liquid-solid reaction [3].

BENEFITS OF OXYGEN ASSISTED COMBUSTION

Increase production without detrimental effect on refractory life, especially on furnace roof and sidewalls.

Undesirable accumulation of solids in the waste heat boiler tubes was not increased in spite of increases in production rate.

Establishing a reducing atmosphere while maintaining a high furnace temperature using molecular oxygen.

Particulate emissions were decreased thus reducing pollution load.

CONCLUSIONS

Oxygen aided combustion established several benefits during the program, which enhanced furnace output without detrimental effect on the refractory.

Operator preference was pre-mixed oxygen-enrichment of preheated air rather than selective oxygen-enrichment lance method as the latter required manipulating lances during the heat.

Oxygen aided combustion is used as a flexible tool for tailoring the flame geometry which resulted in an increased production and a decrease in specific fuel consumption.

ACKNOWLEDGEMENTS

The authors wish to thank the management of INCO Metals Ltd. for permission to publish this paper. The enthusiastic involvement of Dave Stremlaw, Dale Fraipont, Daniel Dobrin and other INCO personnel was critical to the success of the testwork. The contribution of Mr. Lee Cupp, consultant to INCO, is greatly appreciated.

REFERENCES

[1] GIBBS, B.M. and WILLIAMS, A. "Fundamental Aspects on the Use of Oxygen in Combustion Process - A Review".
Journal of The Institue of Energy. 74 June 1983 pp 74-83.

[2] GARRIDO, G.F., PERKINS, A.S. and AYTON, J.R. "Upgrading Lime Recovery with Oxygen Enrichment".
Pulp and Paper Canada 83 : 1 (1982) pp 72-76.

[3] EACOTT, J.G.: "The Role of Oxygen Potential and Use of Tonnage Oxygen in Copper Smelting.
In "Advances in Sulfide Smelting. Vol. 2. "Ed. by H.Y. Sohn, D.B. George and A.D. Zunkel. 1983 pp 583-634.

OXYGEN USAGE OF COPPER SMELTING IN CHINA

Huang Qixing*, Gao Yueze**, Mao Yuebo**, Yan Shiling***

*Central Engineering & Research Institute for Non-Ferrous Metallurgical Industries, China
**Research Institute for Non-Ferrous Metals, Beijing, China
***Daye Non-Ferrous Metals Co., China

ABSTRACT

This paper will describe the use of oxygen in copper smelting in China and discuss some results achieved. After the testing of oxygen enrichment air in blast furnace in Shao Wu Smelter, Tongling Smelter began to use oxygen enrichment air in blast furnace. In comparison with air blast smelting, the feed capacity is increased by 45.86%, SO_2 content in the off-gas is increased by 69.39%, blister copper output is increased by 65.94% and coke consumption is reduced by 36.9% with 30.5% O_2 in air blast. In Daye Smelter a successful test of oxygen/pulverized coal burner has been carried out in a small reverberatory furnace. In Gui Xi Smelter a concept of change to oxygen enrichment air in flash smelting has been studied. In Baiyin Smelter an oxygen plant has been erected and oxygen enrichment air is being tested in the Baiyin furnace.

KEYWORDS

Oxygen enrichment; copper smelting; blast furnace; oxygen/fuel burner; flash smelting.

INTRODUCTION

China has a long history in copper smelting. In recent years technology and equipment are gradually being improved and capacity of blister copper production is also increased rapidly. Some new technology has been adopted, such as flash smelting introduced from abroad and Baiyin Copper Smelting Process (BYCSP) developed by our engineers. On the other hand, the traditional technology, such as blast furnace and reverberatoryfurnace smelting are still operating and made some progress, the fuel consumption in these processes is rather high, the flue gas with low SO_2 content from the reverberatory furnace is not suited for sulphuric acid manufacture and causes environmental pollution, so the modification is required. Oxygen enrichment is one of the best measures for settling these problems. The status of application of oxygen enrichment air in copper smelting of China in recent years can be described as follows:

PILOT TEST OF OXYGEN ENRICHMENT AIR USED IN THE BLAST FURNACE

The blast furnace still finds its application in China , such as in Tong Ling Non-Ferrous Metals Company, Shen Yang Smelter and Zhong Tiao Shan Non-Ferrous Metals Company. Based on the traditional sintering and blast furnace smelting process, the Momoda type blast furnace was developed. The special features of the Momoda type blast furnace are as follows: The pulped concentrate and lump materials, mainly coke and recycled converter slag are charged directly into the furnace top in alternative layers, and furnace gas is exhausted out from both sides of furnace and sent to the sulpuric acid plant. The Momoda type blast furnace is more advantageous than the traditional blast furnace, especially for small plants, but it still has some drawbacks, such as the specific productivity is low (45-50 tonnes of charge per day per m^2 of tuyere reaction area), SO_2 content of furnace gas is low (3-4%), and fuel consumption is high. In order to improve this situation, oxygen enrichment air smelting in Shao Wu Smelter was tested in a Momoda type blast furnace, whose tuyere section area is 1.5 m^2. Test results on the Shao Wu furnace were shown in table 1. These data show that Momoda type blast furnace productivity is increased with the increasing of oxygen content in air blast, the specific productivity is increased by 33.45% for a 24.2% O_2 blast and coke requirement decreased by 28%, i.e. coke rate is decreased from 9.96% to 7.12%. The effect of oxygen content on specific productivity is shown in Fig. 1. It can be seen that increasing of oxygen content of blast can intensify the copper smelting and reduce fuel consumption.

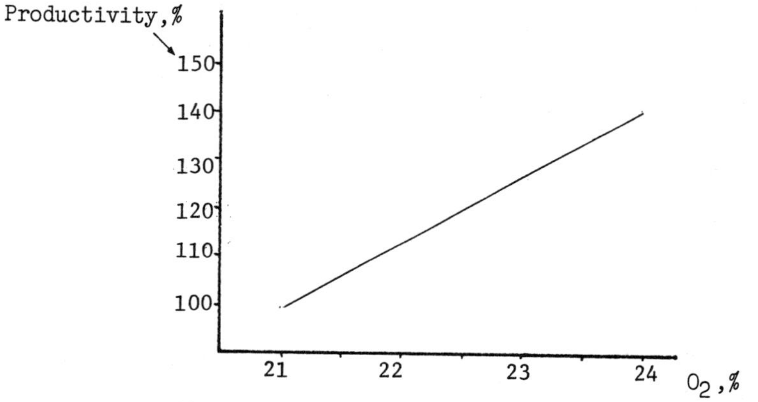

Fig. 1. Effect of oxygen content in blast on the increasing of specific productivity

TABLE 1. Test Results of Momoda Blast Furnace of Shao Wu Smelter

Item	Smelting with air	Smelting with oxygen enrichment blast	
		O_2 22.4%	O_2 24.2%
Specific productivity,%	100	123.91	133.45
Coke rate, %	9.96	9.5	7.12
Cu content in slag,%	0.286	0.27	0.27
Cu in matte, %	32.56	31.67	35.24
Concentrat.rate of matte	3.19	2.99	2.89
SO_2 content in gas leaving furnace	3.5	4.35	5.66

Data in table 1 show that oxygen enrichment of the air blast to the level of 22.4% O_2 results in a 12% increase of SO_2 content of gas leaving furnace, and enrichment to 24.2% O_2 increases SO_2 content of gas by about 16%. It can be inferred that SO_2 content of gas leaving furnace is increased with increasing of oxygen enrichment of air blast. Thus the sulphuric acid manufacture is facilitated.

Test results of Momoda blast furnace in Shao Wu also show that oxygen enrichment of blast to the level of 24.2% O_2 can reduce the operating cost of blister copper by 9.98%. Therefore the oxygen enrichment of blast in Momoda blast furnace is economically beneficial.

COMMERCIAL PRACTICE OF OXYGEN ENRICHMENT FOR THE MOMODA BLAST FURNACE IN THE TONG LING NON-FERROUS METALS COMPANY

After the test of oxygen enrichment in the Momoda blast furnace in Shao Wu, oxygen enrichment (26% O_2) in the Momoda blast furnace of Tong Ling Company was practized since January, 1986. The tuyere section area of the furnace is 10.5 m^2, and operational performances are as follows:

Composition of Copper Concentrates

	Cu	S	Fe	SiO_2	CaO	Al_2O_3	MgO
%	22.28	32.5	26.16	9.11	2.77	1.79	1.28

Composition of Coke

	Fixed carbon	Volatile matter	Ash content
%	79.97	2.21	17.82

Distribution of Coke Size

mm	12	12-40	40
%	5.46	24.36	70.18

Matte Composition

	Cu	Fe	S
%	41.57	28.66	23.79

Slag Composition

	Cu	Fe	SiO_2	CaO	Al_2O_3	MgO	S
%	0.296-0.396	31.75-36.95	32.63-37.62	10.29-12.73	3.1-4.1	1.27-1.39	0.42-0.72

Typical results on a Momoda furnace in Tong Ling are shown in table 2.

TABLE 2. Operational Results of Oxygen Enrichment in Comparison with Air

O_2 vol. % in blast	Specific productivity t/m^2·d	Rate of desulphuri-zation %	SO_2 content in flue gas %
21	42.78	43.34	3.79
30.5	62.4	57.17	6.42

O_2 vol. %	Output: Blister copper t/d	Output: Sulphuric acid t/d	Coke rate % of charge
21	46.77	166	10.2
30.5	77.61	206	6.46

These data show that specific productivity in the Momoda furnace has been increased by 45.86%, rate of desulphurization increased by 31.90%, SO_2 content of off-gas increased by 69.39%, output of blister copper increased by 65.94%, with O_2 30.5% blast the coke requirement decreased by 36.7%. Smelting operation of furnace is normal and stable. According to the results of test and calculation it is shown that 0.1784 ton of coke can be saved for one ton of blister copper produced. It proves that oxygen enrichment in the Momoda blast furnace not only can intensify the smelting, but also reduce energy consumption without change of smelting equipment and violation of furnace running, therefore, the economic benefit is resulted. Thus oxygen enrichment is a better method for the modification of Momoda blast furnace. Although oxygen enrichment can improve the Momoda blast furnace smelting, the specific productivity also closely is related to the percentage of lump materials in feed (coke, fluxes and solidified converter slag). If sufficient lump materials are present in the feed, furnace gas can flow up through the shaft normally and is well distributed, smelting operation will be normal and the furnace charge is increased, therefore, the coke requirement will be reduced. In practice, the percentage of lump materials must be 36.48%. The optimal result of oxygen enrichment can be attained.

Based on results of oxygen enrichment test and production of the Momoda blast furnace in Shao Wu and Tong Ling, Shen Yang Smelter has a plan to adopt oxygen enrichment smelting on its Momoda blast furnace.

TEST OF OXYGEN /PULVERIZED COAL BURNER

In order to modify the reverberatory furnace smelting, we have also studied the oxygen-pulverized coal burner, as one of the alternatives by which the smelters in Canada and Chile have good experiences, but only heavy oil or natural gas are used as fuel in these plants. As the price of oil or gas is expensive in China, so pulverized coal must be used in reverberatory furnace in China instead of heavy oil and natural gas. Test of oxygen/pulverized coal burner was carried out in a small reverberatory furnace in Da Ye Smelter.

The burner is composed of a burner head, a cyclone mixer and a pulverized coal screw feeder. Tests were carried out with two types ofburners. One type is a oxygen pulverized coal direct mixed burner. The pulverized coal is fed into mixer by the screw feeder and in which oxygen is blown. Then the oxygen and pulverized coal mixture through a tube of 1 m long, are sprayed out from the head. The other type of burner is an indirect mixed one, the pulverized coal is fed into the mixer by screw feeder and mixed with compressed air in the mixer, the tube consists of two concentric tubes. Mixture of compressed air with pulverized coal and oxygen flow through inner and outer tube respectively, join together in the burner head and spray out, the burner can be erected in two different manners, namely, horizontal and vertical. Preliminary test was carried out under cold condition, so as to determine optimum pressure and flow rate of oxygen, oxygen-coal ratio and spray velocity, then carried out under high temperature.

Composition of Coal Powder

	C	H_2	O_2	S	N	A	W
%	71.9	4.8	9.5	0.75	1	10.4	1.5

Q=6735 Kcal/kg Size 80% 200 mesh

Test results of direct mixed type burner. The reverberatory furnace was heated up to above 600°C, the operation of oxygen-coal burner was started, oxygen-coal ratio was 0.7-1.2, oxygen pressure was 2-2.5 kg/cm^2, optimum velocity of spray was 30-40 m/s, the flame length was 1.7-2 m, the high temperature area (1800°C) was 500 mm from the burner head, temperature at the end of flame is 1280°C, burning was complete and stable.

Test results of indirect mixed type burner: Test pressure of oxygen and air blast was 2-2.5 kg/cm^2, air/oxygen ratio was 15:40 - 15:20, oxygen/coal ratio was 1 or 0.5, velocity of first air was 30-40 m/s, velocity of mixture spray was 15-20 m/s. The flame temperature and length were the same as in the first type.

Test results of the two kinds of burner show that the burning of oxygen-pulverized coal mixture is complete and stable. If the spray velocity is within above mentioned range, phenomenon of back flame doesn't take place, so that safe operation of the burner is obtained and the ignition point is above 600°C. Both kinds of burner can be used, of which the indirect mixed type is better. Operating cost of oxygen-pulverized coal burner is much lower in comparison with oxygen-oil/gas burner.

OXYGEN ENRICHMENT TEST OF BAIYIN COPPER SMELTING PROCESS

The Baiyin Copper Smelting Process is developed in Baiyin Company during the 1970's. The Baiyin copper smelting furnace is a fixed rectangular furnace in which the hearth is divided into smelting and settling zones by the partition wall. The charge holes are located on the roof of the smelting zone, in which along both sides, there are side blowing tuyeres located and on one side there is a hole for pouring in the converter slag. There is a matte siphon tap hole on one side of settling zone and a slag tap hole on another side. A pulverized coal burner is provided on the end wall and flue gas outlet is located on the opposite end, an auxiliary pulverized coal burner is mounted on the middle of the furnace roof. Mixed with the flux and flue dust, the concentrates are fed into the furnace through the charge holes. Air is blown through the tuyeres to the bath. The melt flows to the settling zone through the hole below the partition wall for separation of slag from matte. The matte can be tapped out through the siphon and transferred into converter for the production of blister copper. Slag is tapped out intermittently and is granulated in water and discarded. The liquid converter slag returns to the smelting zone for recovering copper. The flue gas exhausts from the furnace and is sent to the sulphuric acid plant. In comparison with the reverberatory furnace, capacity is increased from $3.8 t/m^2 \cdot d$ to $8.48 t/m^2 \cdot d$, the SO_2 content of flue gas is increased by 3-4 times, which facilitate greatly the sulphuric acid production. The original process of fluidized-bed roasting-reverberatory smelting was replaced with the Baiyin process by the Company in 1980.

However, after the Baiyin furnace was enlarged, the copper content of slag is increased, resistance to the gas flow and air in filtration is increased, so that SO_2 content of the offgas is lower than that in the testing period. In order to overcome these problems, oxygen enrichment test is being carried out instead of air blast. At present, the Baiyin Company has erected an oxygen plant of 1500 m^3/h. Oxygen enrichment test on the furnace having 44 m^2 in the smelting zone is under testing. The test will be completed in 1987. Production results will be greatly improved.

OXYGEN ENRICHMENT FOR FLASH SMELTING FURNACE IN GUI XI

Gui Xi Smelter, one of the biggest smelters in China, was built based on import flash smelting technology and equipment. The Smelter was started up on December 30, 1985. All the equipments operate normally, and production is continuous and stable. The smelting operates successfully, but during the production in the initial period the energy consumption is high. Since the contract for construction was signed in 1979, the price of energy rose to a great extent, and also the price of oil as fuel is expensive. In order to decrease oil consumption and increase production capacity, the Smelter will adopt oxygen enrichment smelting instead of air. The feasibility study indicates that usage of oxygen enrichment smelting can save the consumption of heavy oil greatly and improve the production results.

CONCLUSION

The preliminary economic results have been obtained in the application of oxygen in copper smelting plants of China. Some old plants are planning to use the oxygen to modify its process. It is foreseen that oxygen can be used more widely in copper smelters of China in the near future.

PRODUCTION OF OXYGEN - KEEPING PACE WITH METALLURGICAL DEMANDS

D. V. Eyre, I. L. Gorup, T. S. Pawulski

Liquid Air Engineering Corporation
2121 North California Boulevard
Walnut Creek, California 94596

ABSTRACT

Oxygen was first produced at the turn of this century by continuous low temperature distillation of air.

At the end of the second world war, Canada led the world in the use of tonnage oxygen for smelting and refining. The Inco Flash Smelting Process provided the first oxygen challenge. This pioneering non-ferrous smelting process required tonnage quantities of oxygen. The challenge was met in 1952, when the world's largest Oxyton was built to supply the process with 300 tons of oxygen per day at Copper Cliff, Ontario. Subsequently, two more Oxytons were installed at this location. Single oxygen plants producing 2500 tons per day are engineering realities today.

In 1911, Canadian Liquid Air pioneered oxygen production in Canada. During the first year of operation, 20 tons of oxygen were produced. The largest oxygen column which is now in operation can produce this quantity in 10 minutes.

KEYWORDS

Discovery of oxygen; production of oxygen; oxygen cost; uses of oxygen; flash smelting; non-ferrous metallurgy.

INTRODUCTION

The intent of this paper is to provide an historical background on the development of the air separation industry. Its relationship to the metallurgical industry has been very close from the beginning and has been one of the main driving forces in its expansion. In addition, the necessity to increase efficiency and decrease costs has been ever present. Over the years, the industry response has been one of continuous improvement, as will be seen, and its ability to meet production demands is more than adequate.

DISCOVERY

When oxygen was first isolated more than two hundred years ago, it was called "fire air" by a Swede and "dephlogisticated air" by an Englishman. We should be thankful that a Frenchman, Lavoisier, gave it an easier name, "Oxygene" -- even if he did have to borrow from the Greek! This discovery of oxygen in the early 1770's initiated an intense period of experimentation, the culmination of which was an entirely new theory of combustion and the final demise of Alchemy.

Alchemy, which had thrived in Egypt around 300 A.D. and had spread as far as China, did not penetrate Europe until the 1100's, when the Arabic texts were finally translated into Latin. It was a strange mixture of medicine, magic and metallurgy and was based, to a large extent, on the accumulated knowledge of the metal-workers of the time. They could do many things with surprising skill, and over the ages brought many improvements, but without understanding why or how.

The break between Alchemy and what we now consider as Chemistry was marked by the publication of the "Sceptical Chymist" by Robert Boyle in 1661, but the old Phlogiston and other theories of combustion lingered on until Lavoisier's time. Not only did he present a new coherent theory of combustion, but he also proved experimentally the law of conservation of matter. Then, around 1780, he wrote the first real equations representing chemical reactions, even though they were not in the neat shorthand with chemical symbols that we now use (these did not come for another 50 years).

Once Lavoisier's ideas were widely accepted, a tremendous spate of activity began. During the 1800's, chemists discovered about half of the presently known elements and ascribed to them reasonably accurate atomic weights. At the same time, many new manufacturing procedures were developed, and existing ones were improved through a better understanding of the principles involved. Not only was it a time for improving the methods, but a new element was injected -- a need for more. The demand for goods was expanding rapidly. After the first phase of just adding more machines or foundries came a second phase of trying to improve efficiency by building them bigger.

Both the ferrous and non-ferrous industries realized that the use of oxygen enriched air could improve thermal efficiencies in their oxidation processes. There were many foreseeable problems with furnace design, but the greatest obstacle was the absence of the industrial scale production of oxygen. However, these "small" difficulties did not deter H. L. Bridgman who was granted a U.S. patent for a flash smelting furnace in March of 1897. His process showed remarkable foresight in suggesting that the smelting of sulphide ores could be achieved by their own combustion in a strongly oxidizing furnace atmosphere without the further addition of heat. Nevertheless, it took half a century for a flash smelter to be realized on an industrial scale!

PRODUCTION

Among the first commercial methods of oxygen production was the Brin process which used barium oxide. Barium oxide was heated in compressed air to form the peroxide which, upon reduction of pressure, released oxygen and reverted to the monoxide. There were many other competing chemical methods, using various formulations of peroxides, chlorates and sulphates, which also had some fanciful names such as oxone, oxygenite, epurite and lavoisite. However, since they only yielded between one and five cubic feet of oxygen per pound of material, the processes were very cumbersome.

The other early method of production was electrolysis. This eventually provided more gas to the commercial market than the various chemical methods, but its high power consumption, around 5,000 Kwh per ton of oxygen, made it unacceptable for large scale production.

Fortunately, other developments were taking place which eventually led to the production of oxygen by physical means rather than chemical and formed the basis for the present-day separation industry. Just as Bridgman showed amazing foresight with his patent on a flash furnace, it can be argued that Jonathan Swift also did when writing Gulliver's Travels. In 1726 he wrote of the fictional Academy of Lagado and its workers: "Some were condensing air into a dry tangible substance, by extracting the nitre, and letting the aqueous or fluid particles percolate." Such thoughts about a "permanent gas" were extraordinary at that time and were only proved reasonable by Cailletet in France some 150 years later, in 1877.

While Cailletet initially produced only a fog of liquid air, others followed soon after with small quantities of liquid oxygen, nitrogen, and even hydrogen in 1884. However, the first to succeed in liquefying air on an almost industrial scale was Carl von Linde in 1897 with what was, at that time, a very sophisticated process operating at 2,800 psig.

The search for an efficient way to produce liquefied gases caught the attention of a young engineer working at the Thompson-Houston electrical company in Paris. He had already, in his spare time, solved the problem of storing acetylene in cylinders, by the method still in use today, and was ready for a new challenge. Georges Claude recognized that the efficiency of the air liquefaction cycle would be much improved if the expansion of the air could be made in a work performing machine. In 1902, after 3 years of work, he succeeded in developing such a machine and had a process that operated at only 280 psig.

The objective was, however, to produce oxygen, not just liquid air. It was not until the double rectification column process was established, around 1905, that one could say that industrial oxygen production had been achieved. This system has remained substantially unchanged to the present time. Meanwhile, the other methods of production were gradually being phased out. In Germany, for example, by the year 1909 there was an annual production of about 12 tons of oxygen by the Brin method, 500 tons by electrolysis and 3,200 tons by the fractional distillation of liquid air. Some rather extravagant claims about liquid air and oxygen were made in the early years, for both their production costs and usefulness. When these did not materialize, there was some disillusionment. But the growth of the industry continued: the first air liquefaction plant was installed in the U.S.A. only in 1907 but by approximately 1930 the annual production had reached 150,000 tons. In Canada, the first plant was installed in Montreal in 1911 and 22 tons were shipped that year; within two years this had risen to 165 tons but never reached the production levels of the U.S.A.

The uses of oxygen in the first years were essentially restricted to welding and cutting, but the goal of large plants to serve the metallurgical industries was still in view. To achieve this end, the cost of production and equipment had to be drastically improved. The theoretical power for separation had been calculated to be about 50 Kwh per ton, but the actual power turned out to be closer to 500 Kwh per ton. This was found to be disappointing, and even incredible -- except to those actually designing the plants.

DEVELOPMENT

Having passed the first hurdle of production, the development of the oxygen producing plants to their present state has taken place slowly over the intervening years. The rectifying sections of the process started out as countercurrent partial condensers and evaporators, termed "Dephlegmators", but it was not long before these were replaced by distillation columns as we now know them. The calculation methods and designs were based heavily on the practices used in the manufacture of alcohol.

Other areas that needed attention were those of purification and heat exchange. It did not take long to realize that the then current methods of water removal by calcium chloride granules and carbon dioxide removal by lumps of solid caustic soda would not be suitable for large scale oxygen production. At the same time the design and manufacture of heat exchange equipment left a lot to be desired. It was not actually until after Nusselt put some order into the science around 1917 that the exchangers could be considered reasonably efficient.

The air prepurification system was eventually changed to use adsorption on alumina for water removal and to use circulation of a caustic soda solution in a packed tower for carbon dioxide removal. However, many small plants could still be found operating in the early 1950's with lump caustic bottles. Even these changes were not enough, and it took the ingenuity of Mathias Frankl, in 1932, to devise his "cold accumulators", or regenerators as we now know them, to open the way to truly low cost oxygen. These regenerators, acting in pairs, are vessels filled with aluminum strip packing; while one accumulates the cold from an outflowing product, the other transfers previously accumulated cold to the incoming warm air. An important added advantage is that the moisture and carbon dioxide contained in the incoming air are frozen out and then evaporated into the outgoing stream in the next half cycle. This simple device opened the doors to low cost oxygen, but by its very nature precluded the production of high purity oxygen due to contamination and to residual air trapped upon switching. This was later overcome by the development of stone packed regenerators with embedded coils to carry the pure product. However, there were problems with dusting, which required special materials and very careful operation.

Several process variations were developed, incorporating the better features of the various systems, and the power consumption was progressively lowered. The power required per ton of oxygen had started out at around 1,000 Kwh at the time of its first production and had decreased rapidly to nearly 500 Kwh by 1915. Progress was relatively slow afterwards, with power only reaching the 350 Kwh level by the late 1940's. Production capacities were growing rapidly, reaching over 600,000 tons annually in the U.S.A. Everyone who was building plants had his own ideas, such that by 1950 one could hear the names Linde-Frankl, Heyland, Kellogg, Elliot, Rescol and Oxyton to describe various cycles. At the same time, the old work producing expansion engine devised by Claude had been replaced by efficient centrifugal machines. With all this interest and effort, the power consumption during this period dropped below 300 Kwh per ton for low purity oxygen. The progress of power consumption is shown in Fig. 1 and reflects the power required in a representatively sized plant of the period, around 2 tons a day initially to 500 tons a day now.

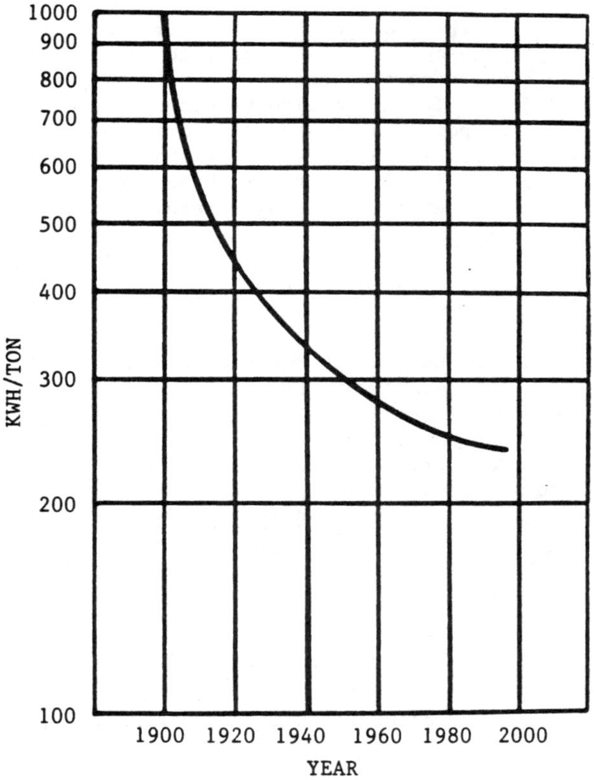

Fig. 1. Power consumption.

One other development around the same time was the concept of reversing exchangers. These resembled regenerators but had the distinct advantage that up to about 40 per cent of the outflowing gases could be clean, pure product, thus permitting simultaneous production of large quantities of both oxygen and nitrogen. This was interesting because, in addition to the production of oxygen, many plants had already been designed exclusively for the production of pure and dry nitrogen. In fact, several of Georges Claude's early plants had been installed in Niagara Falls around 1926 for use in the production of calcium cyanamide -- some were still operable as late as 1975. Some of his other plants had been designed for the production of nitrogen for the synthesis of ammonia.

A more recent development that has occurred, or rather has been applied more extensively in the last two decades, is the use of molecular sieve for the prepurification of air. While this product has been in wide use since becoming available, it was previously restricted to small plants and to plants where all products were required clean because it could not compete, in terms of cost or power, with a plant using reversing exchangers. Improvements brought by Air Liquide and others have now made this system competitive, at least for large plants. The choice now rests more on particular preference or on the degree of air pollution, since by its nature molecular sieve better handles higher concentrations of impurities.

In all of the above discussion, little or no differentiation has been made between pure and impure oxygen. In the early days, one was fortunate to be able to get 98 or 99 per cent purity of oxygen. It was not until the plants became larger with a diversity of uses that any difference in purity was appreciated. In general, the difference in power was in the order of 10 to 15 per cent between 90 and 99.5 per cent pure product. With the advent of an energy conscious economy, this difference is now closer to 5 per cent -- essentially one just pays more in equipment costs for the pure product. It should also be noted that all numbers refer to contained oxygen. Another point to be appreciated is that the gradual reduction in power consumption has not all been due to the ingenuity of the cryogenic fraternity; we have to give credit also to the machinery suppliers, who in the last 30 years or so have improved efficiencies by something in the order of 10 to 15 per cent.

APPLICATION

Although the idea seems obvious that pure oxygen would be useful in all oxidation reactions using air, it took many years to proceed from the idea to industrial reality. The first experiments and applications in the metallurgical field were with oxygen enrichment of blast furnace air in Belgium at Societe d'Ougree Marihaye, around 1913. However, the scale was not large enough and the results were apparently not convincing. It was not until the war years 1939-1945, particularly in Germany where necessity was the mother of invention, that the use of oxygen really developed. The main drive was to increase production from existing facilities, rather than to develop new processes. The applications served a wide range of activities in the metallurgical, chemical and newly emerging rocketry industries.

Prior to the first world war, oxygen was used primarily for maintenance with the oxy-acetylene torch. Between the wars, its use extended into fabrication, particularly with the development of accurate cutting techniques. Its use also extended into the steel industry, but only for improved handling of the products, such as cutting and scarfing. These needs required tonnage oxygen, but nowhere near the quantities we have now. After the second world war, true development of the use of oxygen in processes began. Several sets of tests on the use of oxygen in blast and open hearth furnaces were in progress at various locations in the 1940's, but as late as 1947 no steel company owned its own oxygen plant. The basic oxygen process of steelmaking initiated in Austria was tested in Canada by Dominion Foundries and Steel Co.; and the first of several oxygen plants was installed and in use by 1951, three years ahead of the U.S.A. By the 1960's, it seemed that the flood gates had opened and every steel mill needed oxygen for converters and blast and open hearth furnaces, including the Steel Co. of Canada, Algoma Steel and Dominion Steel. A similar situation occurred in the U.S.A. It was not so much that the oxygen cost had been questioned, but rather that no one had really believed the advantages -- until someone else had tried it.

The non-ferrous industry was making its own sets of tests around the same time. International Nickel started a development program in 1945 for the direct flash-smelting of sulphide concentrates with oxygen. In 1952 they were the first to put on stream a commercial plant with its attendant oxygen plant. The first furnace that treated 500 tons per day of concentrate was followed shortly by one twice the size. The 325 ton per day oxygen plant built by L'Air Liquide was the largest in the world at the time: three times the total capacity of the cylinder oxygen plants in Canada. Further developments for other uses in the overall process led to the requirement for an additional 850 T/D oxygen plant by 1966.

While International Nickel was the pioneer, others were not standing still. Some, such as Outokumpu in Finland, preferred enrichment only, with a lower oxygen requirement. Others, such as Noranda and Mitsubishi, were thinking of more than one step in a continuous process. The long term impact has been such that, to serve only the metallurgical industry in Canada, L'Air Liquide alone has installed nearly 6,000 tons per day oxygen capacity to date, which probably represents only half the country's consumption. The largest single unit is one of 1,400 tons per day in Hamilton, Ontario, commissioned in 1972. The interest in oxygen has since spread to lead smelting with, for example, the QSL process. Additional capacity will have to be provided as experience is gained and perhaps less concentrate is exported.

Over the years, the size of oxygen plants has grown, each one being a function of a particular situation; but what was the largest in 1952 is now commonplace. Unfortunately, the maximum size plant that has been built at any time does not follow any orderly progression. However, an attempt at tracing the growth of plant size is shown in Fig. 2. The curves represent roughly, on a worldwide basis, the maximum size of oxygen plant realized at any time and the average size requested for metallurgical uses. There has to be a need for oxygen and an assurance that the extrapolation taken at each step is not too large. Then, having established a new standard, it may be some time before a greater need arises. We are possibly in that situation at the present time. Probably the largest oxygen

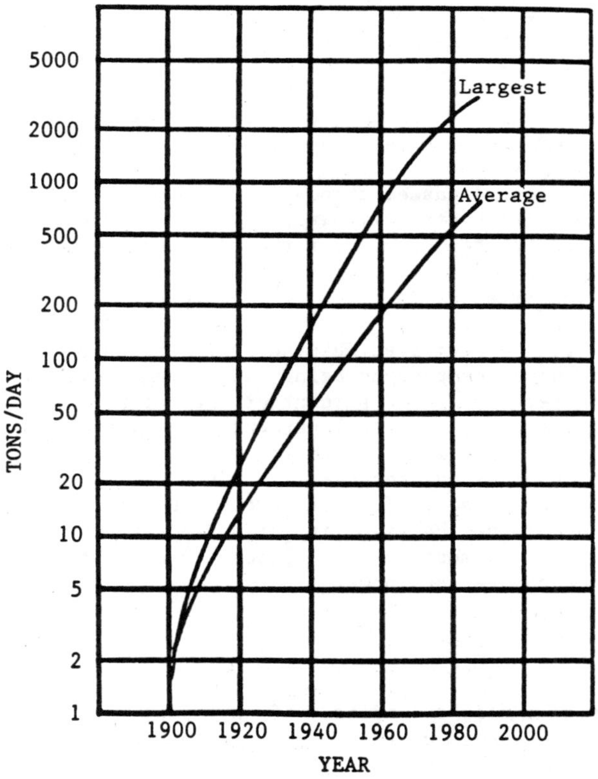

Fig. 2. Oxygen plant size.

plant needed for a smelter at this time is about 850 T/D, similar to the one built for International Nickel back in 1966 and also to the one presently being built for Magma Copper for use in their Outokumpu type smelter. If there were not such a depressed situation, the size for steel mills could well be larger, and units up to 2,000 T/D have already been supplied. The largest plants ever built were the result of a very special set of circumstances in South Africa, where the government felt a need to be self-sufficient in the production of many chemicals, liquid fuels and synthetic gases. This led to a series of twelve units, 2,500 T/D each, followed by another 2,750 T/D unit, all built by L'Air Liquide. We do not see such a large need arising again for some time, but we -- and the industry in general -- would be happy to be pushed even further.

ECONOMICS

It is extremely difficult to talk about the economics of oxygen production without getting mired down in details. It is bad enough just looking at the production alone, but to try and combine this with the economics of its use can be very complex. There are few generalities except that we -- the producers -- are often told that the oxygen is too expensive. In spite of this, we notice that when oxygen is eventually used, it is proclaimed as wonderful and there is never enough!

We were once telephoned by a consultant who would not even give his name but who demanded to know the price of oxygen. He would give no details on quantity, purity, or location, claiming that such considerations were immaterial. Finally we gave him two prices, a low one for bulk supply on the Gulf Coast and something over 100 times higher for cylinders in Alaska, and hung up. This was some years ago, but large price differentials still exist today. However, in spite of this, an attempt to answer his question is shown in Fig. 3. At each period in time, an average set of conditions can be taken -- such as size, purity, utility costs, etc. -- and the costs per ton of oxygen can be plotted in the current dollars.

It can be seen that this curve passes through a minimum in the "good old days" of the 1930's but has been rising steadily since. However, when a smoothed "cost of living" index is applied to this curve over the same period, it appears much more reasonable. Unfortunately, it also appears from this "constant dollar" curve that little has been achieved recently in terms of improvements in costs.

The breakdown of the cost has remained substantially the same for the last two decades -- the two principal components being capital and power -- representing respectively around 50 percent and 30 percent of the total. The labour cost has always been low -- in the order of 5 to 7 percent -- and, with the present use of computer control, will only get lower. The remaining 10 to 15 percent is made up of miscellaneous items such as maintenance, insurance and taxes. Obviously, the cost of money, cost of power, size of plant, product purity and pressure all have their influence. To determine the viability of a project, the particular conditions must be considered, the most suitable process chosen and, if necessary, some trade-offs made between capital and power.

CONCLUSION

It is difficult to write a faithful history of anything: one can never see things through the eyes and have the thoughts of those present at that time. A period of discovery and development always exists, and we are fortunate to be able to see the outcome and not just the beginning. However, we can hope that we too are at the beginning of something new, even though it may not yet be fully recognized. The future of the oxygen industry is inextricably intertwined with both the

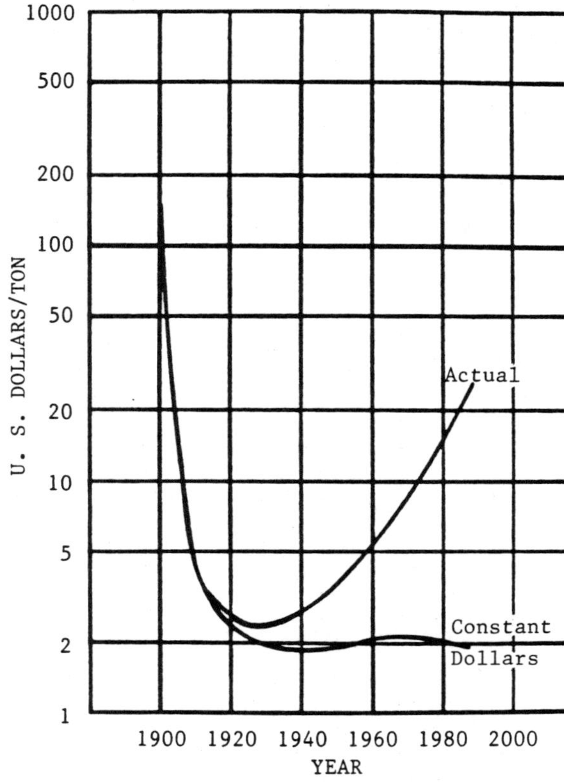

Fig. 3. Oxygen cost.

chemical and metallurgical industries. Is the next development purely in the technology, is it in the incineration of noxious chemical wastes, or is it in the production of metal at every mine by shipping energy to the ore, instead of ore to the energy? Whatever it is, it will surely be as challenging as the past, and our industry can be counted on to play its part.

THE ROLE OF OXYGEN IN THE OUTOKUMPU FLASH SMELTING PROCESS

Y. Anjala*, J. Asteljoki** and P. Hanniala*

* Outokumpu Oy, Engineering Division,
P.O. Box 86, 02210 Espoo, Finland

** Outokumpu Oy, Metallurgical Research Center,
P.O. Box 60, 28101 Pori, Finland

ABSTRACT

The use of oxygen in the Outokumpu flash smelting furnace since 1971 has been one of the most significant developments in the entire smelting process. This has resulted in a widened range of raw materials, increased unit capacity and savings in energy consumption. Additionally this has made possible the developing of new process applications such as Kennecott-Outokumpu flash converting process and the direct blister smelting process, which has been in commercial use since 1978. This paper describes the role of oxygen in the standard Outokumpu flash smelting process as well as in the new developments.

KEYWORDS

Copper; flash smelting; flash converting; Outokumpu; direct blister production; small scale smelter; retrofitting reverberatory smelter.

INTRODUCTION

The Outokumpu flash smelting process was originally developed using preheated air. The most significant advancement in the process development was made in 1971 when oxygen enrichment of the process air was adopted. This resulted in many advantages. The most important of them are:

- Decreased overall energy consumption
- Decreased investment costs
- Several operational advantages
- Decreased pollution and better hygienic conditions

Since the adoption of oxygen enrichment Outokumpu has published numerous papers dealing with this subject. In spite of its clear advantages the idea of using oxygen in smelting was received relatively slowly. However, today it is a proven and well recognized method, which is widely used in the Outokumpu type flash smelters as well as in other smelting processes.

In this paper the essential features in smelting resulting from the use of oxygen and the present practice in flash smelting are reviewed. The new opportunities to apply flash smelting technology including the flash converting process and the direct blister smelting process are described. Interesting new ideas, which are discussed in this paper, are how flash smelting can be usefully applied in low capacity cases and which opportunities the new three shaft flash smelting furnace can offer.

DEVELOPMENT WORK TO APPLY OXYGEN ENRICHMENT

Theoretical Background

The Outokumpu flash smelting process is a combined smelting and converting operation and it utilizes reaction heat from the oxidation of metal sulphides. The degree of oxidation is determined by selecting the desired matte grade taking into consideration the total smelter operation. The matte grade is controlled by the ratio of total oxygen in the process gas to the particular feed. This we call the oxygen coefficient, which determines the degree of oxidation of the concentrate and has nothing directly to do with the degree of oxygen enrichment.

The necessary oxygen coefficient can be accomplished with either straight air or pure technical oxygen or any combination of oxygen enrichment as the process gas. Depending on the nature of the concentrate the pure air approach may require considerable additional heat input in the way of extraneous fuel, while the pure technical oxygen case may require a means of removing the excess heat. Therefore, after establishing the oxygen coefficient to achieve the desired matte grade, the degree of oxygen enrichment is adjusted to the level necessary for autogenous reaction so that neither additional heat nor special heat removal systems should be required in the furnace.

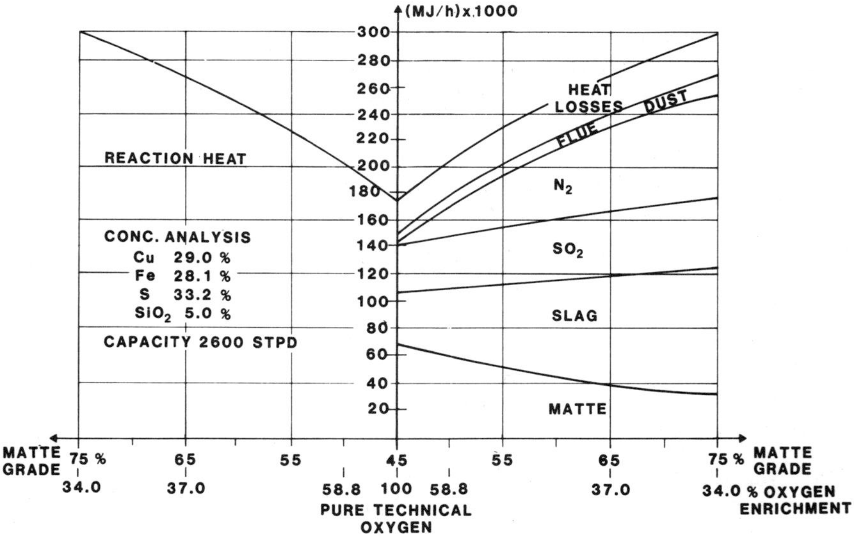

Fig. 1. Outokumpu flash smelting furnace heat balance for autogenous reaction.

The relationship between the matte grade and the oxygen enrichment in the autogenous operation mode is presented in Fig. 1. On the left side is the generated reaction heat for a typical chalcopyrite concentrate and for certain capacity in different oxidation conditions. Matte grade is varied from 45 % to 75 %. On the right side the corresponding heat output is shown. It is divided to different elements (matte, slag, sulphur dioxide, nitrogen, flue dust and heat losses). In the autogenous operation mode the increase of matte grade will not only decrease the matte amount and increase the slag amount respectively, but it will also mean that oxygen enrichment has to be decreased.

In the presented case, the matte grade of 45 % corresponds the use of technical oxygen only. If the matte grade is increased, the additional heat generated from reactions must be removed (if operation is based on the use of technical oxygen) by means of decreasing oxygen enrichment, feeding inert material or removing extra heat from the shaft. Outokumpu has tested special type of reaction shaft in the pilot scale at the Research Centre in Pori and the boiler type of reaction shaft was utilized in commercial scale in pyrite flash furnace in Kokkola.

Development Steps

Historically, the Outokumpu flash smelting process has used preheated reaction and combustion air (about 450°C) and fuel to supplement the heat generated by the exothermic reactions. The fuel was used both in the reaction shaft and in the settler part of the furnace. This approach was selected rather than the use of pure oxygen or oxygen enrichment, because electric power in Finland was expensive and, because furnace capacity was not a primary concern.

Outokumpu investigated and tested the use of oxygen enrichment in the pilot furnace in the late 1960's, in order to increase the furnace smelting capacity. In this stage the purpose was to eliminate the fuel consumption in the reaction shaft and in this way reduce process gas volume. At this stage an oxygen plant was constructed at Harjavalta in 1971 with a capacity of 265 mtpd in order to utilize about 40 % oxygen enrichment on the copper and nickel flash furnaces (1, 2).

Since the installation of the first oxygen plant, the development work of the furnace and the concentrate burner has made possible to further increase the oxygen enrichment and capacity of the existing furnace. At the same time the operation philosophy was changed. The reaction shaft temperature was increased in order to eliminate the expensive fuel consumption in the settler part of the furnace. A new oxygen plant was commissioned in September 1984 with 285 mtpd of oxygen available to the smelter for a total availability of 550 mtpd. Since the commissioning of the new plant, due to the furnace feed containing materials with a low heat value, over 90 % oxygen enrichment has been utilized in the copper flash furnace when necessary (3). The above described development steps are illustrated in Fig. 2 by the areas A, B and C.

- Area A represents the development work at the end of 1960's and the first commercial scale introduction in 1971. The change in the process gas volume was from point 1 to point 2. The reaction shaft was operated without fuel.

- The following stage was to increase oxygen enrichment in order to eliminate the fuel consumption in the settler. This is presented by the area B.

The operation conditions when feeding inert material and at the end part of campaign when heat losses are higher is presented by the area C. It is normal to operate even at the point 5 in the Outokumpu flash smelting process producing high grade matte.

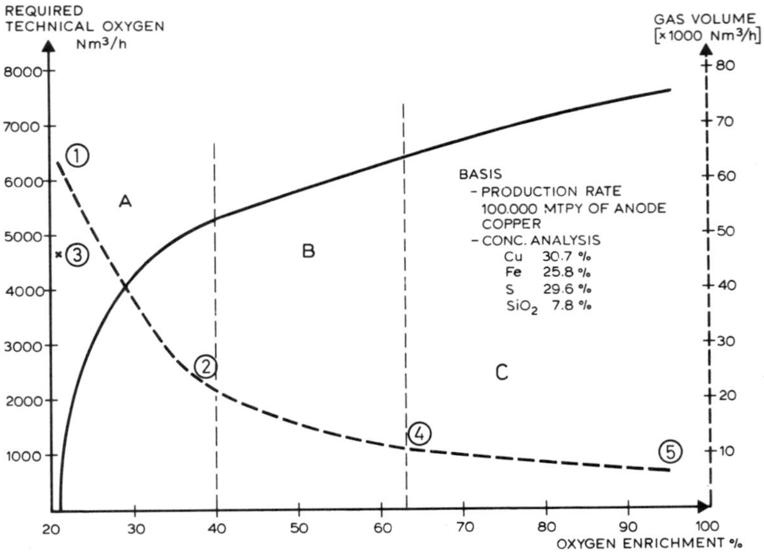

Fig. 2. Process gas volume and required technical oxygen in Outokumpu flash smelting process.

Design Development

The development of the flash furnace cooling system and the concentrate burner during the operational 40 years of the Outokumpu flash smelters has been a fundamental prerequisite for the use of high oxygen enrichment in the flash furnace reaction shaft (4, 5).

The modern cooling arrangement of the reaction shaft and the settler part of the flash smelting furnace is presented in Fig. 3. Cooling of the reaction shaft is made by outside cooling of the steel shell. The settler brickwork is cooled by both horizontal and vertical cooling elements. The elements are copper blocks with cooling water passages inside. The most critical area, on which the most of the development work has been concentrated, is the transition area between the reaction shaft and the settler. The modern design includes special L-shape elements and good overlapping of the elements and the shell cooling. The modern cooling makes long smelting campaigns possible. The record is over 8 years, which is more than 60 times longer than the first 8 week campaigns, which were possible in the beginning with a flash furnace without cooling.

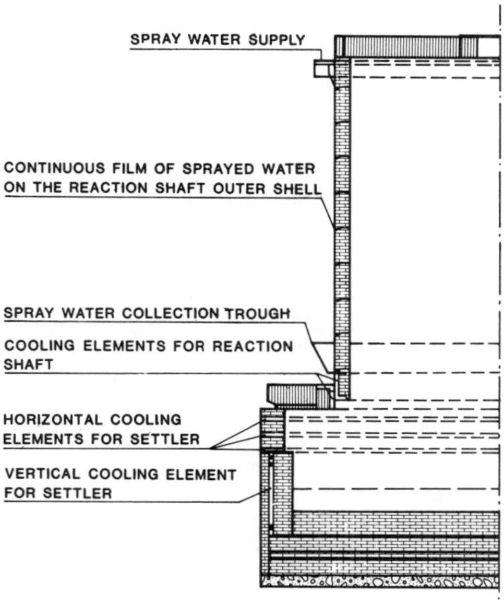

Fig. 3. Principle of flash smelting furnace reaction shaft and settler cooling.

The concentrate burner naturally has been the area of continuous development work. During the early years most of the tested burners were based on the venturi principle. More than 20 modifications of that type of burner were tested during its era, which ended in the late 1970's. Venturi type burner works well with preheated process air and air with low oxygen enrichment. Only when new central jet distribution (CJD) type of burner was developed, the high oxygen enrichment was able to be taken in use with a good utilization of oxygen. The most modern version of the CJD-burner is presented in Fig. 4. Since 1980 the CJD-burner has been a part of the standard design, and several smelters, six until now, have replaced the old venturi burners with the CJD-burner. It also makes the design of one burner for large tonnage possible, as is the case in the latest smelters.

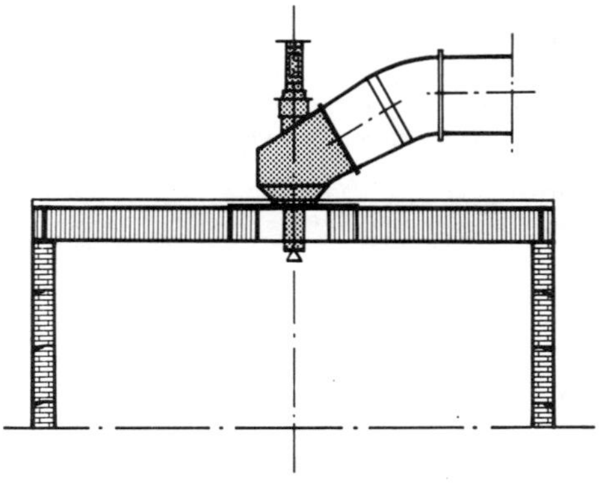

Fig. 4. Central jet distribution burner.

PRESENT PRACTICE

Existing Flash Smelters

The use of technical oxygen in the Outokumpu type flash smelters is presented in Table 1. The number of smelters using or designed to use oxygen enrichment is 25 of total 35. The first smelter designed to use oxygen enrichment in initial operation was KGHM Glogow smelter in Poland. The project started in 1973 and since then altogether 13 smelters have been designed to use oxygen from the beginning. Additionally 12 smelters have increased the smelting capacity by adopting oxygen enrichment. The possibility to increase the capacity of a certain flash furnace by oxygen enrichment is illustrated in the Fig. 2. The additional capacity in a flash smelting furnace is based on the move from operation point 1 to point 2. By oxygen addition the process gas volume is decreased down to one third from 63.000 Nm^3 to 22.000 Nm^3/h, which means an availability of extra capacity in the gas train and a possibility to increase the smelting rate. The gas volume decrease and the respective capacity increase by raising the preheating temperature is given at point 3. The gas volume can be reduced from 63.000 Nm^3/h to 47.000 Nm^3/h by raising the preheating temperature of the process air from 450°C (point 1) to 1050°C (point 3). Figure 2 shows also that the increase of oxygen enrichment over 40 % will reduce the process gas volume very little compared to the change from air to 40 % oxygen enrichment.

TABLE 1. Flash Smelters

		SMELTER START-UP	EXPANSION WITH O_2-ENRICHMENT	DESIGN BASED ON O_2-ENRICHMENT	OXYGEN ENRICHMENT IN OPERATION
OUTOKUMPU OY, HARJAVALTA, FINLAND	COPPER SMELTER	1949	X		60 - 95 %
OUTOKUMPU OY, HARJAVALTA, FINLAND	NICKEL SMELTER	1959	X		85 - 95 %
OUTOKUMPU OY, KOKKOLA, FINLAND	PYRITE SMELTER	1962			
FURUKUWA CO., LTD., ASHIO, JAPAN	COPPER SMELTER	1956	X		60 - 70 %
COMBINATUL CHIMICO-METALURGIC, BAIA MARE, ROMANIA	COPPER SMELTER	1966			
THE DOWA MINING CO., LTD., KOSAKA, JAPAN	COPPER SMELTER	1967	X		30 - 40 %
NIPPON MINING CO., LTD., SAGANOSEKI, JAPAN	COPPER SMELTER	1970	X		25 - 35 %
SUMITOMO METAL MINING CO., LTD., NIIHAMA, JAPAN	COPPER SMELTER	1971			
HINDUSTAN COPPER LIMITED, GHATSILA, INDIA	COPPER SMELTER	1972			
PEKO-WALLSEND METALS LTD., MOUNT MORGAN, AUSTRALIA	COPPER SMELTER	1972			
MITSUI MINING AND SMELTING CO., LTD., HIBI, JAPAN	COPPER SMELTER	1972			
NORDDEUTSCHE AFFINERIE, HAMBURG, FEDERAL REPUBLIC OF GERMANY	COPPER SMELTER	1972	X		40 - 50 %
CHINA QUIXI SMELTER, KIANGZI PROVINCE, CHINA	COPPER SMELTER	1985			
NIPPON MINING CO., LTD., HITACHI, JAPAN	COPPER SMELTER	1972			
WESTERN MINING CORPORATION LIMITED, KALGOORLIE, AUSTRALIA	NICKEL SMELTER	1973	X		25 - 30 %
KARADENIZ BAKIR ISLETMELERI A.S., SAMSUN, TURKEY	COPPER SMELTER	1973			
PEKO-WALLSEND METALS LTD., TENNANT CREEK, AUSTRALIA	COPPER SMELTER	1973			
NIPPON MINING CO., LTD., SAGANOSEKI, JAPAN	COPPER SMELTER	1973	X		23 - 30 %
BAMANGWATO CONCESSIONS LTD., PIKWE, BOTSWANA	NICKEL SMELTER	1973	X		23 - 30 %
HINDUSTAN COPPER LIMITED, KHETRI, INDIA	COPPER SMELTER	1974	X		23 - 26 %
RIO TINTO PATIÑO, S.A., HUELVA, SPAIN	COPPER SMELTER	1975	X		30 - 40 %
PHELPS DODGE CORPORATION, HIDALGO COUNTY, NEW MEXICO, USA	COPPER SMELTER	1976	X		30 - 36 %
KOMBINAT GORNICZO-HUTNICZY MIEDZI LUBIN, GLOGOW, POLAND	COPPER SMELTER	1978		X	55 - 75 %
LA GéNéRALE DES CARRIERES ET DES MINES, LUILU, ZAIRE	COPPER SMELTER			X	50 - 60 %
KOMBINAT NORILSK, USSR	NICKEL SMELTER	1981		X	25 - 35 %
KOMBINAT NORILSK, USSR	COPPER SMELTER	1981		X	25 - 35 %
MEXICANA DE COBRE S.A.. SORONA, MEXICO	COPPER SMELTER	1985		X	35 - 45 %
PHILIPPINE ASSOCIATED SMELTING AND REFINING CORPORATION, PHILIPPINES	COPPER SMELTER	1983		X	30 - 40 %
CARAIBA METAIS S.A., SALVADOR, BRAZIL	COPPER SMELTER	1982		X	50 - 60 %
ONSAN COPPER REFINERY CO., LTD., ONSAN, THE REPUBLIC OF KOREA	COPPER SMELTER	1979		X	50 - 60 %
KOMBINAT G. DAMIANOV, SREDNOGORIE, BULGARIA	COPPER SMELTER			X	30 - 40 %
CORPORATION NACIONAL DEL COBRE DE CHILE, CHUQUICAMATA, CHILE	COPPER SMELTER			X	30 - 40 %
JIWCHUAN SMELTER, GANSU PROVINCE, CHINA	NICKEL SMELTER			X	25 - 35 %
MAGMA COPPER, SAN MANUEL, ARIZONA, USA				X	35 - 45 %
ROXBY MANAGEMENT SERVICES, OLYMPIC DAM, AUSTRALIA	COPPER SMELTER			X	70 - 95 %

The modern design of the flash smelting furnace is based on autogenous operation of the furnace with only oxygen enrichment. This means the operation of furnace in area B (from 40 % to 63 %) in the case presented in Fig. 2. Table 1 shows that the enrichment used in most existing smelters falls in the same area. The high enrichment figures in use indicate that part of the feed must be in the form of inert material like precipitates at Outokumpu Harjavalta. Normal sulfidic concentrate would generate too much heat in this area C of Fig. 2. When blister copper is directly produced in the flash furnace from low heat value concentrate, like in cases of KGHM Glogow and RMS Olympic Dam, a high degree of oxygen enrichment is also needed and is possible to use.

Retrofitting the Reverberatory Smelter

Two of the Outokumpu flash smelter projects started in the 1980's are retrofitting projects of existing reverberatory smelters. The projects are; Codelco, Chuquicamata where the nominal capacity of the furnace is 1920 mtpd of concentrate and Magma Copper, San Manuel with the nominal capacity of 2720 mtpd of concentrate. In the case of San Manuel the flash smelting furnace will handle total feed of the smelter while in the case of Chuquicamata some concentrate is smelted in the reverberatory furnace and the CMT-reactor. As normal in both cases large feasibility studies were prepared to select the smelting method. The final selection was made between different flash smelting processes (6).

The flash smelting furnace of the San Manuel smelter, as presented in Fig. 5 is part of the smelter modernization and air pollution abatement project and it will have about the same smelting capacity as the three reverberatory furnaces theoretically have had.

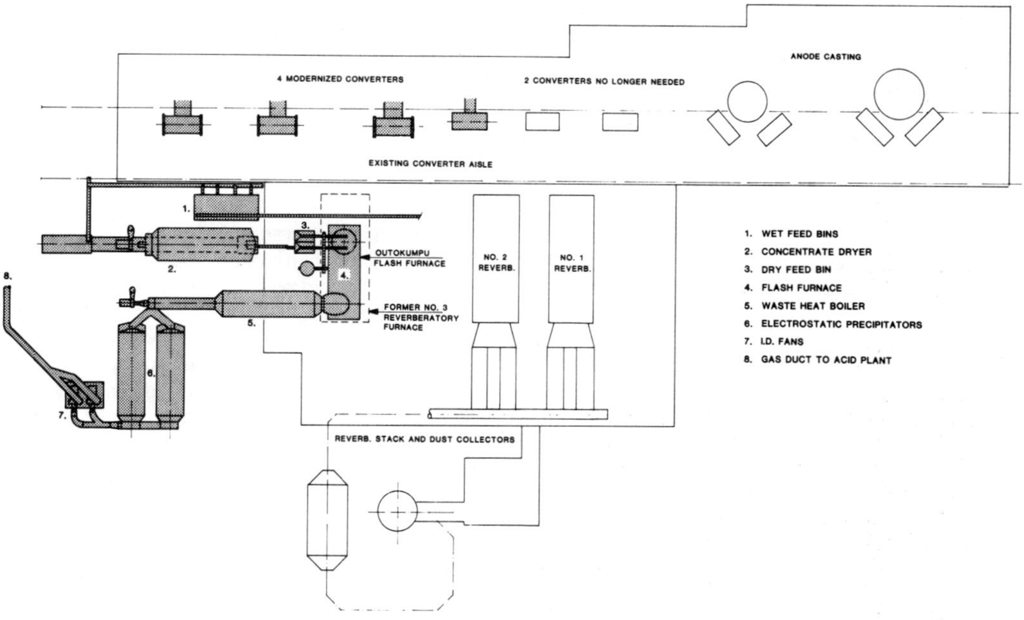

Fig. 5. Retrofit layout of Outokumpu flash smelting process in San Manuel Smelter.

At the same time it will be the largest single unit smelting furnace in the non-ferrous industry in terms of metal production. The flash furnace at San Manuel will incorporate the latest "state of the art" design features and process technology. This contributes to the large production capability with a furnace, which is considerably smaller in physical dimensions than the two or three other flash furnaces operating at similar concentrate smelting capacities. It is also the first Outokumpu retrofit flash furnace practically inside an existing smelter building as is shown in Fig. 6.

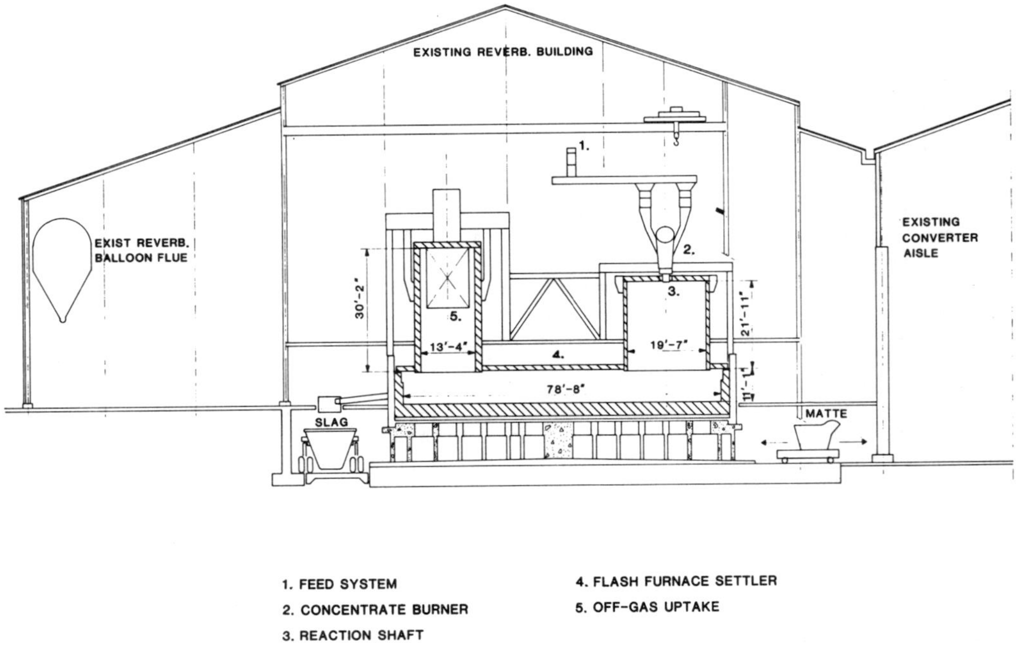

Fig. 6. Outokumpu flash smelting furnace retrofit in San Manuel Smelter building.

Some of the more significant features of the Magma furnace design include the use of one concentrate burner, designed for up to 165 metric tons per hour of total feed to the furnace. A single burner in place of 4 as utilized in former large furnace designs will offer improved combustion and process control and should contribute to longer campaigns. Process calculations for this project were based on the optimum degree of oxygen enrichment of process air to achieve autogenous reactions at the design capacity while producing the desired matte grade of 63 % copper. This results in a great deal of flexibility to alter feed rates or feed compositions, or to adjust matte grades up or down without heat balance or process control problems. Furnace availability and campaign life will be maximized through application of some other new design features. The latest sidewall cooling system design will increase refractory life. The roofs of the settler, reaction shaft and uptake are all totally flat suspended.

Direct Blister Smelting

In principle, blister copper can be produced directly from concentrate as the primary smelting product in one unit using practically any type of sulphidic copper concentrate as raw material. This is an attractive approach and has always been the dream of all copper metallurgists.

However, certain metallurgical facts limit the applicability of this approach (7). Firstly, if the copper content or the copper/iron ratio in the concentrate is too low, the quantity of slag in smelting is high. This means that the recovery of copper into the metal phase is low and that most of copper reports to the slag owing to the high copper content of slag. Depending on the slag cleaning method the circulating copper load becomes extremely high or the slag reduction too costly.

Secondly, the impurity content should preferably be low. The direct blister process has rather poor eliminations for As, Sb and Bi while Pb and Zn don′t give rise to problems. Although the harmful impurities can be eliminated from the blister copper, this results in additional costs.

When designing and constructing a smelter for this kind of concentrate, the advantages of the direct blister process are evident:

- No converting
- One continuous gas stream
- Gas rich in sulphur dioxide
- Low emissions and fugitives
- Reduced labor requirement
- Applicable to automation

By today three licences to use the Outokumpu Direct Blister Smelting Process have been granted. The first one is in Poland, where the smelter has been in operation since 1978. This concentrate is very low in iron. The second licence was granted to Zaire where the copper content of the concentrate is high. The construction of this smelter has been interrupted for the time being, because of local conditions.

The newest example is in Australia where Roxby Management Service Pty Ltd is building a new smelter based on Olympic Dam concentrate, which has a very high copper content. Because of the mineralogical composition of this concentrate, the smelter will operate almost autogenously with pure technical oxygen or very high oxygen enrichment. The Olympic Dam smelter is presented in details in another paper to be presented in this oxygen symposium (8). Smelter will be started mid 1988.

NEW OPPORTUNITIES

In addition to the consequences discussed in the preceding chapters, the use of oxygen has opened new opportunities to apply flash smelting technology in copper smelting. The most important is the flash converting process, which can be used both in existing and new smelters. This process can be utilized in the small scale flash smelter application using same furnace alternately for smelting and converting. For medium capacities the most attractive method is so called three shaft furnace, where smelting and converting take place at the same time in a combined furnace.

Flash Converting Process

A lot of attention has been paid to primary smelting throughout the years so that the modern technologies utilizing the advantages of oxygen are gradually replacing the old copper smelting processes. However, converting has remained at about the same technical level as early in this century when the Peirce-Smith converters were adopted. The PS converting is a batch process which produces fluctuating and high volumes of gas and a lot of fugitive gases.

In the new Kennecott-Outokumpu flash converting process, which has already been introduced in several connections, the benefits of oxygen smelting have been effectively exploited (9, 10). The gas volumes are continuous, small and rich in sulphur dioxide leading to considerable savings in capital and operating costs. Moreover, the process is completely closed and matte is not transferred liquid in ladles. These are factors, which make it hygienic and environmentally attractive.

The basic idea of the flash converting process is to feed the matte solid into the converting furnace. This fact together with a suitable matte grade makes it possible to produce blister copper by using pure technical oxygen or very high oxygen enrichment without additional fuel. Even if external energy is needed to produce oxygen, more is saved in small gas volumes and high sulphur dioxide strengths.

The flow sheet of the flash converting process in connection of flash smelter is presented in Fig. 7. The concentrate is first smelted in the primary smelting furnace to produce high grade matte. The matte is then granulated, ground and dried. The fine-grained matte is finally smelted into blister copper in the second flash smelting furnace. The small amount of formed slag is returned to the primary smelting furnace.

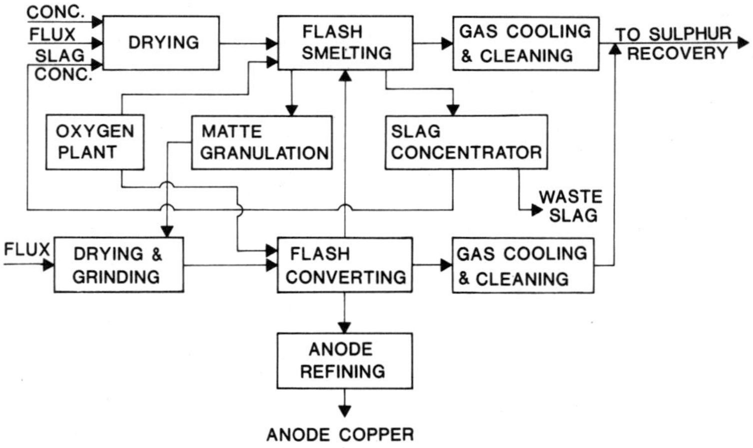

Fig. 7. Flow sheet of flash smelter with flash converting.

The heat balances of Peirce-Smith converting and flash converting and in Fig. 8 and 9 describe well the importance of oxygen and solid matte feed. One can see that in PS converting the sensible heat brought by liquid matte is totally consumed in heating the nitrogen of the blowing air. This is the answer to the frequently presented question: "Why to lose the heat content of the liquid matte?"

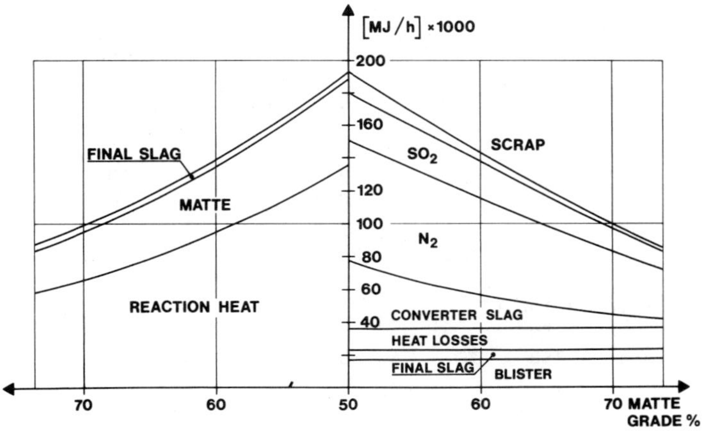

Fig. 8. PS converting heat balance.

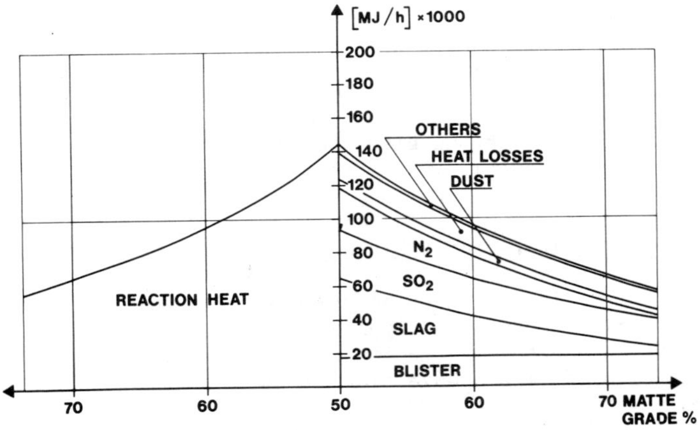

Fig. 9. Flash converting heat balance.

The copper content of the matte to be flash converted can be selected according to the overall economics of the smelter. To flash converting itself the high matte grade is an advantage. Operation is then based on high oxygen enrichment resulting in low gas volume with high SO_2-strength. Also slag amount and circulating copper load are small.

Because of the practically total absence of gangue components in the high grade matte, the type of slag can be freely selected. In Table 2 the slag composition and the recovery of copper into the blister phase is shown as a function of the type of slag and oxidation degree.

TABLE 2 Distribution of Copper in Flash Converting as a Function of Type of Slag and Oxidation Degree.

Matte grade	% Cu	70	70	70	70
Blister copper	% S	0.8	0.1	0.8	0.1
Slag	% Cu	16	23	11	18
	% Fe	32	26	38	30
	% S	0.1	0.1	0.1	0.1
	% CaO	1.5	1.5	20	20
	% SiO_2	25	25	3	3
Distribution of copper into metal phase	%	96	94	98	95

One can see that the copper recovery into the metal phase is very high in spite of the high copper content of the slags. Additionally the copper content of the calcium based slag is typically lower than that of the silica based slag, thus meaning a higher recovery of copper into the metal phase. Also the oxidation degree, which is characterized by the sulphur content of copper, affects the slag composition so that one can notice a difference of 2-3 per cent units in the copper distribution.

The flash converting together with flash smelting is able to eliminate impurities as effectively as PS converting with flash smelting. Lime based slag, which is possible to use in flash converting gives the process even better conditions for removing arsenic and antimony (11).

Flash Converting Process in Retrofit Situation

The Kennecott-Outokumpu flash converting process suits well in a retrofit case to replace old Peirce-Smith converters. Typically the converters and the acid plant are the bottlenecks in smelter, for which situation flash converting is a suitable solution. The benefits of flash converting can be best utilized in smelters that produce high grade matte, typically Outokumpu type flash smelters. However, depending on local conditions lower matte grades are possible, but the profitability has to be carefully checked.

The effect of even gas flow on acid plant design is very worth of consideration. In PS converting the additional unit will increase the design capacity of an acid plant stepwise as presented in Fig. 10. This figure comprises data for the acid plant design capacity as a function of annual copper production. In the PS converting case the flash smelting furnace produces 60 % matte grade.

The comparison shows clearly how the flash converting process will reduce the design values of the acid plant in a green fields case and will make the more efficient utilization of the existing acid plant possible. For example, as presented in Fig. 10, if a smelter with a flash smelting furnace and two PS converters is designed for any capacity in the range of 80-130.000 mtpy of anode copper production using 60 % matte grade, the existing acid plant has the same capacity as is needed to produce 240.000 mtpy of anode copper by utilizing flash converting process instead of PS converters.

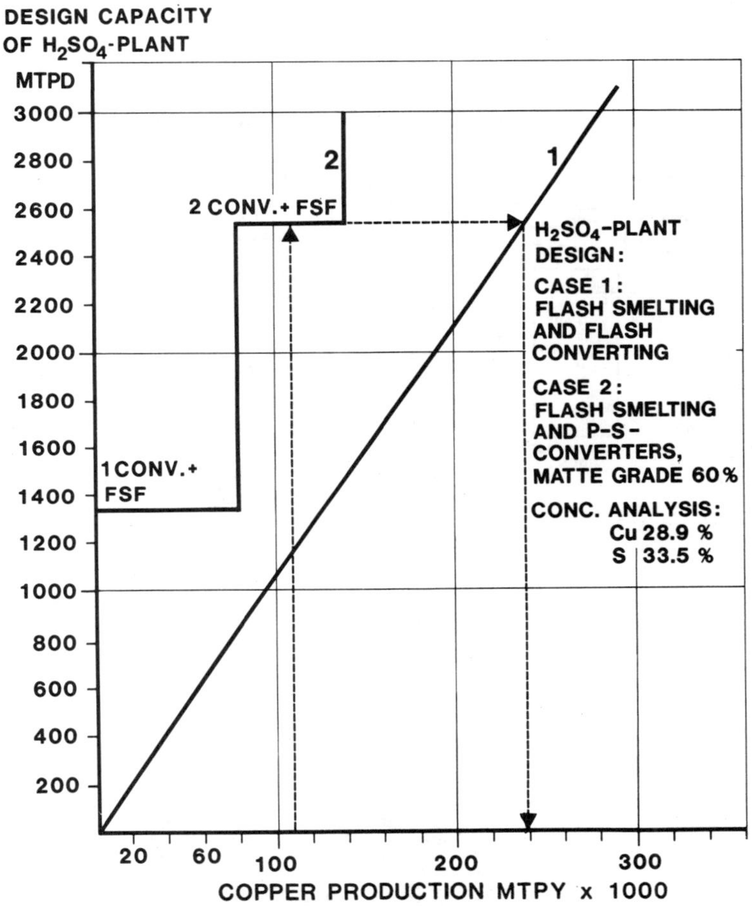

Fig. 10. Copper production and design capacity of acid plant.

Flash Converting Process in New Smelter

When building a new smelter, the combination of Outokumpu flash smelting and Kennecott-Outokumpu flash converting, seems to be the most attractive alternative. Then the design and operating conditions are optimum. The calculations show that the investment cost of this process combination, is considerably (10 to 20 per cent) lower than the corresponding costs for a flash smelter with PS converters. The entire smelter complex, including acid plant, is taken into the comparison of costs.

Figures 11 and 12 show the layout and a section of a flash smelter with flash converting. One can see how compact it is and how little space it requires. No large converter aisle with heavy cranes is needed and manpower can be effectively utilized.

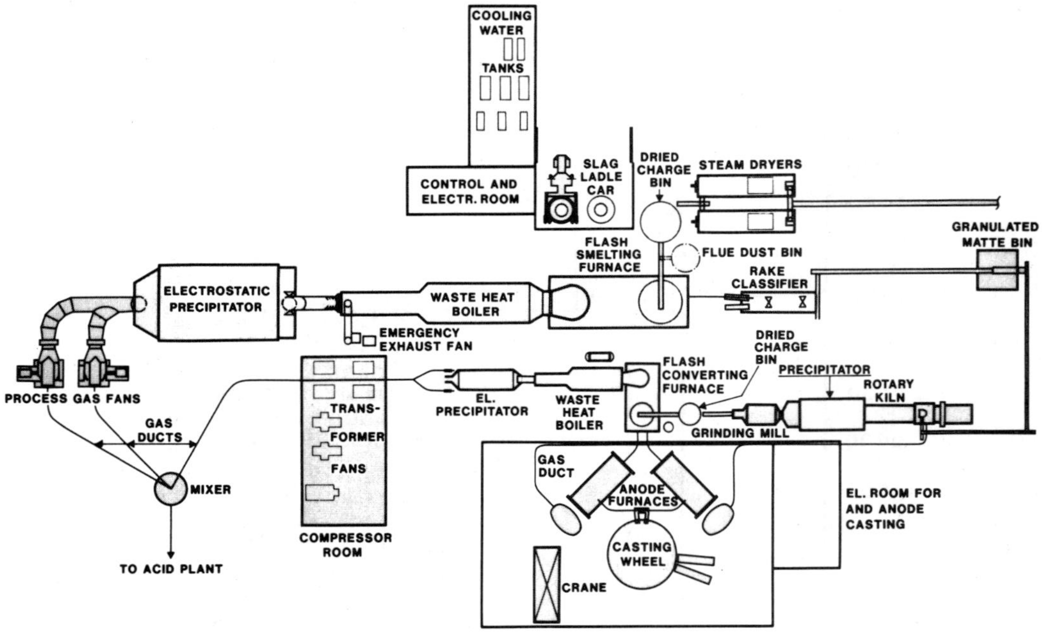

Fig. 11. Flash smelter with flash converting, Layout.

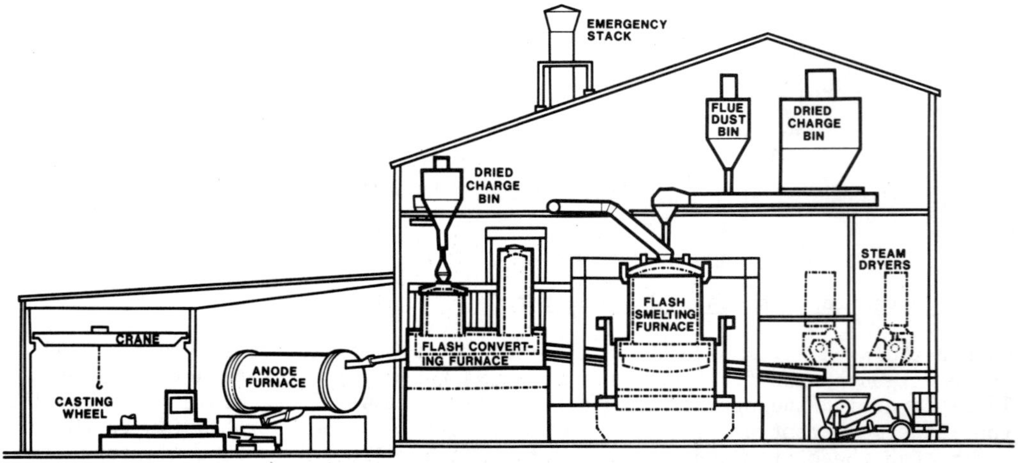

Fig. 12. Flash smelter with flash converting, Section.

Small Scale Smelter

Building of a small scale copper smelter is difficult to justify today, although certain need to do so, exists in several locations. Operation of a small scale copper smelter is characterized by the following features, which make the profitable operation difficult.

- In blister copper production the profitability depends radically on the required investment cost.
- When the production rate is reduced the saving in the capital cost is not in direct proportion.
- The required operation staff can be kept nearly constant in the conventional smelting processes with production rates from 10000 mtpy to 70000 mtpy.
- It is expensive to meet the environmental requirements.
- The smaller the scale of the smelter, the higher the portion of the generated secondary revert.

The use of oxygen in the Outokumpu flash smelting process and practical experience collected in the design and operation of the direct blister process gives a possibility to take a new approach to design a modern copper smelter for low capacity. In our recent studies of a small scale smelter the best alternative to comply the conditions presented above has turned out to be the application of flash smelting and flash converting in same unit by turns. Of course, in this case the furnace itself has to fulfil the requirements of smelting and converting. However, time needed for smelting is longer (about 2/3 of total) than the time needed for converting (about 1/3 of total). The selected process principle is simple and the flowsheets of its periods smelting and converting are presented in Fig. 13 and 14.

In the first period, concentrate, flux and granulated converting slag is dried and smelted to produce high grade matte in the flash smelting furnace. Matte is granulated and transported to storage for further converting. Slag is transported to a cooling area and will be treated in a flotation plant or discarded. The high strength sulphur dioxide gases are conducted to the sulphur recovery e.g. acid plant and/or liquid sulphur dioxide or elemental sulphur production. The gas volume is small and the SO_2-content 70-80 %.

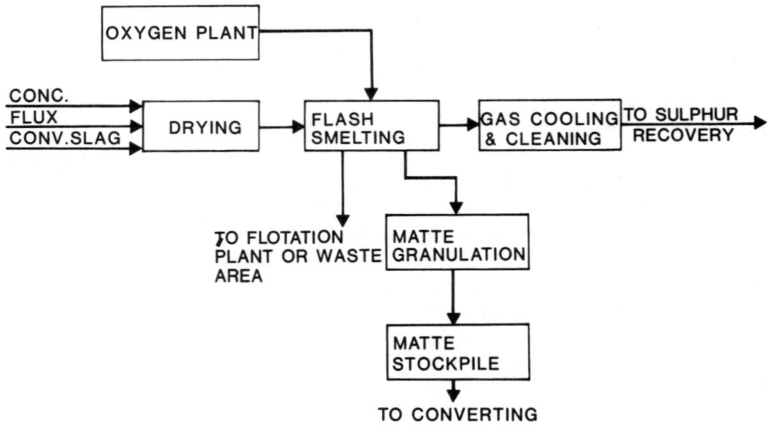

Fig. 13. Process block diagram (smelting).

In the second period the normal flash converting process is carried out in the same main equipment as in the smelting period. Only a grinding mill must be added to the flowsheet. The test results have proven that a part of matte has to be ground. The produced blister copper is laundered to the blister ladle and is transferred by rail to the blister casting machine. The anode refining furnaces and the casting machine may be installed later. Then direct laundering of blister copper from the flash furnace to the anode furnaces is possible. Flash converting slag is granulated using the same granulation basin and equipment as was used for matte granulation in the first period. Slag is transported to storage for further treatment during the smelting period.

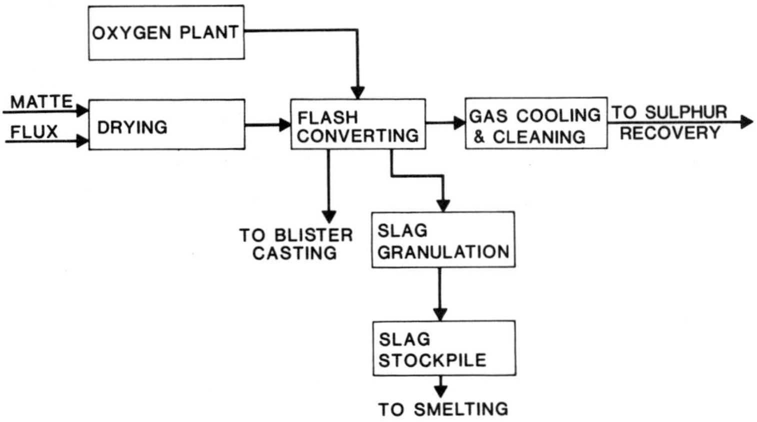

Fig. 14. Process block diagram (converting).

The layout of this small scale application is presented in Fig. 15. The size of the furnace must be kept in mind in comparison purposes. The length of the furnace is about 10-12 m and the gas volume only 3-5000 Nm^3/h for production of 20.000 mtpy of anode copper.

The advantages of using the same unit alternately for smelting and converting in a low capacity smelter are:

- The normal PS converting with heavy converter aisle construction is eliminated totally.
- The gas flow at the inlet of the acid plant is nearly constant, because gas production is the same in smelting and converting stages and overdesign of the acid plant is not necessary. The needed acid plant capacity is considerably lower than with the conventional PS converters.
- The oxygen plant can be operated at a constant production level all the time.
- Operating labor cost, including maintenance, is the main part of cost in normal small scale of operation based e.g. electric smelting with PS converters. This modern oxygen technology based smelter can be operated using the same operating crew for smelting and converting.
- The proposed process provides one of the best solutions to comply with the environmental regulations in low capacity area.
- The increase of the production capacity can be done by adding a smelting unit. The flash converting and anode refining are already

designed for a higher capacity, because converting period is in use only 1/3 of the total time. The slag from converting will be laundered liquid to the smelting furnace after expansion.

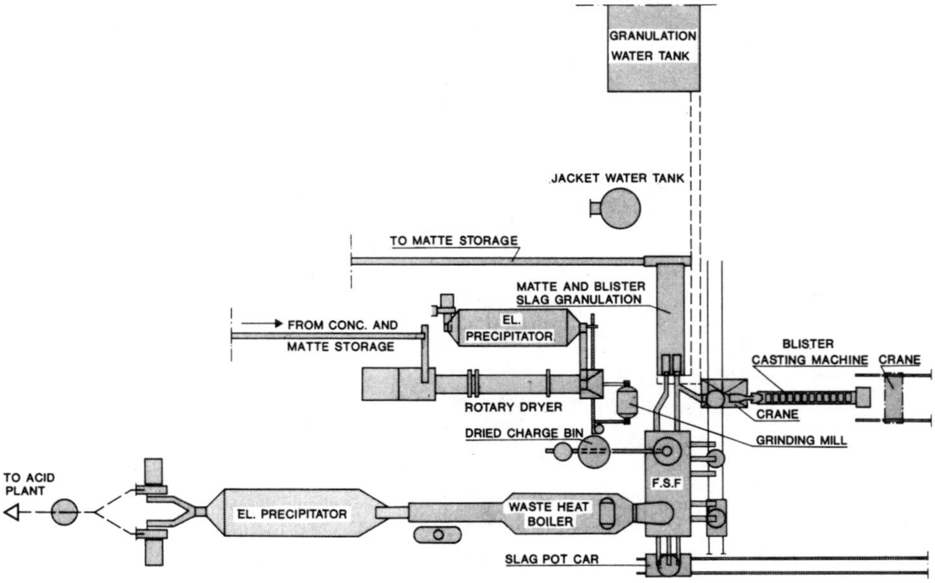

Fig. 15. Flash smelting and flash converting alternately in the same furnace, Layout.

Three Shaft Flash Smelting Furnace

Next phase in the development work on the Outokumpu flash smelting process could be the combining of smelting and converting to the same furnace through separate reaction shafts. This approach will offer certain advantages in addition to those of separate flash smelting and flash converting furnaces (12). The biggest benefits a three shaft furnace will offer in the areas of gas and slag handling. A three shaft furnace will produce only one continuous gas stream and one type of slag, which can be treated in flotation plant with good recovery. Because of the above features, application of a three shaft furnace is in medium size smelters.

Its investment cost will be lower than that of a smelter with separate flash smelting and converting furnaces. The proposed process flowsheet is presented in Fig. 16 and one possible layout of the smelter is in Fig. 17.

CONCLUSIONS

The use of oxygen enrichment in Outokumpu flash smelting process is a part of standard design today. The flash smelting furnace is designed for autogenous operation with oxygen enriched process air only, and oxygen enrichment can vary from air to technical oxygen. Oxygen has made it possible to design an Outokumpu flash smelting furnace to smelt raw materials,which were not possible to treat before adoption of oxygen.

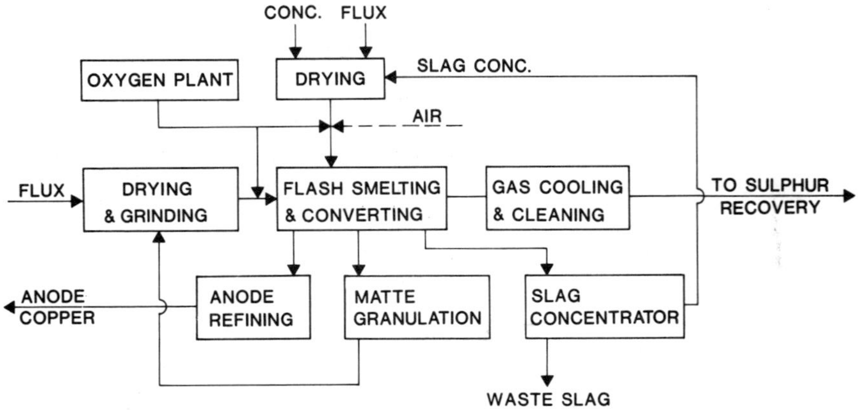

Fig. 16. Process flowsheet for combined furnace.

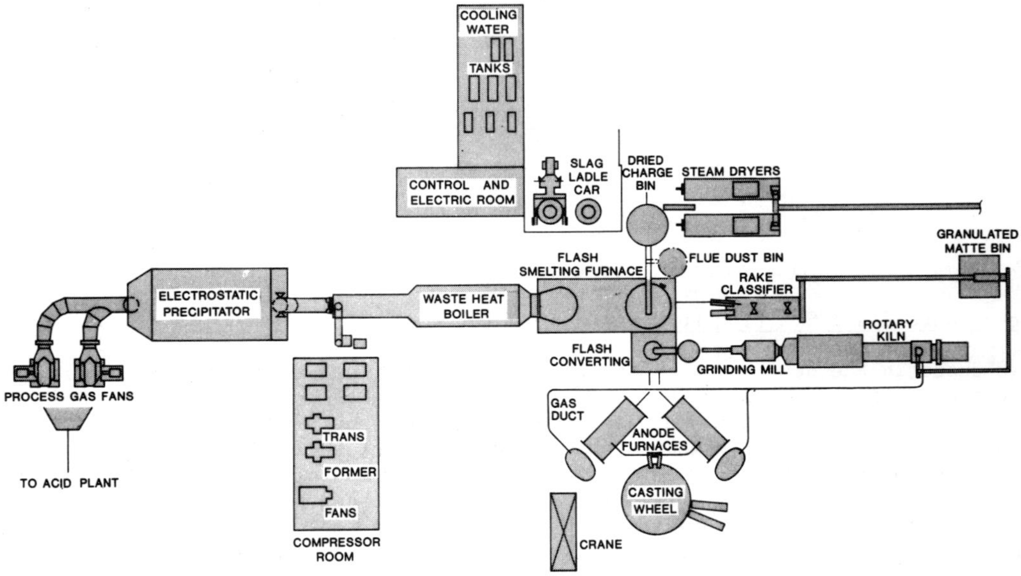

Fig. 17. Flash smelting and flash converting in the combined furnace.

The existing flash smelters have been able to increase the smelting capacity and decrease the operation cost easily by taking oxygen enrichment in use. Using oxygen to design a new flash smelter has resulted in considerably smaller investment costs and lower operation costs as well as easier overall operation of the flash smelter. The benefits are the biggest in the flash smelter where process gases have to be treated and sulphur dioxide captured.

New modifications of the Outokumpu flash smelting process have been possible to introduce after adaptation of oxygen and improvements in the furnace cooling and concentrate burner. The most remarkable applications are the direct blister and flash converting processes. Those inventions result in new solutions for the complete smelter, especially in the small smelting capacity. The following process modifications, presented in Table 3, in the different capacity ranges are proposed to be investigated prior to the final process selection.

TABLE 3 Flash Smelting Process Modifications

Copper production mtpy	Process
10 000 - 25 000	Flash smelting and flash converting in the same furnace alternately.
25 000 - 150 000	Flash smelting and flash converting in combined furnace or flash smelting and flash converting in separate furnaces.
over 150 000	Flash smelting and flash converting in separate furnaces.

REFERENCES

Andersson, B., Y. Anjala, and T. Mäntymäki (1980). Development trends in Outokumpu flash smelting technology with particular reference to the requirements of furnace reaction shaft cooling techniques and refractories. Paper presented at 109th AIME Meeting, Las Vegas, Nevada. (4).

Andersson, B., P. Hanniala, and S. Härkki (1982). Use of oxygen in the Outokumpu flash smelting process. CIM Bulletin, 845, 172-177. (2).

Asteljoki, J., and M. Kytö (1985). Alternatives for direct blister copper production. Paper presented at 114th AIME Meeting, New York. (7).

Asteljoki, J., L.K. Bailey, D.B. George, and D.W. Rodolff (1985). Flash converting-continuous converting of copper mattes. Journal of Metals, 37, 5, 20-23. (9).

Asteljoki, J., Y. Anjala, and M. Kytö (1986). Flash converting. Proceedings of 5th Flash Smelting Congress, Helsinki, Finland. (10).

Asteljoki, J., and M. Kytö (1986). Minor element behaviour in flash converting. Paper presented at 115th AIME Meeting, New Orleans, Louisiana. (11).

Cocquerel, M. A. T., and S. R. Holmes (1987). Retrofitting existing copper smelters with modern flash smelting technology. Paper presented at 116th AIME Meeting, Denver, Colorado. (11).

Hanniala, P. (1986). Method for processing sulphide concentrates and sulphide ores into raw material, U.S. Patent No 4599108. (12).

Härkki, S., O. Aaltonen, and T. Tuominen (1976). High grade matte production with oxygen enrichment by the Outokumpu flash smelting method. Proceedings of symposium on copper extraction and refining, 105th AIME Meeting, Las Vegas, Nevada. (1).

Rodolff, D. W., M. Haani, K. Helne, and B. Andersson (1984). Outokumpu flash smelting an update and retrofit considerations. Paper presented at Arizona Conference AIME 1984 Meeting, Tucson, Arizona. (5).

Rodolff, D. W., Y. Anjala, and P. Hanniala (1986). Review of flash smelting and flash converting technology. Paper presented at 115th AIME Meeting, New Orleans, Louisiana. (3).

Smith, T. J. A., I. Posener, and C. J. Williams (1987). Oxygen smelting and the Olympic Dam project. Paper to be presented at 28th CIM Annual Meeting, Winnipeg, Manitoba. (8).

BULK OXYGEN USE IN THE REFINING OF PRECIOUS METALS

J.G. Cooper, J.W. Matousek and J.G. Whellock

Tolltreck International Limited
7108-M South Alton Way
Englewood, Colorado, U.S.A., 80112

ABSTRACT

The employment of oxygen in bulk quantities for precious metals refining is a relatively new innovation when compared with its application in the iron and steel and copper industries. The impacts of its adoption, however, have been no less dramatic, and oxygen use in pyrometallurgical treatment of precious metals bearing materials is now "state-of-the-art".

In several respects the practical and theoretical bases for oxygen refining of precious metals pre-date those of iron and copper. As in these other industries, the adoption of oxygen had to await the development of the means for producing low cost oxygen in large quantities and of process furnaces that could effectively and economically use this powerful, "new" reagent. Moreover, the trend towards oxygen use has been driven by forces that became particularly evident in the late 1970's -- rising interest rates, rising energy costs, and the general need to decrease in-process inventories and treatment times.

Today the most versatile and efficient furnace for pyrometallurgical refining of precious metals is the top blown rotary converter (TBRC), now commercially available in sizes from 0.5 tonnes to 10 tonnes capacity. Such furnaces are currently either in use or are being considered for use in the smelting and refining of such diverse materials as gravity and flotation concentrates, Merrill-Crowe precipitates, zinc crusts from desilverizing of lead, anode slimes from the electrolytic refining of lead and copper, and secondaries such as electronic scrap, sweeps, and photographic wastes.

Key Words

Oxygen
Precious Metals
Gold
Silver
Cupellation
Top Blown Rotary Converter
TBRC
Refining

INTRODUCTION

This paper is primarily concerned with the adoption of bulk oxygen use to the recovery of precious metals, principally gold and silver. It is appropriate, however, to first review the extractive metallurgy of these metals and some history of the processes involved.

Figure 1 surveys the general methods in use to recover gold and silver from primary, by-product and secondary sources. Some of these have been practiced since before recorded history, and come to us virtually unchanged since their initial discoveries. Of the methods shown for treatment of primary and by-product sources, cyanidation is one of the "newer" processes; its adoption as a replacement for chlorination occurred at the end of the last century. Amalgamation of gold with mercury was a practice known to the Romans; the recovery of silver by amalgamation dates from the 16th century. The use of mercury in gold and silver recovery has rapidly declined over the past several decades in response to environmental and health concerns.

Herbert and Lou Hoover (1950), in notes to their translation of Agricola's De Re Metallica, suggest that the extractive metallurgy of gold and silver began with free-gold recovery from alluvial deposits, progressed to the direct smelting of gold ores, and was followed by the more complex smelting of lead/silver ores and cupellation, the process by which molten lead/silver alloys are oxidized with air or fluxes in shallow-bath furnaces to yield a lead oxide slag (litharge) and relatively pure, silver metal. It is possible that cupellation to separate silver and lead predates methods to separate silver and gold, and is the common ancestor of all oxidative refining processes.

The classical fire assay is also an example of oxidative refining, and shares in common some of the steps for gold and silver recovery as by-products, shown in Fig 1. First a quantity of material is prepared. This corresponds to the concentration step in the flowsheet. Then, depending upon the nature of the charge, a scorification or crucible assay may be performed (Bugbee, 1940). The pyrometallurgical parallel to the first is a combined roasting and smelting operation and to the second, furnace smelting with fluxes. On the commercial scale, it is common to find either roasting and smelting or direct smelting of "green" concentrates in use. In copper metallurgy, electrolytic refining yields slimes that are pretreated before cupellation; lead smelting produces metallic "bullion" that could, if rich enough in silver, go directly to cupellation. This was probably the practice followed by the ancients, discussed by the Hoovers. Such a route would more appropriately be placed under the primary source recovery methods. As practiced today, lead is refined either electrolytically to make slimes that are pretreated, similarly to those from copper, before cupellation, or pyrometallurgically, where the silver and gold are extracted with zinc by the Parkes process into a "crust" that is treated by one of several methods to remove zinc and then cupelled to dore' (a silver/gold bullion).

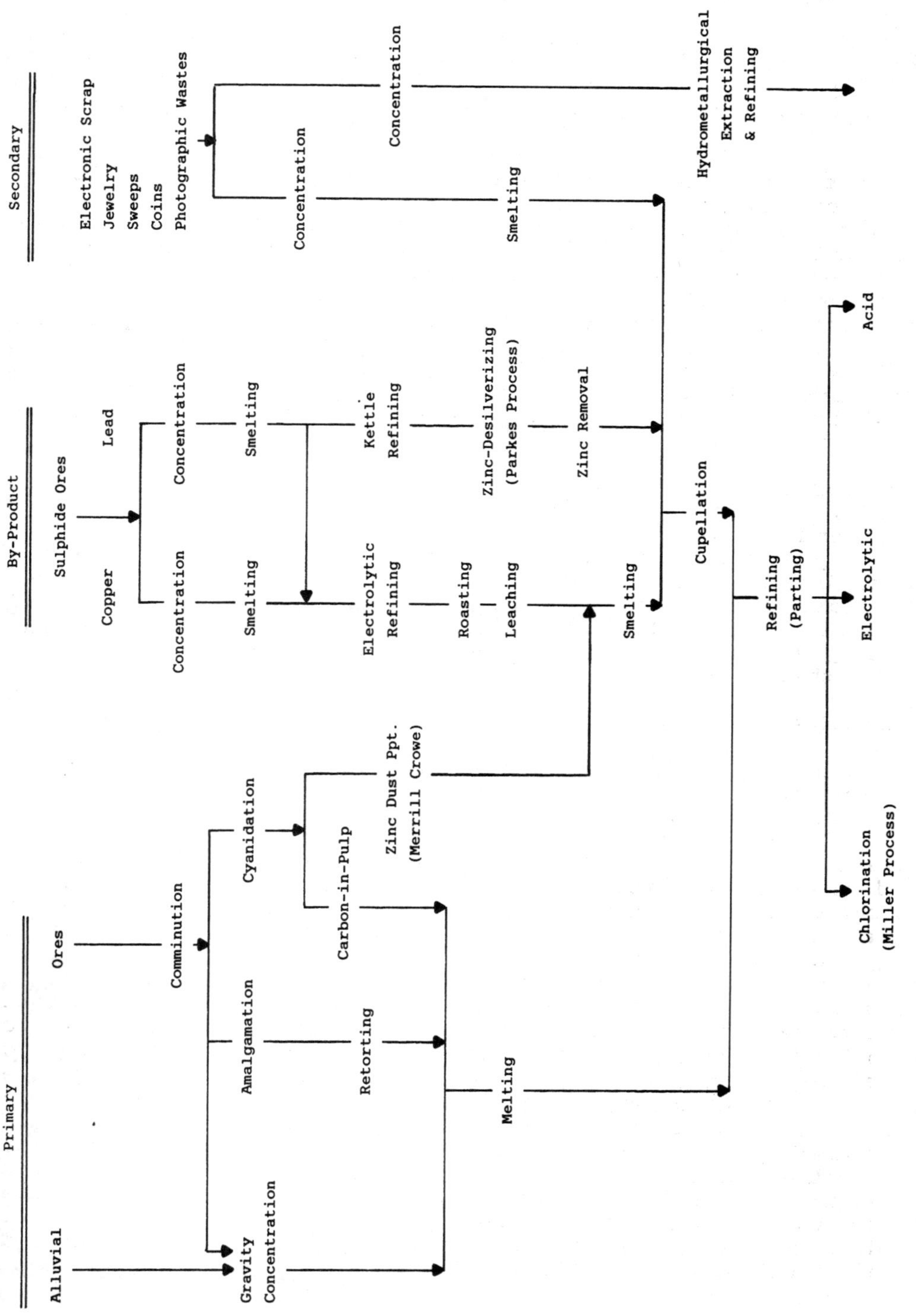

Fig. 1 Extractive Metallurgy of Gold and Silver

Finally gold and silver are "parted" (refined) by one of the three methods shown -- electrolytic, chlorination, or acid. Another method was practiced by the ancient metallurgists, salt cementation. This process is described both in De Re Metallica and more recently by Healy (1979). Briefly, the gold/silver alloy is roasted with fluxes, one of which is common salt. Silver is chlorinated and recovered in the "cement", what we would today term a slag or a dross. The point to be noted here is that coming into the last century, metallurgists would have been familiar with the long history of association between chloride salts and the refining of gold and silver.

OXIDATIVE REFINING: THEORY

The principle of oxidative refining lies in the tendencies of the different metals to form stable oxides. The process is illustrated in the curves of Fig. 2 which show the removal of impurities from "black" copper. An Ellingham (free energy) diagram for the formation of some metal oxides is shown in Fig. 3. The free energy equations for these curves were written for one mole of oxygen gas at one atmosphere pressure and for pure molten metals and their liquid oxides. For a number of these the standard states are supercooled liquids.

$$Me(l) + O_2(g,\ 1\ atm) = MeO_2(l) \qquad (1)$$

The tendency for a metal to react with oxygen and form a stable oxide, as measured by the free energy of formation of the latter, is tempered by the concentration (activity) of the metal in the molten copper. A lowering of the activity of the metal component of the reaction causes the free energy curve to move upwards, towards less negative free energy values. While the Ellingham diagram predicts the order of removal to be Zn:Fe:Sn:Ni:Pb, the test data show the order of zinc and iron reversed and suggest that all of the nickel may never be removed.

Returning again to the discussion of historical metallurgy, oxidative refining of the precious metals in cupellation furnaces was a long, established practice as the metallurgists of the world entered this century, and it would seem that the time should have been right for oxygen use on a more intensive scale. However, when the records are examined, it is found that refining with chlorine gas came first. From the discussion above, this might not come as a surprise. The association of chloride salts with precious metals refining has been noted. In 1838, L. Thompson discovered that molten gold could be refined by passing through it a stream of chlorine (Rose, 1905). In the 1850's, the German chemist, Plattner, experimented with the extractive metallurgy of gold using chlorine compounds, and the process that bears his name, with a salt roast and water leach, was commercialized in the last half of the 19th century. The largest chlorination plant in the world, one based on the reaction of sulphuric acid and chlorinated lime to generate chlorine gas in the presence of gold ores, would eventually be established at Mount Morgan in Australia (Roberts-Austen, 1910). It was probably no accident, then, that the use of chlorine to refine molten gold was first applied on an industrial scale at the Sydney Mint by F. B. Miller.

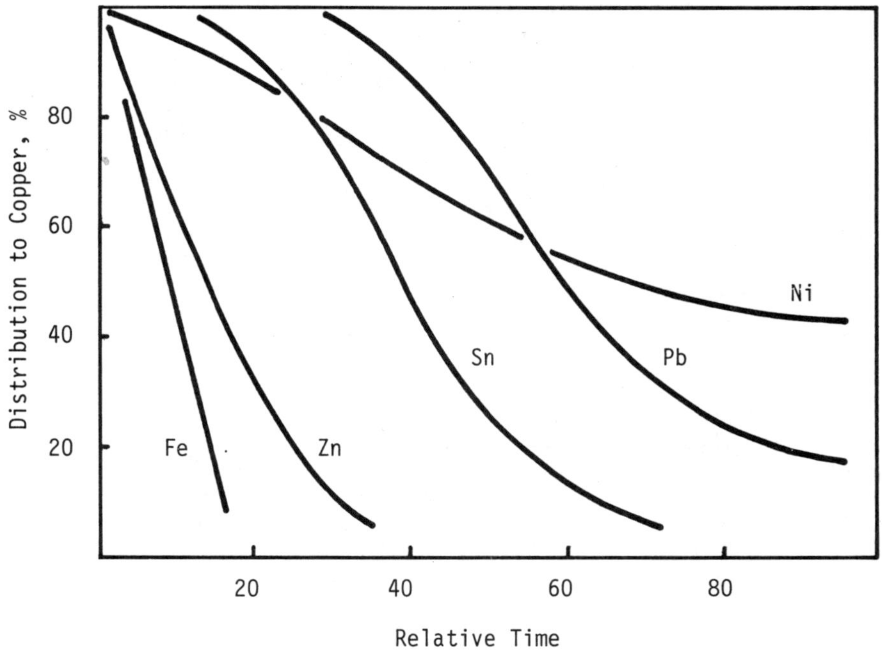

Fig. 2. Oxidative Refining of "Black" Copper

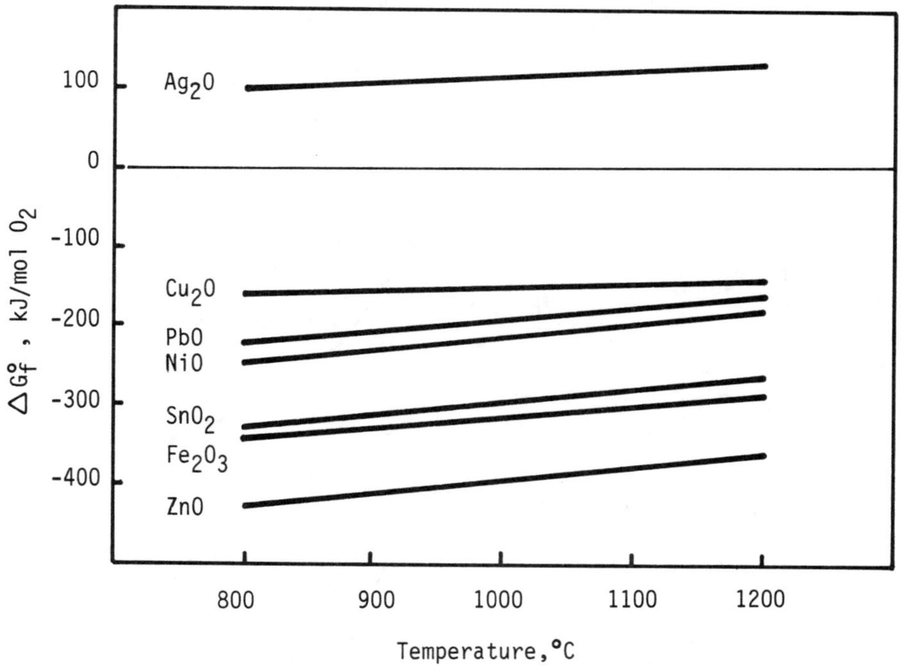

Fig. 3. Free Energies of Formation of Metal Oxides

The Miller patent notes that hydrochloric acid gas in a stream of air or oxygen can be substituted for chlorine for gold refining, and D. Clark, also from Australia, experimented with the purification of molten gold using air and a borax flux. Credit for demonstrating the commercial potential of oxygen in precious metals refining, however, properly goes to T. K. Rose (Rose, 1905).

Figure 4 shows the time progression of the oxygen refining of a gold bullion. The starting material was a synthetic, cyanide process, zinc precipitate (see Fig. 1); oxygen was drawn from compressed gas cylinders. The free energy data of Fig. 3 apply. Rose and Newman (1937) note that in practice the order of impurity removal is generally Zn:Fe:Sb:As:Pb:Bi:Ni:Te:Cu. Atmore, Howat, and Johens (1971) and de Beer (1968) give additional data on refining gold with oxygen.

From the time of its introduction at Sydney in 1867 to the present day, the Miller chlorination process has remained in continuous operation. By 1905, it appears that all of the components had been assembled to also make oxygen refining a commercial success, but nothing happened for 60 years.

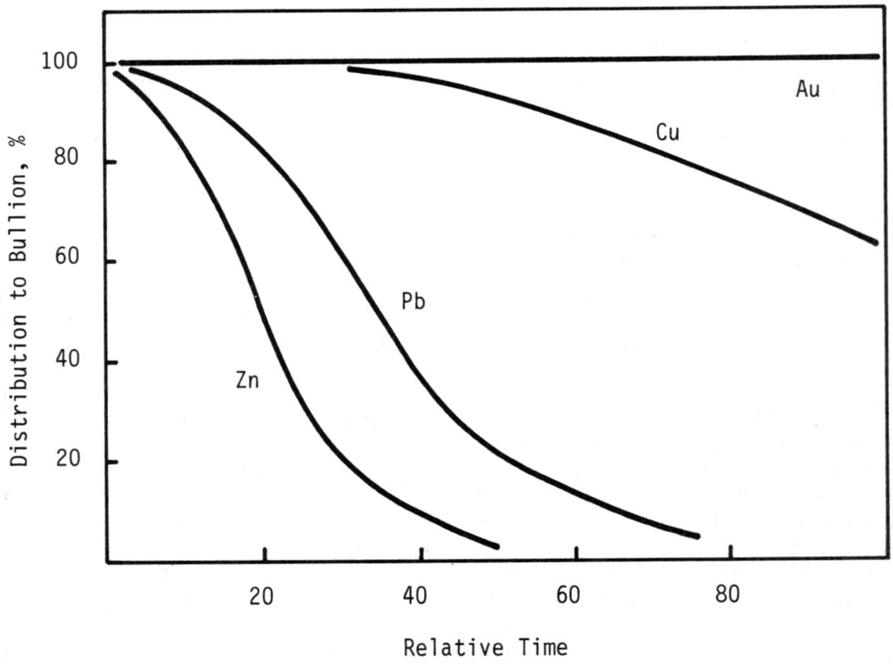

Fig. 4. Oxidative Refining of Gold Bullion

OXIDATIVE REFINING: PRACTICE

Commercial Adoption of Oxygen. All of the reasons for the delay in the commercial adoption of oxygen for precious metals refining are not clear. Certainly there was no shortage of materials that were amenable to pyrometallurgical upgrading. From Fig. 1, the following, in particular, can be listed:

1. high grade flotation concentrates
2. Merrill-Crowe precipitates
3. Parkes process, zinc crusts
4. copper and lead refinery anode slimes
5. precious metals bearing secondaries

These differ considerably in their physical and chemical properties and in their concentrations of precious metals, but each shares in common a need for oxidative refining to remove base metal impurities.

Neither can the unavailability of bulk oxygen be cited as a cause for the delay in technology adoption, at least after the close of the 1940's, and since then cost has not been a factor. From the beginnings of oxygen production in tonnage quantities, it has remained among the least expensive of reagents. Today, at a power cost of $0.05/kWh, gaseous oxygen can be made for $40 to $50/tonne; natural gas, by comparison, costs $275/tonne.

A reason that can be cited for the failure of industry to take up the use of bulk oxygen in precious metals treatment, however, is found in the selling prices of the metals themselves. From the 1930's to the 1960's the price of gold was fixed at $35/troy ounce; silver traded at a fraction of that. With installed plants, cheap labor and fuel, and low interest rates there was little incentive for change. The 1970's altered that situation, probably permanently. Higher precious metals prices and interest rates, higher energy costs, and the increased use of toll smelting brought pressures to bear on the refiners to decrease treatment times and in-process inventories. Then, the results of newer developments in processes, materials and technologies began to be applied to modernizing precious metals smelting and refining.

The Oxidative Refining System. The theory of oxidative refining by the reaction of oxygen with impurities was described above. Not mentioned in that section, but also necessary, is that there be present a solvent that can absorb the oxides of these metals as they form and hold them in a state of low activity. Three systems have been used. The first is based on lead oxide, in what can be called classical cupellation. Lead is oxidized from the precious metals forming a molten, immiscible layer of litharge on the surface of the melt. Other oxides, as they form, are absorbed into this layer. This again is the technique of fire assaying, for which Bugbee (1940) gives the requirements of lead oxide to remove other metals as follows:

	MeO				
	Cu_2O	CuO	ZnO	Fe_2O_3	SnO_2
Parts of PbO required to absorb one part MeO	1.5	1.8	8	10	12

Litharge acts both as a slag (flux) that holds the other metal oxides, and as an exchange medium that supplies oxygen to the melt. In practice, repeated cycles of lead addition, its oxidation, and its removal are used to produce dore' with greater than 98 percent purity.

Second, cupellation can be performed with copper oxide replacing litharge, and at least one secondary precious metals smelter used this system. Higher temperatures are involved than with lead, but, otherwise, the principles are the same.

Finally, oxidative refining can be performed in silicate slag systems, the slag forming components being either naturally present in the charge or intentionally added as fluxes. The first is illustrated by the ferro-silicate and calcium ferro-silicate slags of primary copper and lead smelting. The fire assay may again be taken as an example of the latter with its use of borax ($Na_2B_4O_7$), soda ash (Na_2CO_3), nitre ($NaNO_3$), and silica as fluxes. Exactly the same materials are used to make slags for oxidative refining.

Smelting and Refining Furnaces. The upgrading of the materials in the above list may, then, require both a primary slagging step and an oxidative refining step. Examples are the smelting and cupellation of electrolytic refining copper and lead anode slimes and the fuming of zinc from Parkes process skims under reducing conditions, followed by oxidative slagging of lead and copper. Two furnaces may be used, or, ideally, the two processing stages are performed sequentially in the same unit.

The primary types of furnaces that have been available to the precious metals refiner include:

1. blast and cupola

2. reverberatory (direct-fired)

   a. deep bath - smelting
   b. shallow bath - cupellation
   c. rotary - smelting and refining

3. electric

   a. induction
   b. arc - direct and submerged (resistance)

4. converters

   a. horizontal, cylindrical

      - side blown - Peirce-Smith

b. vertical, cylindrical

- bottom blown - Bessemer
- side blown - Great Falls
- top blown, stationary - BOF, LD
- top blown, rotary - TBRC

The shallow-bath, reverberatory furnace has long been the standard for cupellation, and its use is probably as old as the process itself. It is not a unit well adapted to conducting smelting followed by refining. Oxygen was traditionally provided by blowing air onto the bath surface or by drawing large quantities of air through the furnace. Charging, slag working and product discharge have largely been manual operations. The potentials for inefficiencies are obvious. In general, however, precious metals furnace operations are a small part of a large mine-mill-smelting complex. Inefficiencies in this part of the plant were seldom noted, and if they were noted other concerns with larger dollar values took precedence, or the cost of implementing improvements was larger than the possible savings.

In parallel industries, however, dramatic improvements were being made, especially in the higher tonnage processes, and bulk oxygen was most often at the center of those improvements.

Oxygen Application. In general, there are three approaches to contacting oxygen with a metal bath -- submerged tuyere blowing, submerged top lancing and non-submerged top blowing. The first is widely practiced in the copper industry; it is also the basis for some of the newer, gas-shielded systems being developed. The objections to this method include the need for mechanical punching to keep air paths clear and refractory erosion and corrosion around the tuyeres. The opportunity to increase reaction rates with oxygen enrichment in submerged blowing is limited. With air blowing, nitrogen carries heat away from the tuyeres in addition to agitating the bath and promoting convective and conductive heat transfer. When pure oxygen is used, all of the reactant is consumed in the process, and intense localized heating results. In commercial practice, oxygen enrichment is limited to less than 40 volume percent.

Several attempts have been made to develop practical lances for submerged, top blowing. Testing has included both consumable and water-jacketed types. One area of considerable success is the continuous softening furnaces of the lead industry. The inherent dangers in using submerged, water-cooled lances, particularly with the corrosive slags of nonferrous metallurgy, detract from this method.

The third alternative for contacting oxygen and molten metal is with the non-submerged top blowing lance, as used particularly in the nonferrous metals industry in the top blown rotary converter. The TBRC was revived in the late 1950's by the Swedish steel industry for decarburizing steel using bulk oxygen. Its use expanded to nickel and copper smelting and converting, secondary copper smelting, lead and tin smelting, and slag retreatment. In the past five years, a number of full-scale installations have demonstrated the practicality and versatility of the TBRC in the precious metals processing field. It is effective in precious metals refining for the same reasons that it is used in matte converting. The bath agitation provided by the combination of rotation and surface impingement avoids the local overheating that occurs with oxygen in submerged blowing while still giving acceptable oxygen efficiencies.

The Top Blown Rotary Converter. The advantages of the TBRC have been reviewed in the technical literature (Sealey, 1984; Whellock, 1985; Simmonds and Mills, 1986); the following in particular are noted:

1. improved heat and mass transfer
2. higher fuel and oxygen efficiencies
3. reduced in-process inventories
4. improved environmental control
5. versatility in its ability of smelt (oxidizing or reducing), refine (cupel), and clean slag

In practice it is not possible to separate heat and mass transfer from fuel and oxygen efficiencies. Figure 5 compares the use of oxygen and air for removing lead from a molten bath of silver. The refining reaction is taken as:

$$2Pb\ (l,\ 1000\ °C) + O_2\ (g,\ 1\ atm.,\ 25\ °C) = 2PbO\ (l,\ 1000\ °C)$$

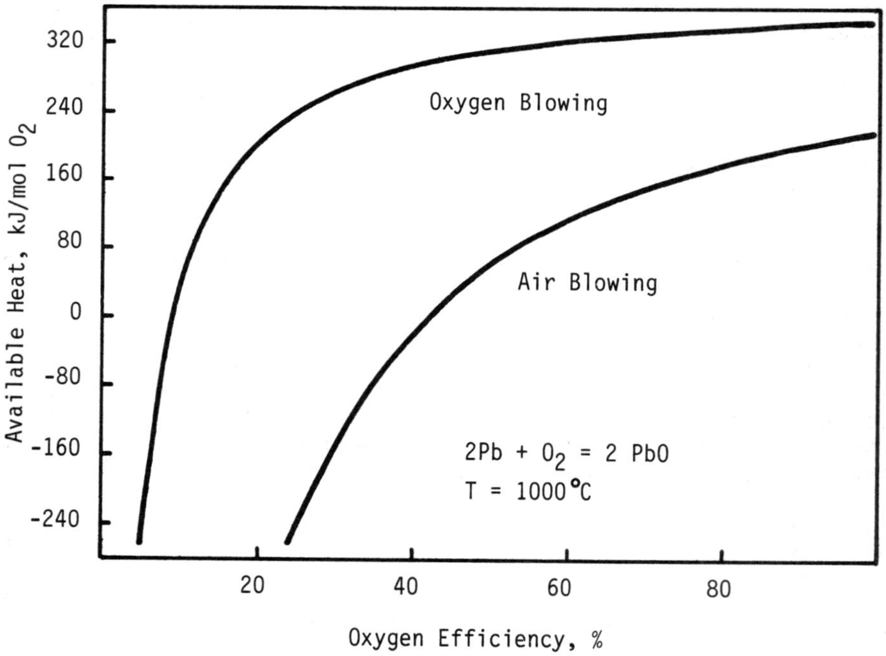

Fig. 5. Enthalpy of Lead Oxidation

Air substitutes for oxygen in the second curve, and N_2 and O_2 appear as reaction products (at 1000 °C) with air blowing and oxygen efficiencies less than 100 percent. The exothermic or available heat from the reaction is shown on the positive scale of the figure. This is the heat that can be used to satisfy all or part of the other heat requirements of the process -- melting fluxes and reverts and making up the furnace heat losses. If the heat demand in the TBRC is taken at a relative value of 250 kJ/mol of oxygen, several times this quantity is needed in the conventional, fuel-fired, air-blown, cupellation furnace. A number of reasons for this can be cited. First, the surface areas of the baths of reverberatory furnaces are large compared with the quantity of metal they contain. This results in containment volumes with large surface areas and, correspondingly, large heat losses. Second, slag layers inhibit reactions, and with relatively quiescent liquid interfaces, only a small area of the total bath can be made available for oxygen transfer; mass transfer coefficients are low. Third, with low rates of oxygen use and, therefore, little heat generation, the need for external fuel increases. With increased quantities of combustion products, the partial pressure of the available oxygen is diminished, and oxygen efficiency suffers.

In contrast, the TBRC operates with oxygen efficiencies approaching 90 percent at the beginning of a cycle, and while this level cannot be maintained as the concentrations of impurities fall, overall efficiencies of 50 percent are achieved. As Fig. 5 indicates, this is sufficient to produce autothermal conditions over a wide range of operating conditions. The increased interfacial areas that are generated by the rotation speed of the TBRC in cascade mode augment mass transfer and reaction rates. Slag layers are no longer inhibiting, and high interphase mass transfer coefficients are obtained.

In pyrometallurgical processes combustion and oxidation efficiencies and slag losses are often listed as the critical economic parameters, but in the precious metals industry in-process inventories may be the most important factor. A typical reverberatory furnace can carry from four to eight days production in the bath and from one to two days production in its refractory lining. Several of these furnaces may be required for a refinery. A TBRC typically processes a day's production in less than 24 hours with no in-process metal left in the furnace at the end of the cycle. The metal lockup in a TBRC lining is less than that in an equivalent number of reverberatory furnaces. This reduction of in-process and lining inventories is usually sufficient to pay a major portion of the cost of replacing an older installation.

The development of the small TBRC has now progressed to the point where more or less standard units are commercially available. Furnaces with capacities of 0.5 to 10 tonnes have been or are being placed into commercial service in precious metals refining. A typical installation is shown in Fig. 6.

Total enclosure of the furnace, feeding system and lances simplifies construction for the containment of dusts and fumes and keeps the quantity of gases to be treated in the trail system to a minimum.

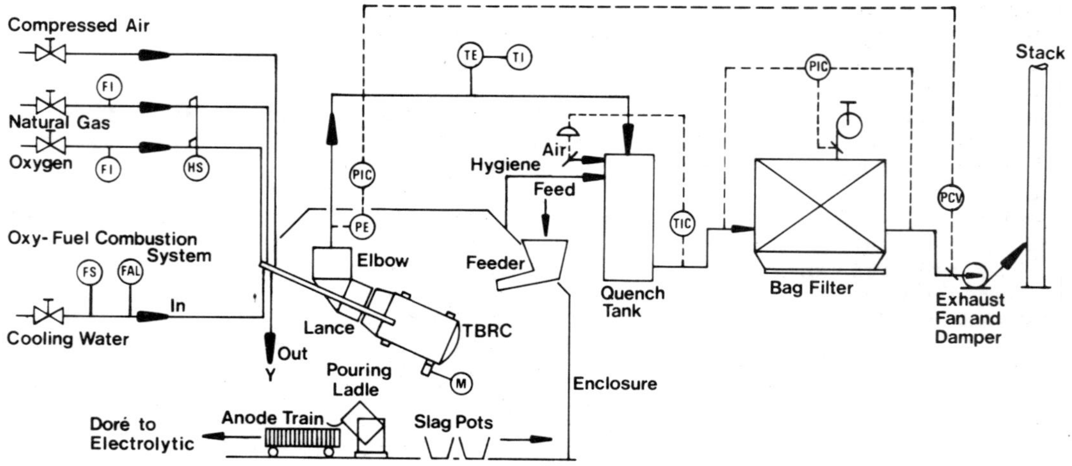

Fig. 6. Typical TBRC Installation

A combination lance capable of operating as either a burner for heating and melting or for oxygen blowing, or separate burners and lances have been installed. The choice depends upon the particular process application and furnace size. Typical turndown ratios for the burners are 5:1. With appropriate engineering and controls, conditions anywhere from highly reducing to highly oxidizing can be generated in the furnaces, giving the capability to perform a variety of activities in a single unit.

CONCLUSION

The oxygen-blown TBRC represents a logical and successful development in the long line of precious metals refining furnaces. It combines high heat and mass transfer rate capabilities with high fuel and oxygen efficiencies in a single vessel. Many of the common metallurgical operations -- drying, melting, reductive and oxidative smelting, fuming, refining, and slag cleaning -- can be performed in one furnace. In conjunction with bulk oxygen, the TBRC provides an advanced solution to processing in the modern precious metals refinery that is unequalled in productivity and flexibility.

REFERENCES

Atmore, R. B., D. D. Howat, and P. R. Jochens (1971). The effects of slag and gold bullion composition on the removal of copper from mine bullion by oxygen injection, J.S. African IMM, August, 5-11.

de Beer, A. G. (1968). The up-grading of gold base bullion. Chamber of Mines Journal, September, 28-31.

Bugbee, E. E. (1940). A Textbook of Fire Assaying. Colorado School of Mines Press (1981), Golden, Colorado.

Healy, J. F. (1979). Mining and Processing Gold Ores in the Ancient World. Journal of Metals, August, 11-16.

Hoover, H. C., and L. H. Hoover (1950). De Re Metallica by Georgius Agricola. Dover Publications, Inc., New York.

Roberts-Austen, W. C. (1910). An Introduction to the Study of Metallurgy. Charles Griffin & Co. Inc., London.

Rose, T. K. (1905). Refining Gold Bullion and Cyanide Precipitates with Oxygen Gas. Trans. IMM, 14, 378-441.

Rose, T. K. and W. A. C. Newman (1937). The Metallurgy of Gold. Met-Chem Research, Inc. (1986), Boulder, Colorado.

Sealey, C. J. (1984). TBRC in Precious Metals Pyrometallurgy. Metal Bulletin Monthly, February.

Simmonds, F. J., and B. E. Mills (1986). TBRC in Precious Metals Smelting and Refining, International Mining, May.

Whellock, J. G. (1985). Systems Approach to the Design of Non-Ferrous Metallurgical Plants. Frontier Technology in Minerals Processing, Society of Mining Engineers of AIME, Chap. 7, pp. 59-74.

CRYOGENIC OXYGEN PLANTS - AN OVERVIEW

T. S. Pawulski

Liquid Air Engineering Corporation
2121 North California Boulevard
Walnut Creek, California 94596

ABSTRACT

Process cycles for the production of tonnage quantities of oxygen have been evolving towards better efficiency in recent years. At the same time progress has been achieved in safety, flexibility and reliability, of the plants.

Typical modern oxygen production cycles are described in this paper and are discussed from the point of view of the above criteria.

The back-up systems to ensure safety of supply are also briefly described.

KEYWORDS

Reversing exchanger cycle; front end purification cycle; plant safety; operating flexibility; reliability; instrumentation; safety of supply; efficient operation.

INTRODUCTION

The purpose of this paper is to present basic information concerning the design and operation of cryogenic plants producing tonnage quantities of oxygen. Essentially there are two process cycles that are currently dominating the field. These two cycles are described in some detail and discussed from the points of view of:

a) Plant safety
b) Operating flexibility
c) The reliability of oxygen producing equipment
d) The plant efficiency

The production plant is normally backed-up by a storage system to ensure security in the event of unplanned shutdowns. Depending on the safe minimum oxygen requirement during emergency conditions, a liquid storage and vaporization system or a pressurized gas storage system may be used. In some cases, a combination of the two is necessary.

DESCRIPTION OF CYCLES

When gaseous oxygen is the only (or the major) product required from the air separation unit, we can safely limit our discussion to two process cycles. These are:

A) Reversing exchanger cycle
B) Front end purification cycle

Reversing Exchanger Cycle (Fig. 1)

This was first introduced some thirty years ago, and now, considerably improved, it is still one of the main systems for the production of gaseous oxygen.

Filtered atmospheric air is compressed in a centrifugal compressor to approximately 620 kPa abs(90 psia). It is then cooled and delivered to the reversing exchangers. These are multi-unit banks of brazed aluminum plate and fin heat exchangers. Their function is to cool the air feed to within a degree or two of the liquefaction temperature of air and to purify the air by freezing out water vapour and carbon dioxide on the heat exchange surfaces. The cooling medium consists of the returning gaseous products of the separation, i.e. gaseous oxygen, waste nitrogen, and, if produced, pure nitrogen. The heat exchange passages for air and waste nitrogen are identical. Every 15 minutes these passages are switched. The accu- mulated water and CO_2 are evaporated and sublimated into the waste nitrogen stream and thus removed from the system. The switching is performed by automatic valves at the warm end of the bank of exchangers operated by an automatic timer. At the cold end there is a set of check valves that follow the pressure changes created by the switching valves.

The air from the reversing exchangers enters the high pressure column. Here it is separated by distillation into relatively pure nitrogen at the top, and a liquid rich in oxygen at the bottom. The nitrogen vapour from the top of the HP column enters the vaporizer-condenser where it is liquified by heat exchange with the liquid oxygen from the low pressure column sump. Part of the liquid nitrogen is removed from the HP column, subcooled, and expanded into the top of the low pressure column as reflux for the second stage of distillation. The remainder serves as reflux for the HP column.

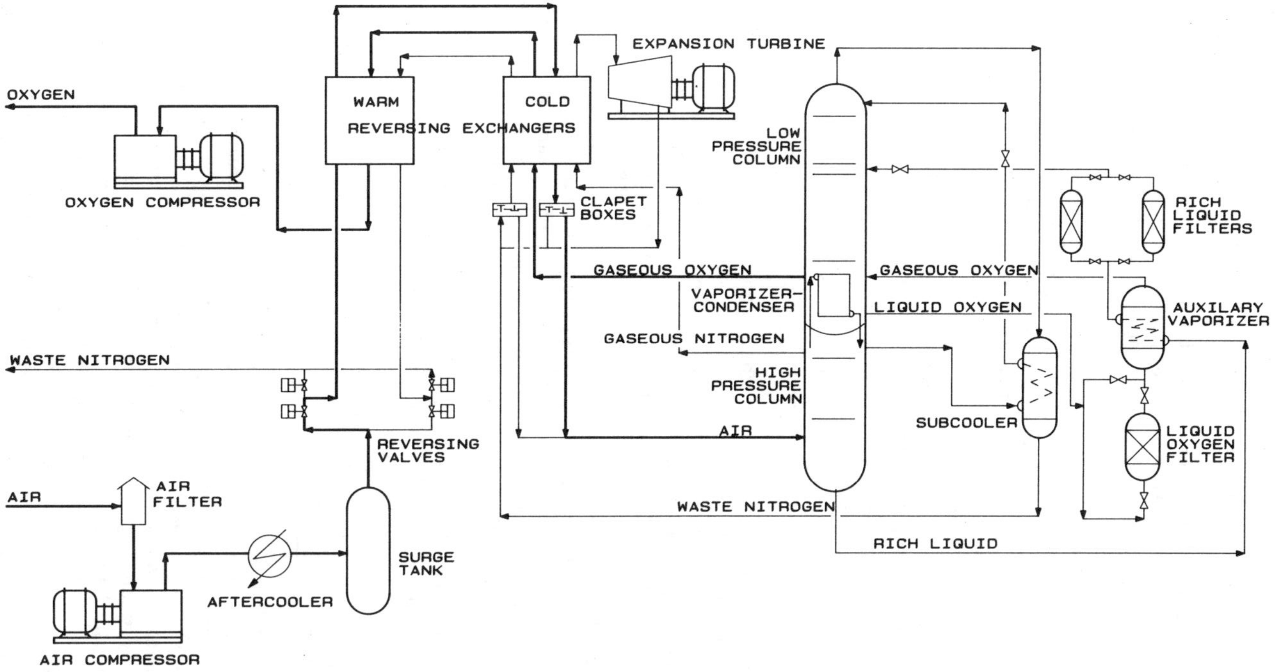

FIG. 1

REVERSING EXCHANGERS CYCLE

The rich liquid obtained from the bottom of the HP column, which contains about 38% oxygen, is subcooled, passed through a rich liquid filter, and expanded to the low pressure column as the main feed.

In the low pressure column which operates at 165 kPa abs (24 psia), the feed is separated into an overhead stream of waste nitrogen and a bottom product of liquid oxygen. Reboil for the distillation as well as for generating oxygen in gaseous form is provided by the vaporizer-condenser, where the liquid is evaporated by heat exchange with gaseous nitrogen from the top of the HP column.

As a safety precaution (further discussed in a later section), we evaporate part of the liquid oxygen in an auxiliary vaporizer rather than in the main vaporizer-condenser. The vaporization of oxygen provides a thermosiphon generating a flow through the liquid oxygen filter. This filter removes impurities that may have broken through the rich liquid filters.

Gaseous oxygen product from the low pressure column is rewarmed back to ambient temperatures in the reversing exchangers, compressed to the required pipeline pressure, and sent to the points of use. Waste nitrogen, after passing through the subcooler, is similarly rewarmed in reversing exchangers. In the reversing exchangers, it picks up the water and carbon dioxide deposited by the air feed. It is then exhausted to the atmosphere.

The distillation process takes place in the temperature range of -170°C to -193°C. The low temperature equipment is enclosed in a "cold box" which consists of a steel casing filled with thermal insulation. Nonetheless, a certain amount of heat from the surroundings is transmitted to the process through the insulation, valve stems, supports, etc. The warm end temperature difference of the main heat exchangers is another source of heat. Since the products leaving the plant are slightly colder than the entering air, some heat is continuously left behind in the process circuit. The refrigeration necessary to maintain the plant in operation, as well as that needed to initially cool down the plant and produce liquids, is generated by the expansion turbine. A stream of nitrogen at 580 kPa abs is taken from the HP column, partially reheated in the reversing exchangers, and expanded to 140 kPa abs. Heat is removed from this stream in the form of power on the turbine shaft. The power, in turn, is removed by coupling the turbine to an electric generator. The expanded nitrogen joins the waste nitrogen stream on the cold end of the reversing exchangers.

Front End Purification Cycle (Fig. 2)

Great progress has been made in recent years in the design of front end purification systems involving adsorption of water vapour on alumina and carbon dioxide on molecular sieves. Double beds of alumina and sieves allow reactivation of these beds at a relatively low temperature of less than 100°C. This eliminates the large power consumption that, in the past, was associated with heating the reactivation gas to a very much higher temperature with electricity or steam. Further, we have found that the heat required can be recovered from the hot, compressed air prior to aftercooling. Another power saving feature, although not entirely new, is the evaporative chilling of water against waste nitrogen gas. The chilled water is then used to bring the air temperature down to approximately 10°C. Previously, mechanical refrigeration units were used for this duty.

As shown in Fig. 2, compressed air is passed through a heat recovery exchanger which heats water to 85 to 90°C.

The water is pumped in a closed circuit through the reactivation heater, heating

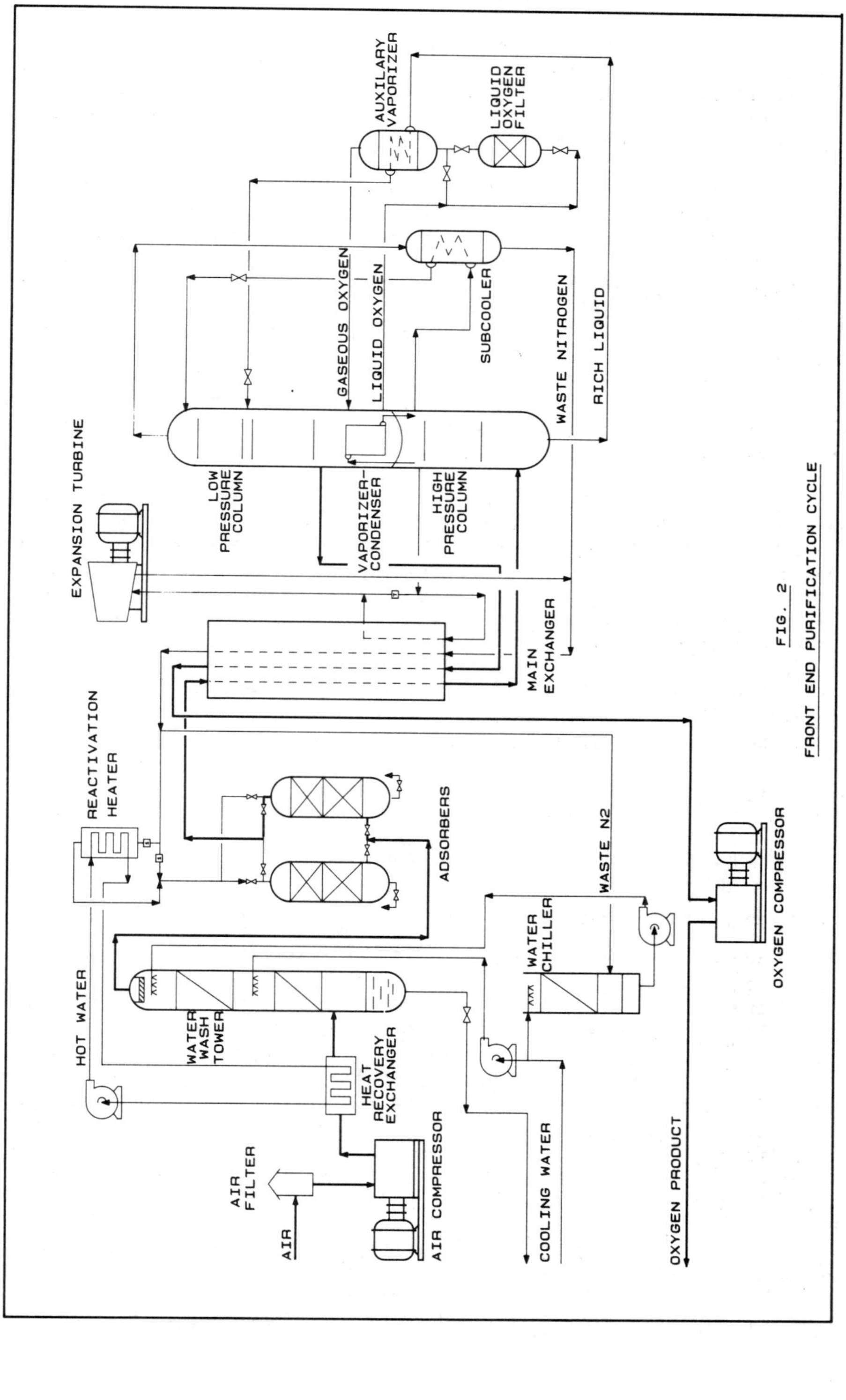

FIG. 2

FRONT END PURIFICATION CYCLE

the waste nitrogen to 80°C which is used for the reactivation of the adsorbents. Alternately, one can use a single heat exchanger for the cooling of air against waste nitrogen.

Air is then sent to a two level water wash tower. In the lower section, it is cooled by direct contact with cooling water flowing over a packing. It is further cooled against chilled water to 10°C in the upper section of the wash tower.

Chilled air is passed through one of the adsorption vessels filled with alumina and molecular sieves. The water vapour is first removed on the alumina, then the dry gas is stripped of its carbon dioxide content by the molecular sieves. It is this absence of water vapour in the molecular sieve section that allows the reactivation at a relatively low temperature. The molecular sieves also remove other impurities such as hydrocarbons and oxides of nitrogen which can cause explosions in air separation units. While the air flows through one adsorption vessel, the other vessel is being reactivated by the passage of heated waste nitrogen flowing in the opposite direction. The vessel is then cooled to the normal process temperature with the waste nitrogen which has bypassed the reactivation heater. The adsorption time can be between two and four hours, after which the vessels are switched over. The switching can be done automatically through timers, or through a computerized control.

Clean, dry air is then passed into the main exchanger where it is cooled down to close to the liquefaction temperature. The exchangers are very similar to the previously described reversing exchangers. However, they are simpler in construction, since they do not need to be built in two sections arranged in a "reversed U" configuration. There are no reversing valves or check valves, and the flow is continuous as there are no impurities to be sublimated into the waste stream.

The distillation process is quite similar to that of a reversing exchanger cycle. The only difference in this part of the flowsheet is the absence of rich liquid filters, since the hydrocarbons are removed on molecular sieves. Liquid oxygen filters are still kept as an extra safety precaution.

The Choice Between the Two Cycles

Choosing between the two cycles depends on a number of factors. The final decision is based on local conditions or, indeed, preferences.

The reversing exchanger cycle has a power consumption equal to or slightly lower than the front end purification system. It has lower investment cost.

The reversing exchangers have suffered from a bad reputation due to the frequency of core failures in the past. The failures have been caused by corrosion of the heat exchanger surface, cracking of partition sheets due to sudden freezing of pockets of water accumulated in parts of the exchanger, or by thermal stresses introduced by the exchanger supports. These failures have led to air leaking into one of the product streams or to outside of the exchanger.

Most of the failures reported in the industry involved leaks of air into the pure nitrogen product stream. It must be realized that nitrogen is normally produced at a very high purity, containing between 1 and 5 ppm oxygen. A small crack in the exchanger partition sheet would increase the oxygen content in the nitrogen to over 10 ppm. In most plants this can be classed as a disaster that leads to an immediate shutdown and emergency repairs. However, in the majority of non-ferrous metallurgical operations, the only product is oxygen, and that is at fairly low purities ranging from 95% to 98% O_2. It would take a massive series of cracks to spoil that purity. Further, where oxygen is used for air enrichment, the system can be operated at lower purities.

In the last ten to fifteen years considerable progress has been made in the industry's understanding of problems and cures for the reversing exchanger failures. The problem is now under control and failures are much less frequent than previously experienced.

PLANT SAFETY

The main source of danger in a cryogenic air separation plant is the hydrocarbons present in the air feed. Even though their concentrations are very low, over a prolonged period of time a considerable amount of hydrocarbons can accumulate in the sump of the low pressure column. Their vapour pressure is much lower than that of liquid oxygen and therefore they will tend to concentrate in the liquid.

The solubility of some hydrocarbons, like acetylene, in cryogenic liquid is very low. This leads to the formation of solid deposits. With time, that accumulation of solid hydrocarbons becomes quite dangerous, as it occurs in the LP column sump where there is a very high concentration of oxygen. All that is needed then for an explosion is the ignition energy. That can be supplied easily by friction, shock, dust particles, or static electricity.

Fires and explosions in the air separation plants used to occur from time to time, some of them with quite serious results. Lately, however, they have become extremely rare, due to a concentrated effort in identifying the sources of hazards, the mechanisms of explosions, and the proper way to design plants.

Reversing exchangers: The heat transfer surface is used for the simultaneous elimination of water and carbon dioxide as well as for heat exchange. Other condensible impurities will freeze out to very low concentrations, and will then be removed by sublimation into the waste nitrogen stream.

However, there is an element of risk in this removal mechanism. One or more of the heat exchanger cores may become unduly warm from time to time through maloperation or upset. Due to this warming up, the hydrocarbons that have accumulated on the cold surface may suddenly be released, re-entering the air stream. The amount of impurities suddenly released may be more than the capacity of the rich liquid filters. If this should occur, the impurities would enter the the sump of the low pressure column and accumulate in liquid oxygen.

This risk is especially great in reversing exchanger systems that have the so called "straight up" arrangement and are subject to frequent and quite drastic temperature fluctuations. That is why we insist on the "reversed U" arrangement for multi-core reversing exchangers.

Rich liquid filters: These are actually two adsorber vessels in parallel, one in operation while the other is being reactivated. They are filled with generous amounts of silica gel. Any impurities that do arrive in the high pressure column eventually end up in the rich liquid. They may consist of acetylene (the most feared component), other light hydrocarbons, oxides of nitrogen, and traces of CO_2. These impurities are adsorbed on the silica gel and removed during reactivation. The only hydrocarbons not effectively removed by silica gel are methane and ethane - these, however, are quite soluble in liquid oxygen and have a significant vapour pressure, and therefore they can effectively escape from the sump with the gaseous oxygen product.

Nitrogen turbine: The classical reversing exchanger cycle, or its predecessors, had an expansion turbine working on the air feed to the column. The expanded air was then admitted to LP column for final recovery of oxygen. At present, however, we favour the use of nitrogen from the top of the HP column as the gas feed to the turbine. The nitrogen is then sent to the waste stream without further processing. In this configuration all of the feeds to the low pressure column are pre-purified; since the top reflux is a product of the distillation it can contain no heavy components. The rich liquid is passed through silica gel filters.

Liquid oxygen filters: This consists of a vessel filled with silica gel to remove any additional impurities. This safety step is a redundant item supplied only as an extra safety precaution. On reversing exchanger plants, the impurities are removed in the exchangers and the rich liquid filters. With the front end purification system, the dangerous contaminants are eliminated very effectively by adsorption on molecular sieves.

Liquid purge: Most plants produce at least a small amount of liquid oxygen for the back-up system. This "blowdown" from the sump of the LP column provides an additional step of continuous removal of contaminants. If the plant operates in purely gaseous mode, a small purge stream should nevertheless be maintained and its composition monitored.

OPERATING FLEXIBILITY

The "cold box" equipment can operate with air input between approximately 55% and 110% of design flow producing from 50 to 110% of design oxygen production. The limitation is imposed by the design of the distillation columns. The trays that we use are a modified version of sieve trays allowing a certain liquid retention capability even at very low vapour velocities through the perforations.

The other potential limitation on operating flexibility of the "cold box" equipment is the design of reversing exchangers. In the conventional "straight up" arrangement of the cores, with the air flowing up from the bottom of the cores towards the top, it is necessary to operate at a reasonably high air velocity. This is to avoid maldistribution of flow which would lead to temperature gradients across the exchanger. With the "reversed U" arrangement the flow of air and of the products can be varied over a wide range of velocities with no detrimental effects.

The air compressor can usually be turned down to provide 70-75% of the design air flow. Even with inlet guide vanes, the flexibility will not be much better, although the efficiency at the low flow will be improved. Below the minimum turndown limit of the air compressor, it is necessary to vent the excess air in order to keep the machine from surging. For installations where it is expected to operate frequently at 50% of the designed oxygen output, one should consider two half size compressors in parallel. A compressor rated at 50% flow and full pressure will generally deliver 55% of flow when the HP column pressure is reduced reflecting the lower pressure drops in the system at a lower oxygen output. We have done this pairing of air compressors quite a few times. In some cases only one 50% compressor was provided for immediate requirements; the second compressor was added a few years later.

Oxygen compressors are just as inflexible as air compressors. It is common practice to provide two or more parallel units.

Expansion turbines, especially when equipped with electric generators as brakes, have a very wide operating range and do not provide a bottleneck for the operation.

RELIABILITY OF AIR SEPARATION PLANTS

Compressors

The compression machinery, both air and oxygen, is very reliable, and can keep the plant on stream continuously for a year or two. It is generally recommended that after 12 months of operation the plant be shut down, thoroughly derimed, and restarted again. The thawing period, from shutdown to completion of restart, usually takes 2 to 5 days. During this time one should also do all the necessary preventive maintenance on the compressors.

Reversing Exchangers

The reversing exchangers, as previously mentioned, can be a source of problems due to the possible failures of exchanger cores. One must avoid the possibility of water freezing in the warm end, which has in the past frequently damaged reversing ex- changers. The exchangers must be designed carefully with strictly defined manu- facturing procedures and proper layout in the cold box. Good operating practice must be established.

The reversing exchangers should be in the "inverted U" configuration. Air flows from the bottom of the unit towards the top in the warm section of the exchanger, and then from the top towards the bottom in the cold section. The products of the separation flow in the opposite direction. this is a more costly solution than the "straight up" arrangement, but it is well worth it.

In the conventional system the cold reversing cores are stacked directly above the warm cores, or, even worse, they are combined into single "jumbo" cores. The air flows from bottom to top. If, due to some maldistribution of flow or plant upset, a core becomes too cold, the air flow in that core will become diminished due to its density increase, and the core will tend to keep getting colder. The return streams, flowing downward, will speed up due to their increase in density, further cooling the core. The core will soon "freeze," possibly cracking the partition sheets.

In the "inverted U" configuration the cold cores are placed in parallel with the warm cores, with the flow direction being reversed. This leads to a flow stability independent of any flow or temperature variation. If air becomes too cold it tends to fall faster through the cold cores, pulling in more warm air through the core, and thus rewarming it. No valves are required to make continuous flow corrections through each core, and there is no need for continuous monitoring and intervention by the operators.

The Non-Reversing Exchangers

These are quite reliable, and can claim an indefinite life without failure. However, care should be taken to avoid any dust carryover from the molecular sieve system to the exchanger thus plugging up parts of the passages. This leads to maldistribution of flow and an increase of the warm end temperature difference.

Expansion Turbine

This is a single stage, radial inflow machiné, operating at speeds between 20,000 and 40,000 rpm. For maximum reliability one should consider a fully installed spare turbine, ready to run at a moment's notice. However, if liquid oxygen is available on site, it may well be enough to have a "liquid assist" line installed from storage to the column. This allows for prolonged operation of the plant without the expansion turbine, the refrigeration being provided from the liquid storage.

Stainless Steel Construction

All welded construction, with stainless steel vessels and piping add considerably to the reliability of the cryogenic unit. All the vessels must be built to Code, and circuits provided with sufficient safety valves. Instrumentation should be generous, with all major variables in the process being continuously recorded. This ensures a long lasting, reliable plant.

INSTRUMENTATION

The plants are generally operated from centrally located control panel in a control room. However they can be designed for remote operation, i.e. from some other control room away from the plant. Alternately, the plant can be designed for unattended operation with occasional visits by an operator.

More and more plants are presently being built with computer control. This allows smooth operation at optimum efficiency under varying demand conditions. However, for a single product plant operating in a narrow range of flow requirements the efficiency savings are rather limited. After all, we expect to recover over 99% of the oxygen present in the air feed with manual operation. For computer control to be truly effective, one would need to integrate the plant operation into the larger system of production planning of the entire facility with feed forward from the oxygen application end, through the air separation unit, to the air compressor output. This should also include considerations of the demand charges for electric power as a function of time of day or of previously established peak demand.

SAFETY OF SUPPLY OF OXYGEN

In spite of all the precautions taken, there may be occasions when the oxygen producing plant is shut down due to a power failure or other unforeseen breakdown. In some cases the loss of oxygen flow will only cause a loss of efficiency of the downstream process. It is then probably best not to install expensive back up systems for oxygen supply.

Gaseous oxygen storage system: This will have to be provided when the oxygen supply is to be maintained continuously at all costs, or at least until other operations can be completed and shut down safely. Figure 3 shows one variant of such a system. Part of the gaseous oxygen product from the pipeline is further compressed to some fairly high pressure, say 4000 kPa, and stored in one or more high pressure vessels. If for any reason the pipeline pressure falls below a preset value, a pressure control valve will start to open to maintain stable supply from

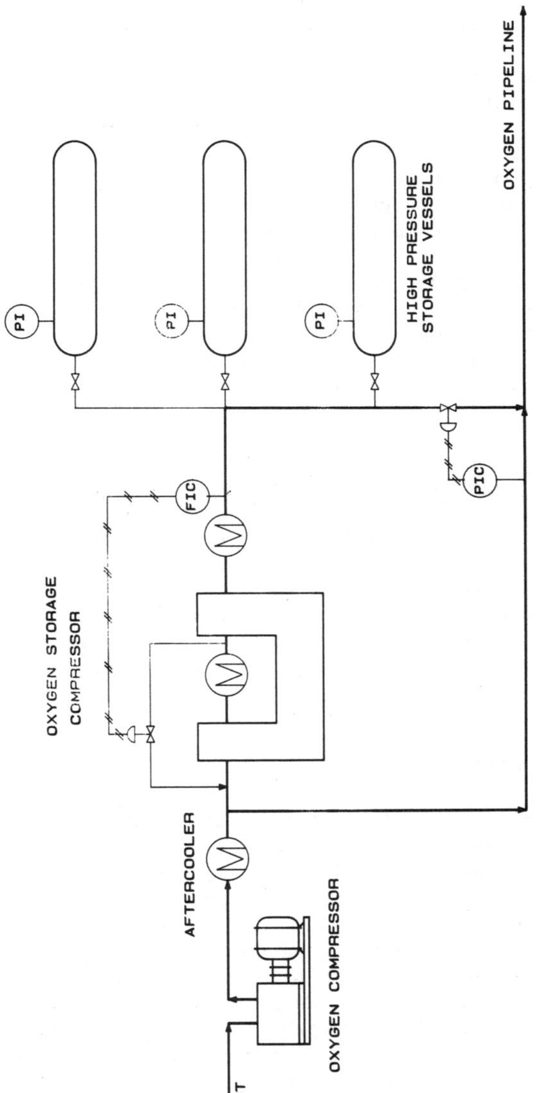

FIG. 3
GASEOUS OXYGEN STORAGE

the storage. This is a most reliable short term back up system. It can only be considered for a short term back up since the high pressure storage tanks are large and very expensive. The capacity of the vessels should be sized so that any critical metallurgical operation can be completed without damage to furnaces and spoilage of the material being processed.

Liquid oxygen storage system: This is capable of maintaining a supply of oxygen at a reasonable rate for a prolonged period of time; even for several days. The supply of oxygen can also be replenished from outside sources by road transport from neighbouring suppliers while the inventory is being depleted in the back up operation.

A much greater mass of oxygen can be stored in a given volume as low pressure liquid than in high pressure gaseous form. Liquid storage tanks of several thousand ton capacity are quite common.

Figure 4 shows a typical liquid back up system. When the oxygen line pressure falls due to loss of production, or maybe due to unusually high demand, the liquid oxygen pump will be started up. Liquid oxygen pressurized to the line pressure is then passed through the vaporizer where it is evaporated and rewarmed to ambient temperature for entry into the main pipeline.

Combined storage system: As can be seen, the liquid storage back up system, while capable of continuous supply of oxygen, cannot ensure an uninterrupted supply. If the plant is shut down due to power failure, we would not be able to even start the liquid pumps. If power is available, time is needed to cool down the pump, start it up, and establish the flow. A combined storage system consisting of a small high pressure gas storage capacity and large liquid oxygen tanks may best ensure both uninterrupted and longer term back up requirements. As an alternative scheme one can install a small high pressure liquid oxygen storage system to bridge the time gap between plant shutdown and the supply of oxygen from the main liquid storage tank.

EFFICIENT OPERATION

Plant efficiency can only be defined in terms of power consumption. The main power user is the air compressor. It has to deliver enough air flow so that the required oxygen can be extracted. It also has to supply sufficient pressure to enable the condensation of nitrogen in the high pressure column.

Modern air separation plants are designed to recover over 99% of the oxygen present in the air fed to the columns. This is ensured by adjustment of reflux flow to the top of the low pressure column as a function of the oxygen content in the waste nitrogen stream, i.e. of the oxygen lost. It is a very simple and straight forward operation. A slightly better recovery of oxygen from the air can be achieved if oxygen product is made at lower purity, but the effect is really quite small.

Air compressor pressure requirement is fixed by the design of the cryogenic equipment and there is very little one can do in operations to reduce it. However, if the plant is operated at a lower purity of oxygen, the oxygen will boil at a lower temperature, allowing nitrogen to condense at lower pressure. This will be a definite power saving on the air compressor. If, for example, we reduce oxygen purity from 99.5% to 95%, the pressure reduction will be 22 kPa, allowing a power saving of 1.9%.

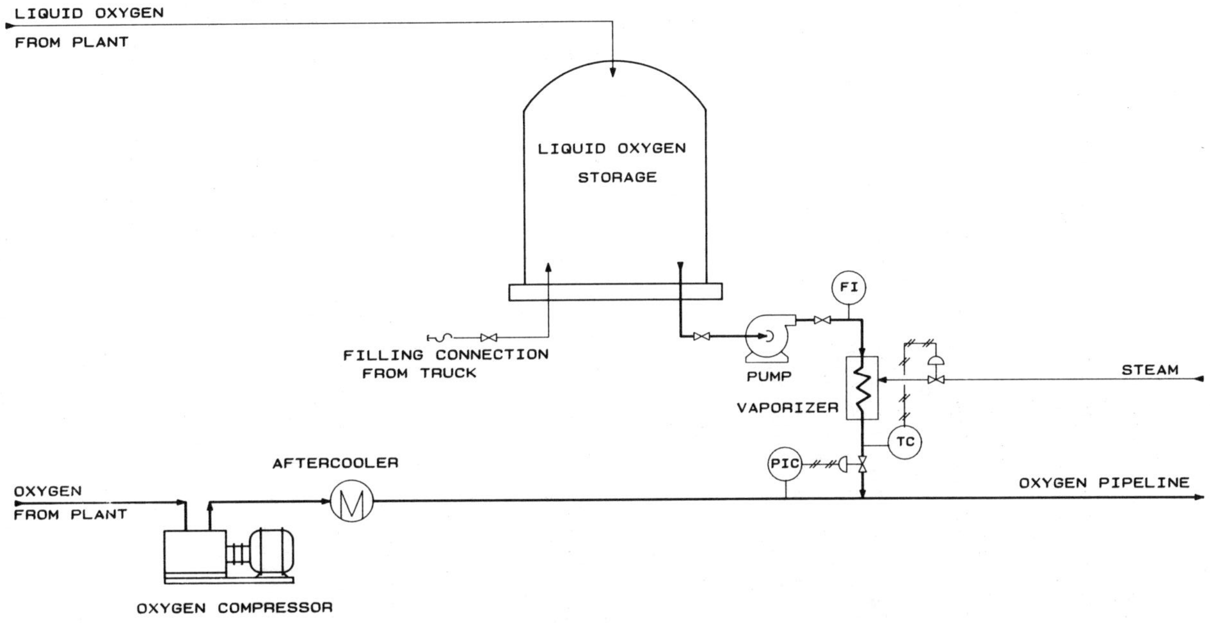

FIG. 4

LIQUID OXYGEN STORAGE

If the plant is designed for lower purity only, there will be fewer distillation trays installed. The pressure drop of the system will be lower than in a comparable high purity plant, and this may give further saving in power of 1.5% to 2%.

However, for air enrichment, blending of oxygen product with air will require a higher tonnage of oxygen if oxygen is at a lower purity.

ACKNOWLEDGEMENT

The author wishes to thank the management of Liquid Air Engineering Corporation for permission to present this paper.

BIBLIOGRAPHY

Elmore, G. T. L., and T. S. Pawulski (1983). Reversing Heat Exchangers 1973-1983, A Decade of Progress. 16th International Congress of Refrigeration, Paris.

Grenier, M., I. Y. Lehman, P. Petit, and D. V. Eyre (1984). Adsorption Purification for Air Separation Units. Cryogenic Process and Equipment, The American Society of Mechanical Engineers.

Kerry, F. G. (1975) . Technical and Economic Considerations of Very Large Oxygen Plant. The Metallurgical Society of AIME, A75-15.

Pawulski, T. S. (1978). Pyrometallurgical Support Facilities-Oxygen Plants. The Metallurgical Society of AIME, A78-15.

APPLICATION OF OXYGEN IN PYROMETALLURGICAL PROCESSES OF COPPER PRODUCTION IN POLAND.

J.Czernecki,Z.Śmieszek,J.Botor,S.Gizicki,J.Bystroń

Institute of Non-Ferrous Metals, ul.Sowinskiego 5, Gliwice,Poland.

W.Cis

Copper Mining and Metallurgical Complex,ul.M.Sklodowskiej-Curie 49, Lubin, Poland

1. Introduction

Two methods for copper production are applied in Poland. The first of them consists in blister copper production by means of converting the copper matte obtained from the shaft furnace. The other method consists in obtaining copper directly from the flash furnace.
In this paper the results of tests on the application of oxygen-enriched air blow at various stages of shaft furnace practice, copper matte converting, as well as the flash smelting process are presented.
The effect of oxygen content in air blow on the intensification of these processes, energy problems and quality of metallurgical products are presented and discussed here.

2.Methods of copper production in Poland

Copper is produced in Poland from sulphide concentrates whose main component contents are given in Table 1.

Table 1. Main component contents in sulphide copper concentrates.

Cu	17 - 30 wt%
S	7 - 11 wt%
Fe	2 - 5 wt%
C_{org}	5 - 7 wt%
Pb	1 - 2.5 wt%
As	0.05 - 0.3 wt%
Sb	< 0.01
Bi	< 0.002

Copper occurs mainly as chalcocite and partly as bornite and chalcopyrite. The average content of copper is on the same level as in concentrates commonly produced in the world. The contents of sulphur and iron are, however, several times lower. A specific quality of Polish concentrates has a high organic carbon content.
The contents of lead and arsenic are also high at trace contents of antimony and bismuth. A typical chemical composition of three types of copper concentrates is shown in Table 2. The gangue consists mainly if silica, aluminasilicates, calcium carbonate and magnesium carbonate. Their proportions make the addition of the slag-forming agents unnecessary.

Table 2. Chemical composition of typical copper concentrates

Type of conc.	Cu	Fe	S	Pb	As	Zn	SiO_2	CaO	MgO
I	17.6	5.0	9.7	1.9	0.24	0.35	22.5	8.4	4.9
	Al_2O_3		C_{org}						
	6.6		5.5						
II	23.0	2.1	8.2	1.2	0.18	0.40	17.4	10.4	4.8
	5.7		5.6						
III	28.2	2.5	10.0	1.8	0.06	0.50	16.9	7.9	4.9
	5.8		6.3						

The above concentrates are processed by two methods: rich matte smelting in a shaft furnace and a direct method of blister copper smelting in a flash furnace.

Fig. 1 gives the shaft process flow-sheet, including:

A. Briquetting the concentrate and dusts from the shaft furnace gas dedusting system, with binder being the waste sulphite liquor.

B. Smelting the concentrate briquettes and converetr slag, with coke in shaft furnace of 70 - 85 Mg/ 24 hm^3 capacity.

From the shaft furnace the following are obtained:

i copper matte of copper content of 58 - 65 wt %
ii waste slag of copper content below 0.6 wt %
iii slimes from the gas scrubbing system, containing 38 - 40 wt % Pb.
iv gases of CO contents ranging between 8 and 14 vol.%, which are used for the concentrate drying and steam production.

C Copper matte converter process, with the following products:

i blister copper of Pb content of below 0.35 wt%
ii converter slag recycled to shaft furnace
iii dusts of Pb content of 45 - 55 wt.%
iv gases of SO_2 content of 6 - 12 vol.%, used for the sulphuric acid production.

D. Fire refining in stationary anode furnace.

In 1978, the direct process for the copper production in a flash furnace by the Outokumpu method, was applied at Glogow 2 copper smelter. The process flow-sheet shown in Fig.2 comprises the following:

A. Concentrate drying to achieve H_2O content of below 0.2 wt%,

B. Smelting the concentrate in a flash furnace, using oxygen-enriched air blow of 55 - 75 vol.% O_2 , in result of which the following are obtained:

i blister copper of Pb content of below 0.3 wt% and oxygen content of 0.3 - 0.5 wt.%
ii slag of Cu content of approx. 14 wt%
iii dusts recycled to the flash furnace
iv gases of SO_2 content of 9 - 14 vol.%, used for the sulphuric acid production

C. Slag decopperizing in an electric furnace, in result of which the following are obtained:

i copper alloy of 10 - 18 wt% Pb and 3 - 8 wt.% Fe
ii waste slag containing below 0.6 wt.% Cu
iii dusts contain 35 wt.% Pb and 12 wt.% Zn

Fig.1 GŁOGÓW I COPPER SMELTER

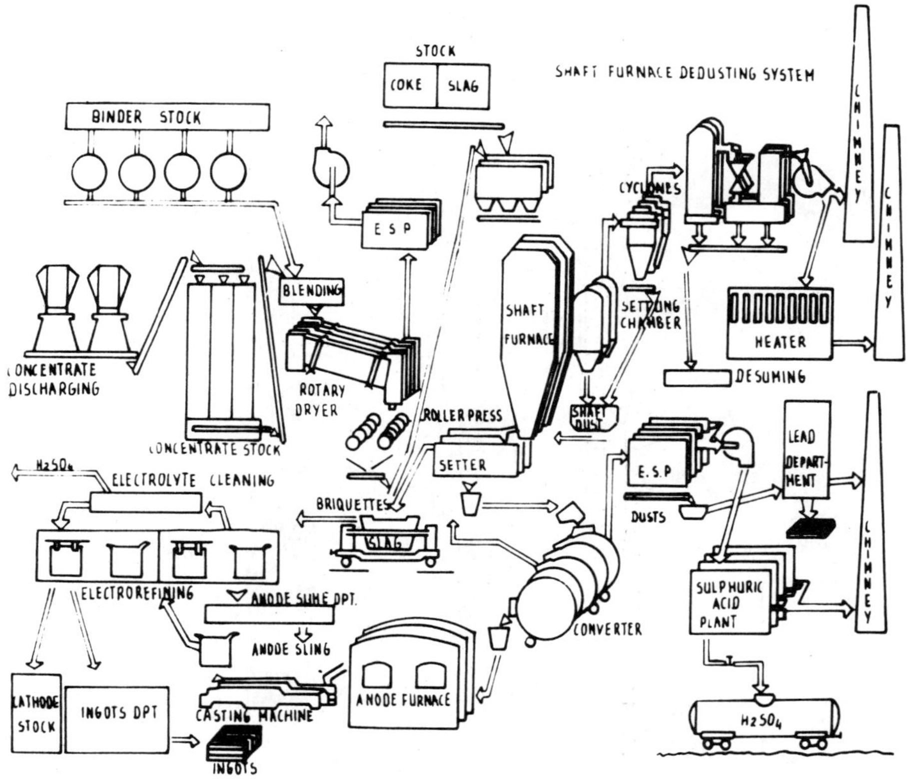

FIG. 2. GŁOGÓW II COPPER SMELTER

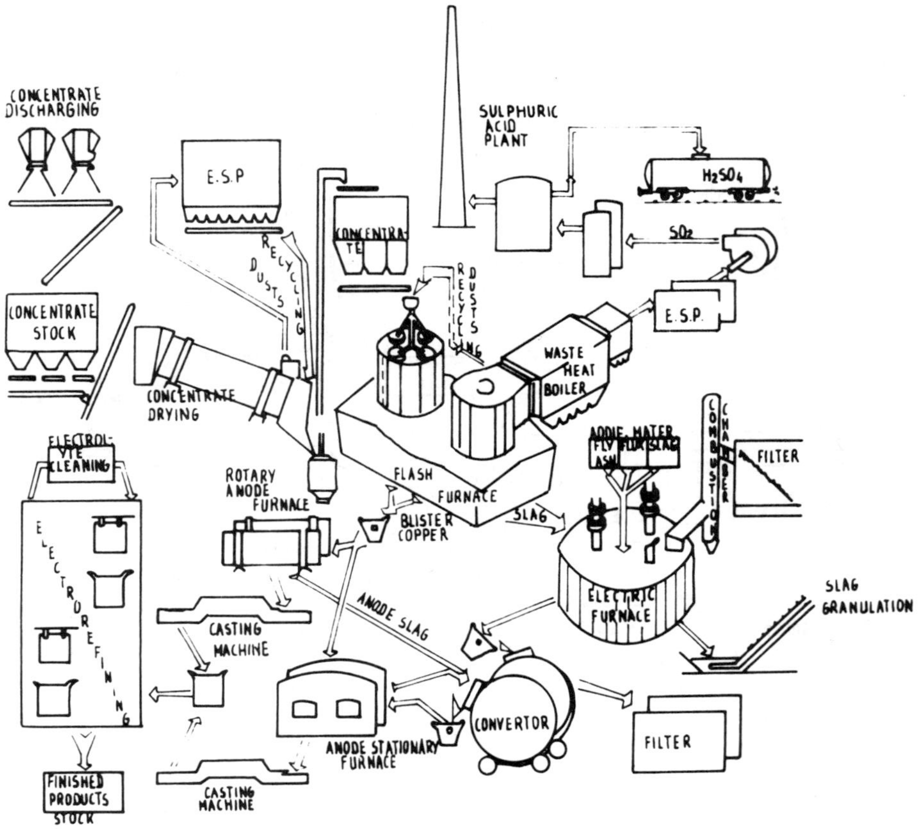

iv gases of the concentration of up to 40 vol.% CO

C Converting the CuPbFe alloy to obtain copper of Pb content of below 0.3 wt%

D Fire refining of blister copper from the flash smelting process and of converter copper in stationary and rotary anode furnaces.

3. Direct flash smelting process with oxygen-enriched air blow.

Oxygen-enriched air blow of 55 - 75 vol.% O_2 is used in the flash smelting process. It proceeds in a typical Outokumpu flash furnace which consists of a reaction shaft, settler and gas uptake. The reaction shaft is made of steel and is 7.7 m in diameter and 7 m high. It is lined with magnesite-chromite refractory materials and with ceramic materials, and it is cooled by water. The settler 216 m^2 in area is also lined with magnesite-chromite materials and cooled by a water jacket. The concentrate dried to water content of less than 0.2 wt.% is transported to four burners located in the reaction shaft roof. A schematic diagram of the burner is shown in Fig.3

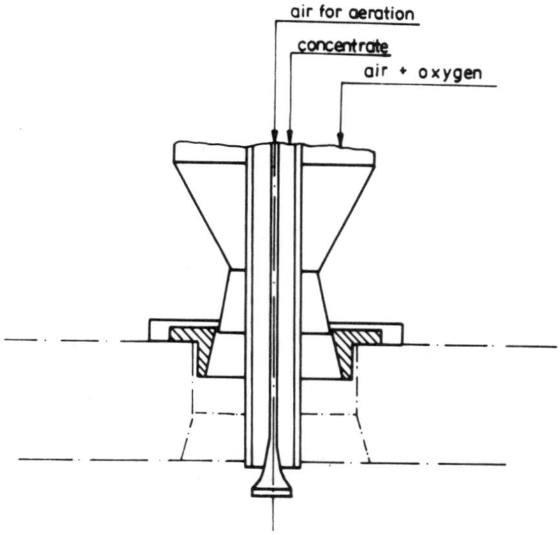

Fig.3. Schematic diagram of a burner in the flash smelting furnace

The concentrate batching and weighing systems are individual for each burner. The gravity-fed concentrate is atomized at the burner outlet and mixed with the oxygen-enriched air. Additionally, small amount of oil is used. In result of oxidation in the reaction shaft, metallic copper and cuprous oxide are formed. These products together with the unreacted copper sulphide pass to the settler. The estimated amounts of copper forms in the reaction shaft products are as follows: 20% of metallic copper, 20% of copper in form of sulphides, and 60% of copper in form of cuprous oxide. The reactions between sulphide and oxides proceeed in the settler, as a result metallic copper is formed and the liquid products such as blister copper and slag are separated. Temperatures of copper and slag are close to each other and range between 1550 - 1570 K. The process gases of 1570-1590 K pass through the gas uptake to the waste-heat boiler and after cooling to 660 - 680 K are directed to electrostatic precipitator.
The composition of Polish concentrates, especially their low iron content allows their application in the direct flash smelting process for blister copper production. Low sulphur content is the factor which determines the necessity of oxygen application in this process.
In Table 3 the composition and amounts of gases produced in the flash smelting process have been compared, at the application of oxygen--enriched air blow, and at smelting capacity of 70 Mg/h.

Table 3. Amount and composition of gases and fuel (oil) consumption in the flash smelting process of the capacity of 70 Mg/h

No.	Parameter	Oxygen content in blow 21 vol.% O_2	70vol.% O_2
1.	gas quantity (Nm^3/Mg)	1770	550
2.	gas composition (vol.%)		
	SO_2	3.8	12.7
	CO_2	14.2	34.5
	H_2O	8.3	12.0
3.	Oil consumption	59	14

At application of the oxygen-enriched blow, the amount of gas produced is about 3 times lower, and sulphur dioxide content is 3 times higher in comparison to the air blow method.
That enables gas utilization with the application of the cooling system, gas scrubbing and sulphuric acid production equipment of much smaller dimensions.
Flash smelting process proceeds under conditions, assuring high oxidation of the concentrate. This results from the high lead contents the Polish concentrates.
Data included in Tables 1 and 2 have indicated that the Polish concentrates are characterized by a high lead content. The lead concentration in copper is determined by thermodynamic equilibrium of reaction:

$$[Pb] + (Cu_2O) = 2[Cu] + (PbO) \qquad (1)$$

whose free enthalpy change in temperature function is determined by relation:

$$\Delta G_T^\circ = -10240 + 2.7\,T \qquad (2)$$

Cuprous oxide concentration in slag determined basing on eqs.(1) and (2) is as follows:

$$X_{(Cu_2O)} = \frac{X_{(PbO)}\ \gamma_{(PbO)}\ a^2_{[Cu]}}{K_{P(1)}\ X_{[Pb]}\ \ \gamma_{[Pb]}}\ \frac{1}{\gamma_{(Cu_2O)}} \qquad (3)$$

where

X molar fraction

γ activity coefficient

a activity

Parentheses denote that a component is the slag phase, and square brackets denote that a component is in the alloy.

As results from the lead concentration in blister copper dependence on the copper concentration in slag (Fig.4), the copper oxide concentration in slag should be higher than 13 wt.% if the lead content in copper is to be less than 0.3 wt.%. Thus there is a necessity to apply oxygen excessive to the stoichiometric demand to ensure the required copper oxidation in reaction shaft. The oxidation degree depends mainly on the concentrate grain size. As the coarse grain

Fig 4 Relation of lead concentration in blister copper and copper concentration in flash smelting slag

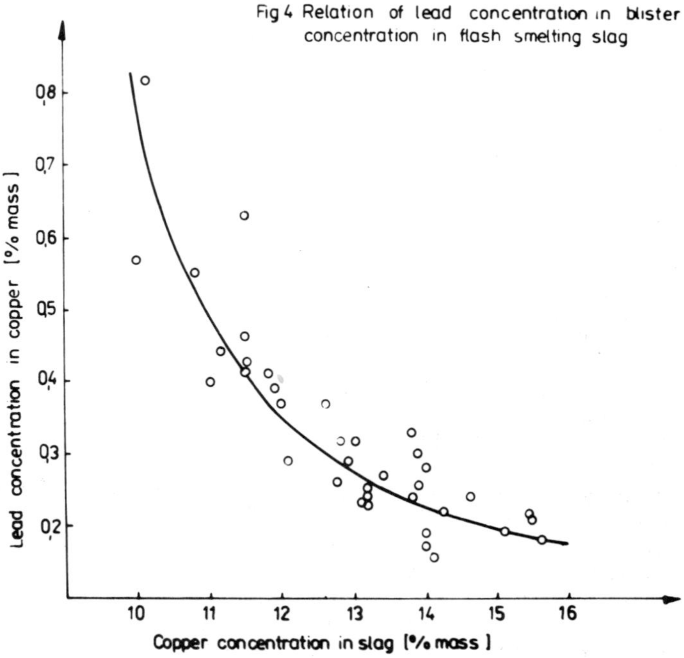

share in the concentrate becomes reduced during the process, the oxygen consumption in the process is also reduced. This results from a more complete oxygen utilization in the reaction shaft at the same product oxidation degree. In consequence, besides the lower oxygen consumption, also its lower concentration in the product gases is achieved.
Thus the SO_2 to SO_3 transformation is limited, which in turn reduces the gas cleaning system corrosion caused by the aggresive gases and dusts. Moreover, certain oil savings (approx.200 kg/h) are achieved.
The oversize concentrate particles share has been reduced gradually from 10 - 14% to 2 - 4%, achieving the reduction of the oxygen concentration in process gases from 10 - 16 vol.% to 6 vol.%. This tendency has been illustrated by the diagram in Fig.5.
The change of oxygen concentration in process air applied is ranges between 55 - 75 vol.% (generally 65 - 75 vol.%). It is the function of concentrate chemical composition, especially carbon content in concentrate, process capacity and thermal losses of the flash furnace.

Fig 5 Effect of coarse-grained structure of concentrate on oxygen concentration in process gases

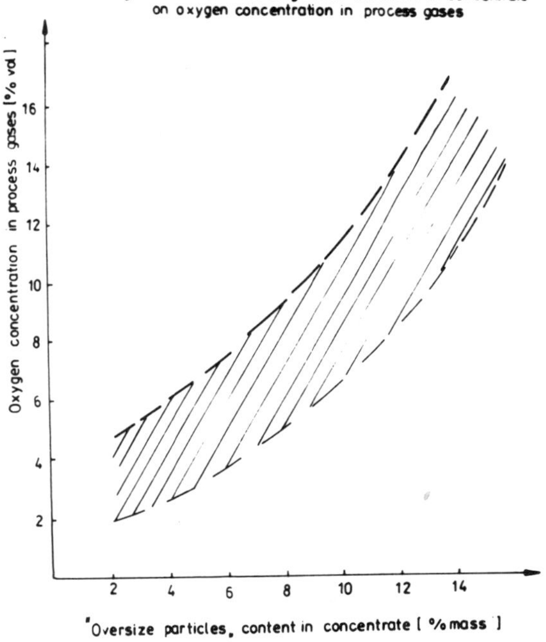

In order to achieve stable process parameters,mainly the temperature and chemical composition of copper and slag, and quantity of heat taken by the process gases, a mass and heat balance is being made. In that way the process air quantity, its degree of oxygen enrichment and quantity of oil fed to the reaction shaft are determined. The exemplary heat balance of the flash smelting process , 85 Mg/h in capacity is shown in Fig. 6.

Fig 6 Heat balance of the flash smelting process [85Mg concentrate/h]

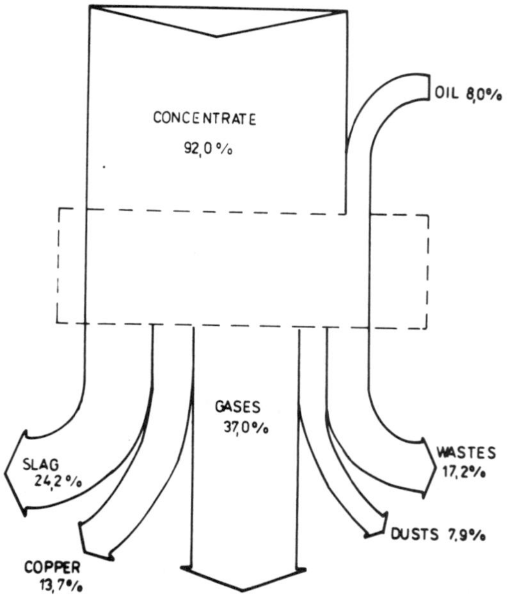

By applying the oxygen-enriched air of the oxygen content of 70vol.%, oil consumption lower by 4 times than in case of air is achieved. Therefore the process is autogenic. Input of thermal energy produced by oil and concentrate in dependence on the process capacity has been shown in fig. 7. This input does not exceed 10% if the smelting capacity exceeds 75 Mg/h. The highest amount of heat is carried away by the process gases. There is a necessity of maintaining constant amount of energy, taken by gases; this amount ranges between 100 - 125 GI/h. This is connected with assuring steady operating conditions of the cooling-dedusting system. This is achieved by applying variable oxygen contents in the process air, in dependence on the concentrate chemical composition and the process capacity.
Provided that all components of the heat balance change little in time, heat losses undergo considerable changes. This is connected with a high wear of the refractory lining during the flash smelting process. These losses range from 25 - 75 GI/h. The increased heat losses are compensated by a higher oil consumption and a higher de-

Fig 7 Oil consumption and oxygen concentration in process air as process capacity function

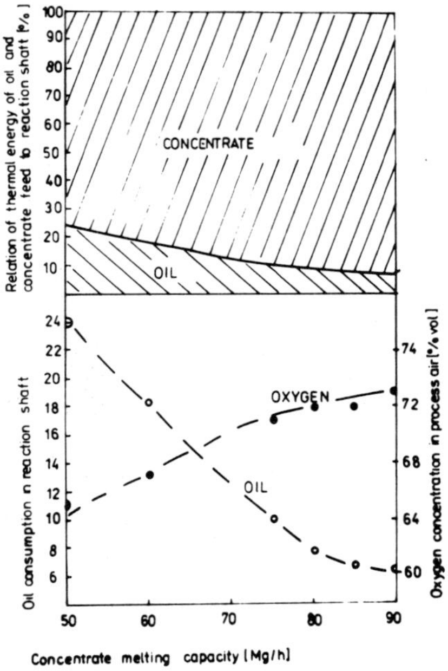

gree of blow enrichment with oxygen.

4.Matte smelting in shaft furnace with oxygen-enriched air blow.

By starting with the flash smelting process in Poland copper concentrate smelting was done entirely in shaft furnaces of cross-section area on the tuyeres level being 10 m² and 20 m². In this process coke is applied in the amount of 10 wt.% of the charge; with air blow of 3 - 3.5 10^3 Nm²/hm², smelting capacity is 70 - 75 Mg/m² 24h. To attain intensification of this process some tests on the effect of oxygen enrichment of the blow on matte smelting capacity have been carried out. The test were conducted in a pilot shaft furnace of the cross-section area on the tuyeres level being 1.5 m². The furnace has been equipped with a tank of capacity 10 Mg of liquid oxygen and an oxygen gasification plant having the capacity 600 Nm³/h with instrumentation. Coke addition and the amount of blow have been identical as those of industrial furnaces, i.e. 9-10 wt.% and 3.5 10^3 Nm³/hm², respectively.
Two series of experiments have been conducted at constant coke amount of 9 wt.%, oxygen content in air blow was changed in the following way: 21,23,25 and 27 vol.% O_2. In the second series , at constant oxygen concentration of 27 vol.% O_2, coke addition to the charge was reduced to 10, 7.5, 3 and 0 wt.%.
The results of tests on the effect of oxygen content in air blow on the charge smelting capacity with a constant coke addition of 9 wt.%, have been shown in Fig. 8. The data obtained indicate that each percent of oxygen concentration increase (ranging between 21 - 27 vol.%) causes the increase of the process capacity of 7.5 Mg/m²/24 h.

Fig 8. Melting capacity of copper concentrate briquettes in shaft furnace as function of oxygen concentration if coke addition is constant [9% mass]

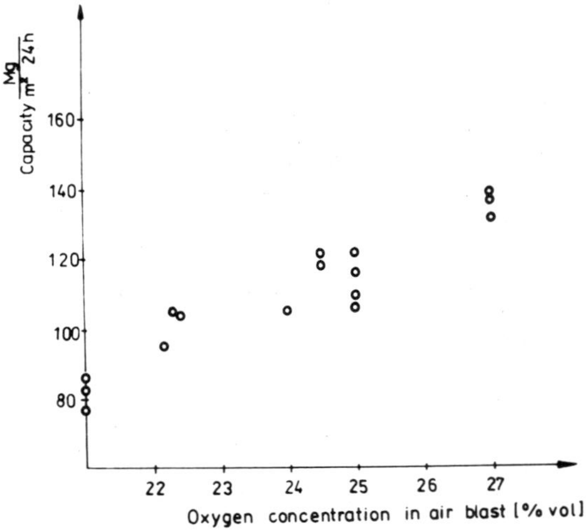

As oxygen enrichment proceeds, shaft furnace operation changes favourably. They result from: reduction of the combustion zone at tuyeres and temperature increase in that zone. In result of these, there is a possibility of shaft furnace operation at lower charge level and obtain higher temperature of the furnace products which in turn causes more stable operation. Apart from all these advantages tests have proved that application of the oxygen-enriched blow enable the use of concentrate briquettes and coke of lower mechanical strength.

In Fig. 9 the results of tests on the effect of coke amount on the charge smelting capacity with a constant oxygen concentration in blow being 27 vol.% O_2 are presented.

A 1% reduction of coke addition in relation to charge causes a proportional increase of the smelting capacity(about 8.5 Mg/m^2 24 h. At coke addition of 3 wt.% of charge, high smelting capacity, approx.170 Mg/m^2 24h has been achieved. However, in that case a decrease of the shaft furnace stability has been noted. When the process was conducted without coke addition there were some symptoms of charge freezing in the furnace. Tests have shown that a maximum reduction in coke addition is permissible at the level of 3 wt.%. Such a low coke consumption is also possible because, as the coke addition decreases, the utilization degree of organic carbon in concentrate increases. In this process coke is not only the energetical agent, but it also stabilizes the gas flow through the column of solid charge in the furnace shaft.

Change of the oxygen concentration in blow is reflected in chemical composition of the process gases. The data are given in Table 4.

Fig 9 Melting capacity of copper concentrate briquettes in shaft furnace depending on coke addition if oxygen concentration in air blast is constant [27% vol]

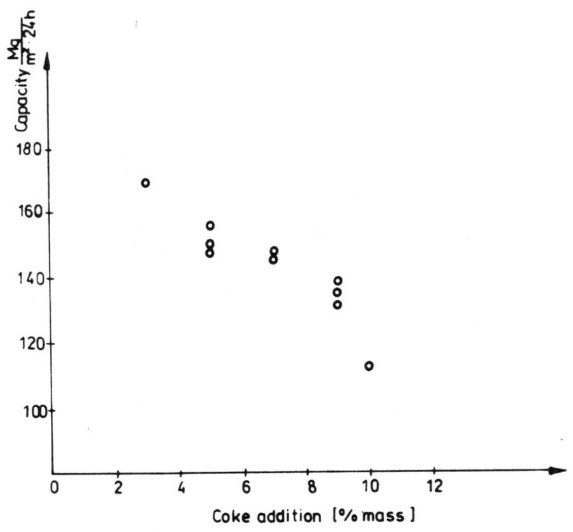

Table 4. CO_2 and CO concentrations in process gases in dependence on the amount of coke addition and oxygen concentration in blow.

O_2 content in blow (vol.%)	Coke amount (wt.%)	Concentration (vol.%) CO_2	CO
21	9	14.0	17.0
23	9	15.0	16.0
25	9	16.0	15.8
27	9	18.0	16.0
27	7	19.0	15.5
27	5	20.0	15.0
27	3	22.0	11.5
27	0	25.0	8.5

At low coke additions, in the range of 3 - 7 wt.%, the increase of the oxygen content in throat gases over 2 vol.% has not been observed.

The results obtained in the pilot shaft furnace indicate that increasing the oxygen concentration in blow from 21 to 27 vol.% and decreasing the coke addition from 9 to 3 wt.%, will enable an even 100% increase of the smelting capacity. Despite of the high process capacity copper content in slag does not increase, which might be due to the temperature of the bath leaving the furnace higher by 100 - 200 K.

The application of oxygen-enriched blow in shaft furnaces of 20 m² cross-section area on the tuyere level is expected. It is assumed, that the oxygen-enriched blow (to 24 vol.% O_2) will be applied and a decrease of coke consumption is expected. It is connected with the necessity of keeping the proper calorific value of throat gases which enables an effective coke combustion and assures the melt intensity stabilization at lower quality of the charge. Coke consumption reduction and heat recovery from throat gases will compensate for the production costs and oxygen feed to shaft furnaces.

5.Converting copper matte by oxygen-enriched blow.

Oxygen-enriched blow is one of the fundamental elements of the intensification of matte converting process. It results from the nature of the process which consists in oxidation of metal sulphides, mainly: FeS in 1st stage and Cu_2S in 2nd stage. The reaction rate is defined mostly by the amount of oxygen fed into liquid matte volume in time unit. Therefore the application of oxygen-enriched blow reduces the duration of the process with a simultaneous decrease of converter gases volume. This leads to the improvement of the process heat balance thanks to the reduction of heat losses in gases and to the atmosphere. Excess heat obtained in 1st and 2nd stages of the process enables melting anode wastes from electrorefining or converter copper ingots.

These materials are melted usually in anode furnaces which causes a prolonged furnace operation and the increased fuel consumption. At air enrichment to the content of 4 vol.%, the amount of nitrogen and argon fed to the converter is decreased, according to the following relationship:

$$\Delta V = \frac{m\, V_z}{\alpha} \left(\frac{79}{21} - \frac{100 - Y}{Y} \right) \quad / \; Nm^3 / \; cycle \qquad (4)$$

Reduction of the converter process time with the blow is expressed by equation:

$$\Delta t = m \frac{V_z}{\alpha / 1 - \beta / \frac{V_d}{60}} \left(\frac{1}{0.21} - \frac{1}{0.01Y} \right) (min) \qquad (5)$$

where

- m — copper matte weight, Mg/cycle
- V_z — required amount of oxygen necessary for oxidation reaction of sulphides in matte, Nm^3/Mg of matte
- α — oxygen utilization coefficient in oxidation reaction of sulphides
- β — coefficient of blow losses at tuyeres
- V_d — blow intensity, Nm^3/h

The amount of thermal energy saved in the process due to the application of oxygen-enriched blow, results from the decrease of thermal energy in converter gases and reduction of the process duration. It can be expressed in the following:

$$\Delta Q = \Delta V \Delta H^\circ_{(N_2 + Ar)} + q \Delta t \qquad (kJ/cycle) \qquad (6)$$

where

- $\Delta H^\circ_{(N_2+Ar)}$ — enthalpy of N_2+Ar mixture of composition the same as in air, kJ/m^3.
- q — converter unit heat losses, kJ/min.

Assuming that the saved amount of heat ΔQ will be used for copper smelting, there is the possibility of smelting the following amount of solid waste:

$$M = \Delta Q / \Delta H^{\circ}_{Cu} \qquad (kg) \qquad (7)$$

where

ΔH°_{Cu} enthalpy of liquid copper, kJ/kg

Calculations ahve been made at the following process parameters:

- copper matte mass (m) 147 Mg/cycle
- blow intensity/V_d/ at 1st stage 26000 Nm³/h
- blow intensity/V_d/ at 2nd stage 22000 Nm³/h
- oxygen efficiency coefficients α 0.90
- oxygen efficiency coefficients β 0.015
- oxygen concentration in blow 22 - 30 vol.%
- heat losses (q) at 1st stage 64000kJ/min
- heat losses (q) at 2nd stage 66000kJ/min.
- converter gases temperature at 1st stage 1470K
- converter gases temperature at 2nd stage 1490K

In Table 5, the theoretical requirement of oxygen for copper matte converting of 60 wt.% and 65 wt.% copper contents have been listed.

Table 5. Theoretical oxygen requirements for matte converting

wt.% Cu	O_2 requirement /Nm³/Mg of matte	
	1st stage	2nd stage
60.0	66.6	106.7
65.0	46.9	116.3

Using the data enclosed in Table 4, and based on the assumed converter process parameters, the mass of solid copper additions to be smelted in the converter, in dependence on the oxygen concentration in blow, has been calculated from relations (4) and (7).
The results have been presented in Fig.10. The relations obtained indicate that there is a possibility of smelting considerable amounts of solid copper additions both in the 2nd and 1st stage of the converter process while applying the oxygen-enriched blow.
In order to confirm the above calculations industrial tests were conducted in a Hoboken Q-80 converter. This converter 3.23 m in diameter and 7.6 m long has chromomagnesite lining. These tests have been conducted using identical process parameters as in the case of air blow application. They are as follows:

- amount of matte charged in (m) 147 Mg/cycle
- copper content in matte 64.0 - 65.(wt.%
- primary amount of solid charge 10 - 15 Mg/cycle
- amount of flux (93 wt.% of SiO_2) 8 - 12 Mg/cycle
- blow intensity (V_d) at 1st stage (24-26)/10³ Nm³/h
- blow intensity (V_d) at 2nd stage (19-22)/10³ Nm³/h
- blow pressure 800 - 1200 kPa
- number of tuyeres 42 mm dia. 44

Applying the blow enriched with oxygen to 30vol.% (average content is approx.25 vol.% O_2),the reduction of converter operation time with blow has been achieved; in the 1st stage: from 80-100 min. to 65-75 min., and in the 2nd stage from 245-280 min to 200-230 min. The additional amount of smelted solid copper additions was 7 - 9 Mg at 1st stage, and 35 - 45 Mg at 2nd stage of the converter process, which is in agreement with the calculated values given in Fig.10.

Fig10. Mass of additions of solid copper in I and II stages of converting copper mattes of Cu content 60 % and 65 % depending on overage oxygen concentration in air blast

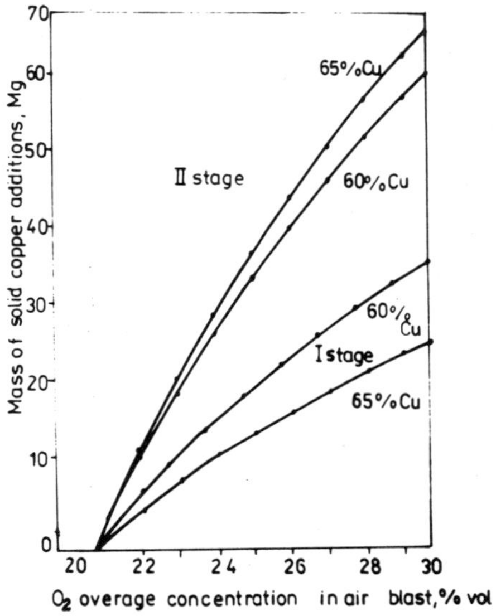

Copper matte converter process with the application of oxygen-enriched blow with a simultaneous smelting of solid charge is prepared to be introduced in Poland.
This method has already been applied at Caraiba Metais S.A. in Brazil, and partly at Khetri Copper Smelter in India.

6. Summary

The application of the oxygen-enriched blow brings significant technological advantages both in flash smelting of blister copper and in copper matte smelting in shaft furnaces.
In the flash smelting process, the oxygen-enriched blow of up to 75 vol% is applied, which enables a minimum fuel consumption, and a considerably reduced amount of gases of relatively high sulphur dioxide content. The blister copper obtained is characterized by low content of metallic inclusions, mainly lead, which enables the elimination of the converter process.
The application of oxygen-enriched blow (to 27 vol.%) enables a considerable intensification of the copper matte smelting process, reduction of coke consumption and reduction of requirements in relation to the quality of charge material.
In the converter process of copper matte the application of oxygen

enriched blow (to 30 vol.%) content improves the process heat balance which subsequently enables smelting of anode wastes from the electrorefining process or converter copper in blocks.

OXYGEN UTILIZATION IN HYDROMETALLURGY:
FUNDAMENTAL AND PRACTICAL ISSUES

E. Peters
Dept. of Metals and Materials Engineering
The University of British Columbia
Vancouver, B.C., V6T 1W5, Canada

ABSTRACT

Molecular oxygen, because of its poor solubility in aqueous solutions, presents major kinetic problems of utilization in leaching systems. These problems are discussed theoretically in terms of:
(a) mass transfer kinetics at the gas-liquid interface
(b) homogeneous reactions between dissolved oxygen (ore its surrogate, formed at the gas interface) and reduced leach products, and
(c) rate limitations at the mineral-solution interface, including chemical or electrochemical kinetics and mass transfer. The effective rates of these elementary processes are estimated from known data, and compared with the observed leaching rates for Sherritt's zinc pressure leach, and for Anaconda's oxygen ammonia (Arbiter) leach, as well as several other applications.

KEYWORDS

Hydrometallurgy, Oxygen, Oxidation, Mass Transfer, Leaching, Sulphides, Mechanisms

INTRODUCTION

Oxygen has been an important reagent in the non ferrous metals industry for a rather long time. Wherever air had been used, as in blast furnaces, roasters, flash smelters, or converters, experiments with oxygen enriched air followed from the availability of surplus oxygen from air separation plants and from electrolytic hydrogen plants originally designed to support ammonia synthesis. Today, we see on-site oxygen plants dedicated to supplying the needs of these pyrometallurgical operations. The driving forces for oxygen enrichment in pyrometallurgy have been (1) SO_2 enrichment, to improve ease of capture, (2) reduction in supplementary fuel requirements, (3) increasing furnace throughput and (4) achievement of higher temperatures. In fact, oxygen enrichment of air in these operations is usually limited by refractory losses occasioned by excessive temperatures.

Many hydrometallurgical operations that involve oxidation also benefit from the use of oxygen. As in pyrometallurgy, the most important of these involve the oxidation of sulphide minerals. Some of the advantages of oxygen in sulphide leaching processes are similar to those for pyrometallurgy: (1) it is much easier when using oxygen to utilize exothermic reaction heat to maintain leaching temperature, and (2) reactor throughputs are much higher than with air, if conditions of leaching are optimized. It is also easier to reach higher temperatures.

This paper deals essentially with sulphide mineral oxidation. Under hydrometallurgical conditions, sulphur may be converted to elemental sulphur, in which case oxygen utilization is little more than stoichiometric with sulphur (i.e., 1/2 tonne of oxygen will oxidize 1 tonne of sulphur to the elemental form from, say, ZnS[1]. If all the sulphur is oxidized to sulphate, the consumption of oxygen is a least 2 tonnes per tonne of sulphur oxidized. The oxygen that is consumed must first dissolve (or create a surrogate oxidant) at the gas-liquid interface, then the dissolved oxygen or its surrogate must be transported by mass transfer processes to the suspended mineral particle surface, where it is consumed. For a commercially viable process, the rate of oxygen consumption must be of the order 7.50 g - moles (240 grams) of oxygen per cubic meter of slurry per minute, or faster[2]. To achieve these high oxygen consumption rates, the leaching reactors have been designed with very high sparging and agitation rates (sometimes using flotation cell type agitators) or elevated pressures and temperatures, or both. Such reactors are beginning to resemble the high intensity operations more often encountered in pyrometallurgical operations. The fundamentals associated with th oxygen transfer falls into about 4 distinguishable processes[3]:

(1) The creation of large gas-liquid interfaces, through sparging and agitation.
(2) The actual gas-liquid interface processes, which may be dissolution of molecular oxygen, or oxidation at the interface of a reducing agent whose product is a surrogate oxidant.
(3) Homogeneous reactions involving dissolved oxygen or the surrogate oxidant with dissolved reducing agents created at the mineral particle surface.
(4) Reactions at the mineral particle-solution interface that lead to mineral dissolution (leaching).

(1) <u>The creation of large gas/liquid interfaces</u>. The gas-liquid interface area can be defined in units of m^2/m^3 or m^{-1}. The mass transfer rate across this interface is always proportional to this area, and its value is maximized by sparging and by gas-pumping agitators. The steady-state interface area is then established by the gas pumping rate, $\frac{dV_g}{dt}$, the volume average bubble size, $\bar{d}_g$, and the average holdup time, $\bar{t}g$, i.e.,

$$A_{g/\ell} = k_v(\bar{d}_g)^{-1}\,\bar{t}_g\,\frac{dV_g}{dt} \tag{1}$$

where k_v is the shape factor for bubbles (= 6 for spheres).

The factors that affect the mean bubble size, and the holdup time are not too well understood. Westerterp (1963) has studied this for water under ambient conditions, and found that bubbles formed at an agitator tip coallesce to a constant average size with increasing pumping rate, up to a fractional gas hold-up volume of 0.4. Above this extremely high agitation rate the average gas bubble size decreases.

It is reasonable to expect that average gas bubble size should be smaller for lower interfacial tensions, because the driving force for coallescence decreases. Also, smaller bubbles should have longer holdup times, and be more evenly dispersed throughout the vessel.

The gas pumping rate dV_g/dt, is identical to the sparging rate if the reactor agitator is incapable of pumping gas. For oxygen absorption, such gas pumping is

---

[1] The oxygen consumption increases somewhat if iron is present in the mineral.
[2] In volume units, this is 188 liters of oxygen at ambient temperature and pressure per m^3 slurry per minute.
[3] There may also be gas phase reactions, such as the oxidation of NO in HNO_2/HNO_3 catalyzed systems.

quite important, because sparger bubbles are "one pass" bubbles, i.e., if all the oxygen is not absorbed by the time the bubble breaks the surface, it will report in the purge gas. On the other hand, a gas-pumping sparger will recycle gas from the surface in the form of new bubbles.

Reith (1968) has studied the value of the interfacial gas/liquid area as a function of rotation speed and agitator diameter, for a simple turbine type agitator, at a single gas sparging rate (superficial velocity of gas = 4.7 cm/sec.). The data are plotted (Fig. 1) as area ($cm^{-1} \equiv cm^2/cm^3$) vs. N^3d^2 where N is RPM and d is the agitator diameter. On a log-log plot, most of the data fall on a straight line of slope ≈ 2/3, for tanks between 19 cm and 1.2 m in diameter.

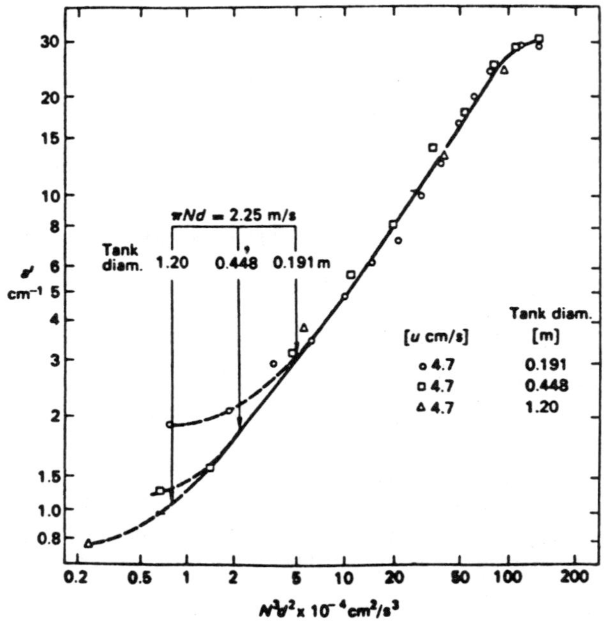

Fig. 1: Interfacial Area in Agitated Absorbers (Reith)

At agitator tip speeds below π Nd = 2.25 m/s, there is a levelling off of interfacial area, as if the agitation intensity no longer affects the area, now limited to a value that is due to sparging alone. Aparently, Reith (1968) did not take into account the possibility that the agitator may pump more gas from the surface than the sparger is providing. In the smallest tank (D = 19.1 cm) the sparger appears to be capable of generating interfacial areas in the range of 2 cm^{-1} and this falls to less than 0.8 cm^{-1} in a 1.2 meter diameter tank. The highest interfacial areas, at $N^3d^2 \simeq 10^6$ cm^2 sec^{-3}, are about 30 cm^{-1} for the full range of tank sizes. If this last value corresponds to Westerterp's (1963) hold-up volume of 0.4, it would represent a mean bubble size (in equation (1)) of .08 cm for spherical bubbles.

There are three parameters that will ultimately determine the gas/liquid interfacial area: (1) gas pumping rates, in which the agitator may supplement sparging (2) the bubble generation size, which is determined by shear forces and surface tensions at the sites of bubble entry into the slurry (sparger holes and agitator tips) and (3) coallescence (driven by surface tension and offset by local turbulence.

2. Reactions at the gas-liquid interface. Oxygen may react at the gas/liquid interface if a sufficiently reactive reducing agent is presented to the oxygen bubble. If the bubble is not pure oxygen, there is a gas-side mass transfer limit, to the reaction rate. In the absence of gaseous oxidation products (as in oxygen pressure leaching), the gas-side mass transfer resistance is not significant. Typical cases involve oxygen bubbles saturated with water vapour (in acid systems) or ammonia and water (in ammoniacal systems). Even when air is used, as in the Sherritt Gordon ammonia pressure leach, it is unlikely that the mass transfer resistance for oxygen is significant, even though gas bubbles are up to 90% nitrogen.

The Arbiter process for ammonia pressure leaching (Kuhn, et al, 1974), as practiced by Anaconda in the recent past obtains perhaps the most rapid oxygen absorption rates in hydrometallurgy. Almost all the sulphur in copper concentrates is oxidized to sulphate, requiring around 120 Kg oxygen per m^3 of slurry over a period of ~ 5 hours. The oxygen was supplied at only 0.35 atm. partial pressure (to avoid explosive NH_3-O_2 ratio's) and flotation cell agitators were used to obtain high gas/liquid interfacial areas. At the steady-state, the leach liquor is believed to contain fairly significant concentrations of cuprous ammine, ($Cu(NH_3)_2^+$), and this species is known to react very rapidly with oxygen[4]. The rate-determining step is believed to be diffusion of the $Cu(NH_3)_2^+$ species to the oxygen bubble surface, i.e.

$$R_1 = k_i A_{g/\ell} \{C_i^b - C_i^o\} \simeq A_{g/\ell} \cdot C_i^b \tag{2}$$

where $A_{g/\ell}$ is the gas liquid interfacial area,

C_i^b is the $Cu(NH_3)_2^+$ (bulk) solution concentration,

C_i^o is the $Cu(NH_3)_2^+$ concentration at the interface ($\simeq 0$),

and k_i is the liquid-side mass transfer coefficient of $Cu(NH_3)_2^+$.

If this is the rate-determining step for ammonia pressure leaching of copper or nickel concentrates, it would mean that the oxidation chemistry at the bubble surface is

$$4Cu(NH_3)_2^+ + O_2 + 2H_2O \rightarrow 4Cu(NH_3)_2^{+2} + 4OH^- \tag{3}$$

The cupric ammines ($Cu(NH_3)_2^{+2}$ and species that complex further with ammonia) then become the surrogate oxidizers at the mineral surface, where they are reduced back to $Cu(NH_3)_2^+$ by an electrochemical reaction driven by mineral oxidation.

If there is no aqueous reductant at the bubble surface capable of almost instantaneous reaction with oxygen, the oxygen will dissolve. The rate of dissolution is given by

$$R_2 = k_{O_2} A_{g/\ell} \{C_{O_2}{}^* - C_{O_2}\} \tag{4}$$

---

[4] In a laboratory test, deGraaf (private communication) once showed that a solution containing between 0.5 and 1.0 molar $Cu(NH_3)_2^+$, prepared by dissolving metallic copper in a 1.0 molar $Cu(NH_3)_4^{+2}$-containing solution, reacted completely at ambient temperature with oxygen at 0.7 MPa (100 psig) in about two minutes.

where k_{O_2} is the mass transfer coefficient for oxygen passing through the liquid-side boundary layer, $C_{O_2}^*$ is the solubility of oxygen at the given partial pressure in the bubble, and C_{O_2} is the steady-state oxygen concentration in solution. The dissolved oxygen reacts, either at the mineral interface, or with a homogeneous reductant in solution such as Fe^{+2}, whose oxidation product then becomes the surrogate oxidant at the mineral surface. Equation (4) is very similar to Equation (2). The value of $A_{g/\ell}$, the gas-liquid interfacial area is identical, while k_{O_2} will be very similar[5] (within a factor of 2) of k_i. Since $C_{O_2}^*$ is close to 1 g mole/m^3.atm., in water, it is evident that the value of $\{C_{O_2}^*-C_{O_2}\}$ is usually much smaller than C_i^b in Equation (2), so oxygen dissolution is a fairly slow process. It has been studied repeatedly at ambient temperatures in a model system consisting of sodium sulphite (.005 to 0.8 molar) with a cobalt catalyst ($\simeq$ 5ppm = 8.3×10^{-5} molar), in which the reaction produced sodium sulphate, i.e.

$$SO_3^= + \frac{1}{2}\,O_2(aq) \xrightarrow{Co^{+2}} SO_4^= \qquad (5)$$

under these conditions, the steady state oxygen concentration is negligible ($<$0.3ppm $\simeq 9 \times 10^{-6}$) while the oxygen solubility at 0.21 atm. (air) is about 2×10^{-4} molar. Thus, the sodium sulphite oxidation rate is an accurate measure of maximum oxygen dissolution rates at the gas-liquid interface.

Recent work by deGraaf (1984) shows that in this model system, on a 2000 liter scale, air bubbles are approximately 20 to 35% depleted in oxygen when they break the surface. If the mean bubble size is about 0.08 cm. and the mean residence time is 0.8 sec., the liquid side mass transfer coefficient works out to about about 1.2×10^{-3} m./sec. This is a rather large number, compared to mass transfer from particles. There are large discrepancies in the technical literature relating to mass transfer from bubbles, and different models have been used to explain this. Danckwerts (1951), for example, adopted a "penetration" model for mass transfer on the liquid side of a gas interface, based on a convective renewal of the interface (no stagnant liquid film as on solids). This leads to a mass tranfer coefficient that is proportional to $D_{O_2}^{1/2}$ (rather than to D_{O_2}[6], as in the film model)* and also to higher mass transfer coefficients than would be reasonable in a film model. Besides this, there is always the contribution to mass transfer the bubble nucleii, before they detach from the sparger or agitator, or before they coallesce to the observed size. Smaller bubbles or attached bubbles will make a contribution to mass tranfer that may be substantially out of proportion to the gas volume represented in this form.

---

[5]In general, k_i or k_{O_2} represent mass tranfer coefficients that are linear functions of the molecular diffusion constants D_{O_2} or D_i. Except for the diffusion constants of H^+ and OH^-, these diffusion constants fall into a rather narrow range of aqueous solutions.

[6]Where $D_{O_2}^*$ is the diffusion coefficient of dissolved oxygen.

De Graaf (1984) also studied mass transfer due to gas pumping by the agitator alone. He came to the conclusion that the agitator functions in several different ways:

(a) Below a critical agitator tip velocity

$$v^* = C_{Ag}\ (2\ g\ h)^{1/2} \tag{6}$$

the agitator does not pump gas, and above this velocity, gas pumping increases linearily with tip speed v[7] (or with R.P.M., since the agitator diameter is fixed).

(b) The power number of the agitator, Np, decreased with gas pumping rate, in much the same way as it does with gas sparged below the agitator (Michelle et al, 1962) (see Fig. 2 and Fig. 3).

(c) The oxygen mass transfer rate is approximately linear with agitator tip speed for a given agitator (see Fig. 3).

It is clear from this work that gas can be dispersed without spargers, but that there remains inadequate knowledge with respect to optimizing agitator configurations for such a case.

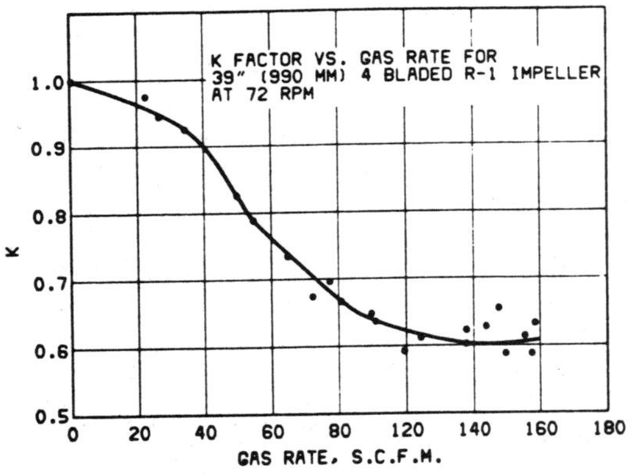

Fig. 2: Relative Power Consumption for an Agitator in a Gas-Sparged Vessel.

---

[7] where C_{Ag} is a constant, dependent on the agitator design (= 1 for an "ideal" agitator),

g is the acceleration of gravity (= 9.81 m^2sec^{-1})

and h is the agitator imersion depth

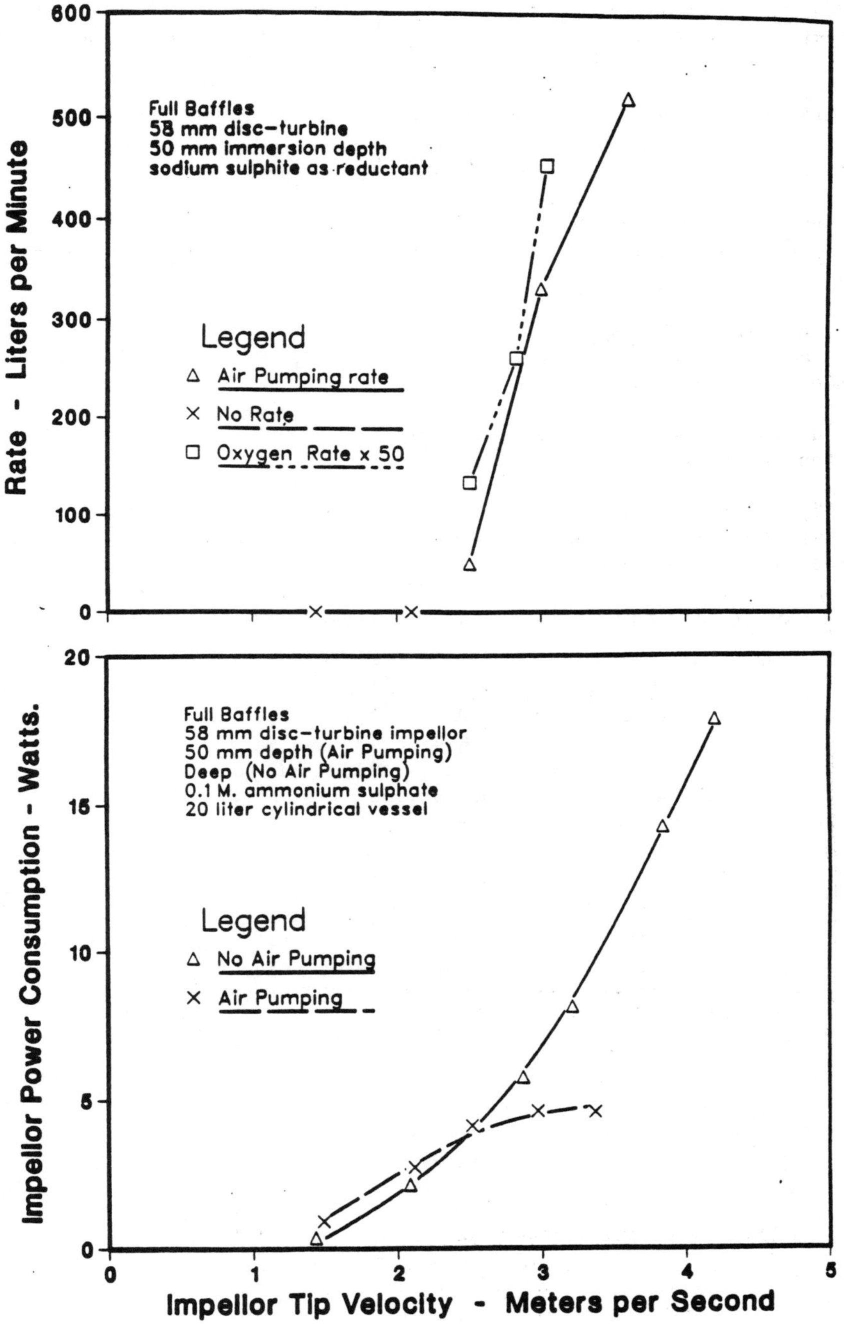

Fig. 3: Effect of Impellor Tip Velocity.
(a) On air pumping rate and oxygen absorption rate by dissolved Na_2SO_3
(b) On power consumption

When equation (4) is examined, it is clear that the rate of oxygen absorption, R_2, can be increased by three independent parameters, k_{O2} (the mass transfer coefficient), $A_{g/\ell}$, and $C_{O_2}^*$. Most studies on gas absorption have established values for $k_{O2} \times A_{g/\ell}$, i.e. the product of these two parameters. The remaining variable, $P_{O_2}^*$, can be substantially enhanced by using oxygen instead of air, preferably at high pressures. Another alternative would be to use a much more soluble oxidant than oxygen, but this is remarkably expensive.

3. Homogeneous processes involving dissolved Oxygen. For a sparger-agitator system to continue the process of dissolving oxygen from bubbles, there must be a sink for the dissolved oxygen. In the zinc pressure leach, the dissolved oxygen reacts with Fe^{+2} ions, according to the equation:

$$4Fe^{+2} + 4H^+ + O_2 \rightarrow 4Fe^{+3} + 2H_2O \tag{7}$$

The rate of this reaction when studied alone has been reported as:

$$\frac{-d\,[Fe^{+2}]}{dt} = k_8\,[Fe^{+2}]^2\,[O_2] \tag{8}$$

where $[O_2]$ is the actual dissolved oxygen concentration. This equation is over simplified, because the reaction is known to be catalyzed by copper (McKay and Halpern, 1958), and inhibited by acid (Mathews and Robins, 1972). It is also catalyzed by $SO_4^=$ (Dreisinger, private communication, 1987). Most of the work has been done at temperatures below the zinc pressure leach conditions, but recent work at 150°C, and in solutions 2 M. in $ZnSO_4$ and 0.4 M. in H_2SO_4, there are indications that the rate constant for equation (8), k_8, is about 48 liter mole^{-2} sec^{-1}. The solubility of oxygen has never been measured in this particular solution, but there is evidence (MacArthur, 1916) that solutions similar to this composition have only about half the oxygen solubility of water, or about 5×10^{-4} moles liter^{-1} atm^{-1}. At a partial pressure for oxygen of about 8 atm., this leads to an approximate value of $C_{O_2}^* = 4 \times 10^{-3}$ moles liter^{-1}, and the rate of Fe^{+2} oxidation at 0.1 M. Fe^{+2} is then about 1.92×10^{-3} moles liter^{-1} sec^{-1}, approximately twice the rate normally observed in the first compartment of the Cominco autoclave (Martin and Jankola, 1985). Thus, the ferrous oxidation process is fast enough to support observed rates of the zinc pressure leach, provided that the steady-state oxygen concentration is at least 50% of the saturation value.

4. Reactions at the mineral interface In the Anaconda-Arbiter process, (Kuhn et al, 1974), dissolved oxygen is non-existent under normal conditions, and cupric ammines ($Cu(NH_3)_n{}^{+2}$) are the surrogate oxidants at the mineral surface. In the zinc pressure leach, the surrogate oxidant is Fe^{+3}. However, if dissolved oxygen is present, it can react directly at the mineral surface. The rate at which it reacts is determined by either the heterogeneous reduction kinetics (an electrochemical process whereby the rate is determined by the potential of the mineral surface), or by the limiting mass transfer rate.

The cathodic reduction of oxygen on mineral surfaces has been studied by Biegler et al (1977). Polarization curves for oxygen reduction on various surfaces, including metals, are shown in Fig. 4.

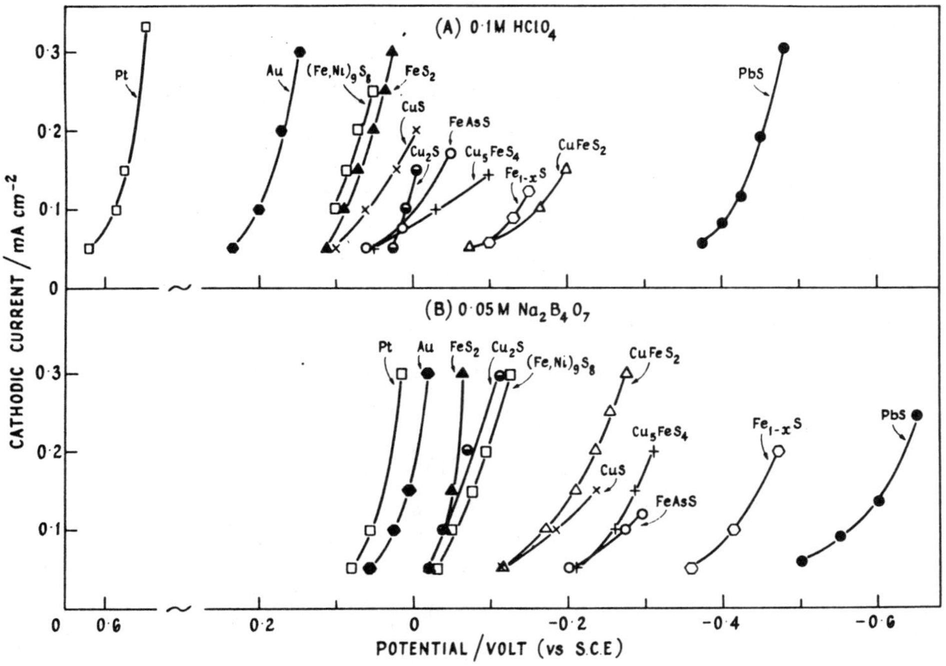

Fig. 4: Activation-controlled currents at the foot of the O_2 reduction wave on various electrodes in O_2-saturated acid (A) and alkaline (B) solutions (Biegler et al, 1977).

For leaching reactions to take place at reasonable rates (say, 0.1 μm per minute minimum, so that a 50 μm particle can dissolve completely in 5 hours), it is necessary to have an anodic current density of 1 to 5 ma/cm^2, depending on the molar volume of the mineral and how it oxidizes. The curves in Fig. 4 indicate that, at room temperature, only platinum will reduce oxygen at a measurable rate at the potential of Fe^{+3}/Fe^{+2} (about + 0.5 volts on the S.C.E. scale of Fig. 4). Since significant rates of oxygen reduction require cathodic potentials below the reversible Fe^{+2}/Fe^{+3} values there is no hope of oxygen acting directly (inorganically) on any mineral at room temperature at a useful rate. However, bacteria can catalyze the reactions of oxygen with inorganic sulphur containing species. In the presence of such bacteria as thiobacillis thio-oxidans, sulphide minerals can be decomposed by oxygen at useful rates.

Polarization curves have not been made for oxygen reduction under autoclave conditions. However, Bailey (1976) measured mixed potentials on pyrite in the presence of oxygen, and showed data that can be interpreted as a shift in the FeS_2 curve in Fig. 4 (a) from 0.1 volts at room temperature to about 0.8 volts at 130°C and 10 atm. oxygen pressure. The Fe^{+3}/Fe^{+2} potential measured in this same system was 0.85 volts (1M.$HClO_4$). Since FeS_2 is one of the most noble of the minerals in Fig. 4, it is quite clear that autoclave conditions to 130°C will not lead to significant oxygen reduction on the surface of sulphide minerals, even if dissolved oxygen is present at appreciable levels due to high pressure. The oxygen reduction on the surface of these minerals is simply too slow at normal mixed potentials to support practical mineral decomposition rates, or to compete with ferric ions if these are present.

Under zinc pressure leaching conditions, the limiting mass transfer coefficient for oxygen on pyrite is probably around 10^{-4} m/sec, based on Harriott's equation (Harriott, 1962) for 50 μm particles under free settling conditions:

$$k_t^i = \frac{D_i}{\mu} \{2 + 0.6\ N_{Re}^{0.5}\ N_{Sc}^{0.33}\} \qquad (9)$$

where μ is the mean particle diameter
D_i is the molecular diffusion constant of i
N_{Re} is the Reynolds number ($\mu v \rho/\eta$)
and N_{Sc} is the Schmidt number $\eta/\rho\mu$)

In this equation, η is the solution viscosity and v is the Stokes Law settling velocity:

$$v = \frac{g\mu^2(\rho s - \rho \ell)}{18\ \eta} \qquad (10)$$

The mass transfer rate for oxygen is given by the equation:

$$R = k_t^i\ (C_i^b - C_i^s) \qquad (11)$$

where i is oxygen, and the superscripts b and s represent the bulk and surface concentrations. If $C_i^b = 4$ moles/m^3 and $C_i^s = 0$, the value of R is 4×10^{-4} moles meters^{-2} sec^{-1}. This rate is fast enough to cause 50 μm particles to shrink[8] at a rate of about 2.3 μm/min., or more than fast enough to account for the 1 μm (min. observed reaction rate. However it has already been stated that dissolved oxygen is not even remotely reactive enough to react on a ZnS surface, and additionally, the observed reaction rate is far slower for "pure" (low iron) ZnS in the absence of dissolved iron.

The mass transfer rate for oxygen supplied to bacteria that are decomposing sulphides by organic sulphur oxidation is no different than an inorganic case, except that oxygen is usually supplied as air (0.2 atm. O_2) at low temperatures (20-40°C). In an agitated system, maximum oxygen transfer (if bacterial populations are not rate-limiting) would be about 10^{-5} moles meter^{-2} sec^{-1}, if the mineral is under free settling conditions, and this would lead to a leaching rate for, say $CuFeS_2$, of 0.007 μm per minute. It would take about 5 days for a 50 μm particle of $CuFeS_2$ to decompose completely. This is not normally regarded as a satisfactory leaching rate. Faster observed rates must involve a contribution from inorganic Fe^{+3} leaching, or more finely ground mineral.

The first compartment of the zinc pressure leach operates at about 0.1 molar Fe^{+3} in the steady state. If Fe^{+3} is identified with i in equation (11), the result at $C_i^s = 0$ is that $R = 10^{-2}$ moles meters^{-2} sec^{-1}, and this is enough to dissolve 50 μm ZnS particles at a rate of 14.5 μm/minute, or about 15 times the observed rate.

---

[8] Particle shrinkage in μm min^{-1} is ZRV_{ZnS} (1.67×10^4) where Z is the moles ZnS leached per mole O_2, R is the leaching rate, and V_{ZnS} is the molar volume of ZnS.

It is therefore clear that the reaction between Fe^{+3} and ZnS is not fast enough over the whole ZnS surface to permit the surface concentration of Fe^{+3}, C_i^s, to be close to zero; the reaction rate is chemically, rather than diffusionally controlled.

In ammine systems, the mineral surface reaction is equally unlikely to be controlled by mass transfer of oxygen. The Anaconda-Arbiter process operates at only 0.35 atm. oxygen pressure, and this would lead to a maximum dissolved oxygen concentration of only about 0.2 moles O_2 per m^3 solution. The maximum rate of leaching would be 2×10^{-5} moles $meter^{-2}$ sec^{-1}, which translates to a mineral particle size reduction rate of .015 μm/min, or less than 15% of the observed rate. In the case of ammine leaching, Cu^{+2} is probably the surrogate oxidant at the mineral surface. If it reacts at its limiting rate, as determined by mass transfer kinetics (equation (11) with $C_i^s \approx 0$), then it is possible to oxidize chalcopyrite particles at the observed rate of the Anaconda-Arbiter process when the $Cu\,(NH_3)_n^{+2}$ concentration is about 5 moles per m^3 (.005 M or 0.3 g/ℓ Cu^{II}). This concentration is present even in Sherritt's ammonia leach for nickel. The necessary chemistry (for chalcopyrite) would be expressed by the equation (Peters, 1986):

$$3Cu(NH_3)_n^{+2} + CuFeS_2 \rightarrow 4Cu(NH_3)_m^{+} + Fe^{+2} + 2S^{\circ} \tag{12}$$

and

$$2S^{\circ} + 2OH^{-} \rightarrow 1/2\ S_2O_3^{=} + HS^{-} + 1/2\ H_2O \tag{13}$$

In this system, $Cu(NH_3)_n^{+2}$ is also the surrogate oxidant for homogeneous oxidation of Fe^{+2} and intermediate valence states of sulphur. In fact, it is probable in this system that 80% of the oxidant formed at the gas/liquid interface is consumed by homogeneous reactions, and not more than 20% is reduced at the mineral interface by reactions such as equation (12).

SUMMARY AND CONCLUSIONS

The utilization of oxygen in pressure leaching processes such as the zinc pressure leach or the ammoniacal leach for nickel or copper concentrates calls for mineral oxidation at a solution-mineral interface by dissolved oxygen, or by a dissolved species defined as surrogate oxidants because they have been created by the consumption of oxygen. Oxygen itself is a relatively insoluble gas, having a solubility of 0.8 to 1×10^{-3} moles $liter^{-1}$ atm. in water, and as little as half the above value in solutions commonly used in these leaches. The transfer of oxygen from gas bubbles to the mineral interface is the simplest model for this leach (Fig. 5(a)). This process, while slow, is fast enough at elevated oxygen pressures to account for the kinetics of the zinc pressure leach. However, dissolved oxygen arriving at the mineral interface is not reactive enough under inorganic conditions to be reduced at the observed leaching rate, because the electrochemical oxygen reduction rate is too slow, even if the oxygen concentration is at a saturation level. The bacterially catalyzed process may operate at the mass transfer-limiting rate, but this rate is further reduced by the practice of using air at ambient pressure.

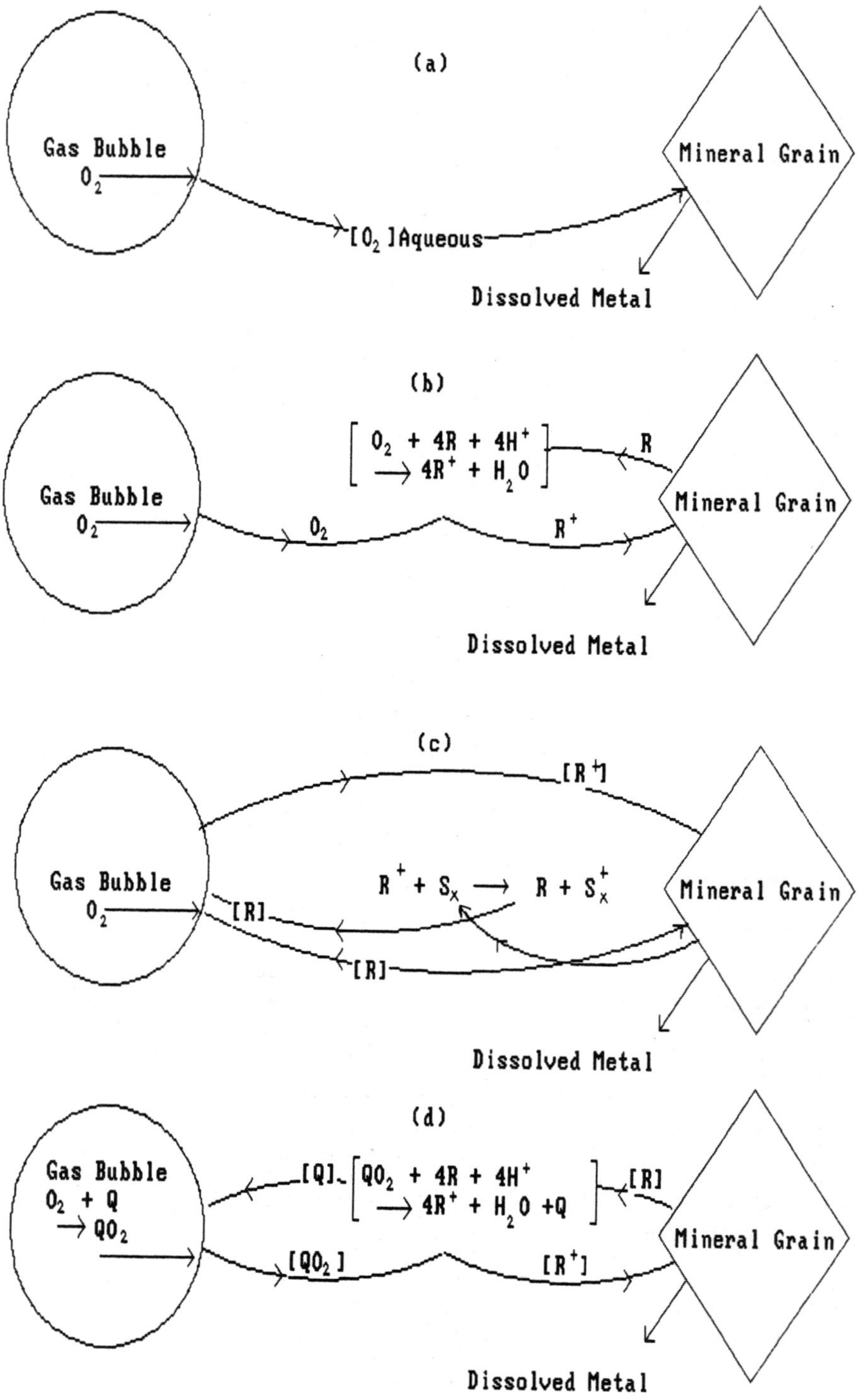

Fig. 5: Models for Oxygen Absorption During Sulphide Mineral Leaching.
(a) Transfer of dissolved oxygen, (b) Homogeneous reaction of dissolved oxygen, (c) Heterogeneous reduction of oxygen (d) Gas phase conversion

Dissolved oxygen can also react with reducing agents present in solution by a homogeneous reaction, and create a secondary oxidant. In the case of the zinc pressure leach, Fe^{+2} ions are oxidized to Fe^{+3}. The Fe^{+3} ions so produced can then be delivered to the mineral surface at a rate that is proportional to its concentration, which is normally much higher than that of dissolved oxygen. The Fe^{+3} ions, being created by the homogeneous consumption of dissolved oxygen, act as a surrogate for oxygen in oxidizing the mineral. The reduction of Fe^{+3} at the mineral surface is almost reversible, and so the rate-determining step defaults to the mineral decomposition process. Such a model is represented by Fig. 5(b), where the Fe^{+3} is represented by the more generalized oxidant R^+. The reaction rate for the model in Fig. 5(b) is intrinsically faster than that for the Fig. 5(a) model, because the oxidant $R^+$is available at the mineral surface at much higher concentrations than dissolved oxygen. Furthermore, under bacterial leaching conditions, free-floating bacteria can catalyze the reaction between Fe^{+2}ions and dissolved O_2.

A reducing agent R generated by the mineral decomposition reaction might be far more reactive to oxygen than Fe^{+2} ions, and so may react at the gas/liquid (bubble) interface. This is almost certainly the case for $Cu(NH_3)_2^+$, which is known to oxidize extremely rapidly, even at room temperature. Under these conditions there is no significant dissolved oxygen, and the surrogate oxidant ($Cu(NH_3)_n^{+2}$) is the agent for both mineral oxidation and for any homogeneous reactions with even stronger reducing agents released by the decomposing mineral. Ammoniacal leaching is accompanied by the presence of reduced sulphur species, which are almost certainly first formed by the non-oxidative leaching of sulphur (see equation (13)). The resulting reaction model (Fig. 5(c)) is (in principle) capable of being faster than the earlier model (Fig. 5(b)) because a reducing agent such as $Cu(NH_3)_2^+$ can be transferred to the oxygen-containing bubble faster than oxygen can be transferred to solution across the gas/liquid interface, provided that the steady-state cuprous concentration is substantially higher than the solubility of oxygen.

In principle, the fastest oxygen transfer rate is achieved if a highly soluble surrogate oxidant can be generated in the gas phase. This model is depicted in Fig. 5(d), where the surrogate oxidant is identified as QO_2. Some nitric acid leaching processes would fall into this class, although the oxidant, NO_2, is not exactly dissolved; it reacts with water to form a mixture of nitric and nitrous acid:

$$2NO_2 + H_2O \rightarrow HNO_2 + HNO_3 \tag{14}$$

Although nitrous acid is a very active oxidant (it reduces rapidly at a mineral surface at a potential well above that of Fe^{+3}), nitric acid is slow. The maintenance of high mineral oxidation rates is dependent on keeping the slow equilibrium of reaction (15)

$$3\ HNO_2 \rightleftarrows HNO_3 + 2NO + H_2O \tag{15}$$

well to the left, a condition that is greatly assisted by high NO partial pressures in solution. Because NO is generated from HNO_2 at the mineral surface, the reaction beginning with HNO_3 is found to be autocatalytic.

Because of the high oxidation potential that can be exerted by HNO_2, the above system is useful for processes in which sulphur must be oxidized rapidly. It is the basis of the Arseno process for opening gold ores (Beattie et al, 1987) as well as for the NITROX process, (Van Weert et al, 1986).

REFERENCES

Beattie, M.J.V., Raudsepp, R., and Peters, E. (1987). Hydrometallurgical Arsenopyrite Process, U.S. Patent No. 4,647,307, March 3, 1987; Canadian Patent No. 1,219,132, March 17, 1987.

Biegler, T., Rand, D.A.J., and Woods, R. (1977). In J.O.M. Bockris, D.A.J. Rand, and B.J. Welch (Eds.), Trends in Electrochemistry. Plenum Press, N.Y. and London, pp. 291-302.

Chen Feng, K. (1979). Thermodynamic and Kinetic Studies of the Iron-Sulfite System by Electrochemical Means. Ph.D. Thesis, Columbia University.

Danckwerts, P.V. (1951). Ind. Eng. Chem., 43, (1460-).

DeGraaf, K.B. (1984). An Investigation of Gas/Liquid Mass Transfer in Mechanically Agitated Leaching Systems. M.A.Sc. Thesis, The University of British Columbia.

Harriott, P. (1962). Mass Transfer to Particles: Part 1. Suspended in Agitated Tanks. A.I.Ch.E.J., 8, (93-102).

Kuhn, M.C., Arbiter, N., and Kling, H. (1974). Anaconda's Arbiter Process for Copper. C.I.M. Bulletin, 67, February (62-73).

Martin, M.T., and Jankola, W.A. (1985). Cominco's Trail Zinc Pressure Leach Operation. C.I.M. Bulletin, 78, April (77-81).

Mathews, C.T. and Robins, R.G. (1972) Proc. Aust. Min. Met., 242 (47-).

McKay, D.R., and Halpern, J. (1958). J. Trans. Met. Soc. of A.I.M.E., 212, (301-309).

Michelle, B.J., and Miller, S.A. (1962). A.I.CH.E. J., 8,

Peters, E. (1986). The Leaching of Sulfides. In P. Somasundaran (Ed.), Advances in Mineral Processing, Society of Mining Engineers of A.I.M.E., Chap. 26, pp. 445-462.

Pray, H.A., Schweickert, C.E., and Minnick, B.H. (1954). The Solubility of Hydrogen, Oxygen, Nitrogen and Helium in Water at Elevated Temperatures. Ind. Eng. Chem., 44, pp. 1146.

Reith, T. (1968). Physical Aspects of Bubble Dispersions in Liquids. Ph.D. Thesis, Delft University, The Netherlands.

VanWeert, G., Fair, K.J., and Schneider, J.C. (1986). Prochem's NITROX Process. presented at the 88th General Meeting of the C.I.M., Montreal, May 11-15.

Westerterp, K.R. (1963). Chem. Eng. Sci., 18, p. 495.

ROLE OF OXYGEN IN DUMP LEACHING

J. Brent Hiskey
University of Arizona, Tucson, Arizona

Roshan Bhappu
Mountain States Research & Development, Tucson, Arizona

ABSTRACT

As invaluable as oxygen is in many hydrometallurgical processes, dump leaching of copper is no exception. Without the availability of oxygen in the dumps, pyrite cannot be oxidized to provide the required ferric iron and acid for leaching. Moreover, without oxygen, the primary and secondary copper minerals cannot be solubilized. Finally, without oxygen, bacterial activity in the dumps would be retarded resulting in poor copper extraction. Accordingly, the presence of oxygen is essential in dump leaching and every effort must be made to provide this essential ingredient in the dump leaching recipe.

This paper presents the important role that oxygen plays in dump leaching of copper. The theoretical and practical aspects of pyrite and sulfide copper minerals are presented. Similarly, the role of oxygen in bacterial leaching is discussed. Finally, efforts have been made to present the engineering, technical and economic aspects of dump design and operation for providing and sustaining maximum oxygen in the dump leaching system.

KEYWORDS

Copper dump leaching; oxidation of sulfide minerals; bacterial leaching; iron oxidation and hydrolysis; dump design.

INTRODUCTION

The dump leaching of lower-grade copper ores has been practiced since the last 50 years as an integral part of the open-pit mining operations in the Western United States. The recovery of copper by this technique has been and will continue to be one of the major sources of copper. Each year, several hundred million tons of copper wastes are added to the existing dumps of over 10 billion tons. In recent years, copper production from dump leaching and in situ extraction amounted to about 15 - 30 percent of the total copper production . These dump leaching practices have also contributed to the production of a small quantity of uranium (Brooke 1977) and could be a potential source of other by-products such as aluminum and molybdenum (Chase and Potter 1982) and strategic metals like cobalt (Hiskey 1986). Besides the United States, heap and dump leaching operations are carried out in other countries such as South Africa, Zambia, Zimbabwe, Japan, Australia, Chile, Bulgaria, Mexico, India, Portugal, Spain and the USSR.

The recent acceleration in dump leaching can be attributed to:

1) Availability of large tonnage of lower-grade ores dumped at newer mines as well as those resulting from increased stripping ratios at older mines.

2) The relatively small capital investment required for dump leaching facilities.

3) Low labor and supervision required for dump leaching operations.

4) Advances in solvent extraction-electrowinning (SX-EW) technology.

5) Attractive environmental considerations.

Historically, heap leaching is the original leaching method established at Rio Tinto, Spain (Taylor and Whelan 1942) more than 300 years ago and is still utilized for leaching oxide and sulfide ores that are too low in copper content to be economically concentrated and smelted, but economical enough not to be discarded as waste. Heap leaching can be carried out on crushed or uncrushed ore and usually heaps are deliberately designed and constructed to provide maximum extraction of copper within a relatively short period (30 to 100 days). On the other hand, dump leaching is usually carried out on uncrushed waste rock containing marginal or submarginal material dumped in the most convenient areas adjacent to the mine and leached over a long period (2 to 4 years or more). In practice, both heap and dump leaching involve application of acidic lixiviant to the top of the heap or dump surface by sprinkling or flooding with the capture of solution that percolates to the bottom by gravity flow. The dissolved copper from this pregnant solution is usually recovered by cementation of copper on scrap iron or by solvent extraction followed by electrowinning of the metal. The barren solution is then recycled to the heap or dump as is or after addition of sulfuric acid as needed. In recent years, attempts have been made to introduce the leach solution, especially in dumps, through injection wells to obtain more uniform solution flows in dumps which can vary from about 30 meters to more than 200 meters.

In the past, the mine operators have not been concerned about efficiency in dump leaching and accepted whatever copper recoveries that were obtained without much effort. However, because of the importance of dump leaching currently as a source of lower-cost copper, an operator today cannot leave dump leaching to chance. What appears to be a simple, empirical technique, is in reality a very complicated process involving a large number of critical parameters that encompass several scientific and engineering disciplines. Accordingly, it is essential that the modern operators be cognizant of these issues and utilize them in conjunction with their past practical experiences to obtain maximum recovery from their dump leaching operations.

One of the most important parameters in dump and heap leaching is the behavior and control of iron. This topic was discussed by Bhappu (1986) last year at the Symposium of Iron Control in Hydrometallurgy. Of equal importance is the behavior of oxygen in the dump leaching operation and it is the primary objective of this paper to elucidate the role of oxygen in such hydrometallurgical processes. The paper covers the chemistry of oxygen in dump leaching operations including oxidation and dissolution of copper sulfide minerals; and oxygen distribution in the dumps.

Other related topics covered include the role of iron-oxidizing bacteria in dump leaching practices, iron oxidation, hydrolysis and precipitation; alternate drying and wetting of dumps; oxygen injection and engineering design of dumps to obtain optimum oxidation.

LEACHING CHEMISTRY

The Role of Oxygen

To examine the role of oxygen in the oxidation of sulfide minerals in waste dump leaching environments, one must first examine the dissolution of oxygen in aqueous solution. The dump leaching system involves the countercurrent flow of air and water through piles of rock as depicted in Fig. 1.

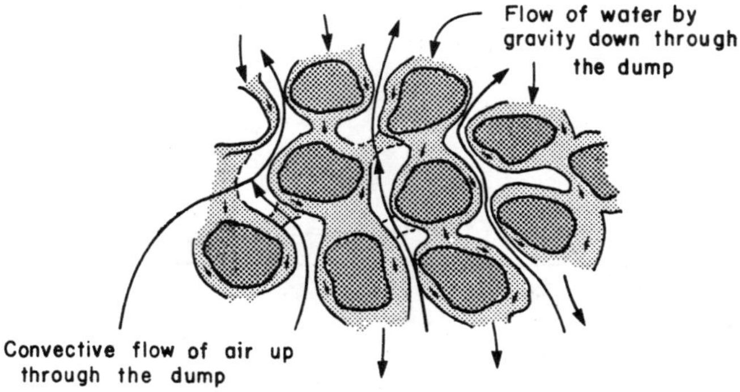

Fig. 1. Schematic diagram showing the counter-current flow of air and water through a leach dump. (After Cathles and Apps 1975)

Oxygen enters the aqueous phase by simple dissolution as follows:

$$[O_2]_{gas} = [O_2]_{aq} \qquad (1)$$

The solubility of oxygen is explained in terms of Henry's law which states "the mass of gas dissolved by a given volume of solvent, at constant temperature, is proportional to the pressure of the gas with which it is in equilibrium." It falls directly from reaction (1) that:

$$[O_2]_{aq} = K_h \; P_{O_2} \qquad (2)$$

where $[O_2]_{aq}$ is the concentration of dissolved oxygen, P_{O_2} is the partial pressure of oxygen in the gas phase, and K_h is the equilibrium (Henry) constant. Fig. 2 shows the effect of temperature on the solubility of oxygen in pure water for one atmosphere of air and pure oxygen. Both gases exhibit a decrease in oxygen concentration with increasing temperature. As expressed in equation (2), at constant temperature an increase in oxygen pressure (0.21 atm → 1.0 atm) produces a corresponding increase in oxygen concentration. At 25°C the solubility of oxygen in pure water in equilibrium with air is 2.6×10^{-4} mol/liter. Correspondingly, the equilibrium concentration for pure oxygen at 25°C is 1.26×10^{-3} mol/liter. In a strict sense, an activity term should be used in equation (2). In dilute solutions the use of concentration is a reasonable approximation. However, in the high ionic strength solution produced by the recycle of dump leach liquors, there is an associated decrease in oxygen solubility.

The effect of ionic strength on oxygen solubility is illustrated in Fig. 3. Oxygen solubility at 25°C for ionic strengths up to 3 M are shown for various electrolytes (Linke 1965). Sodium sulfate solutions have the most pronounced effect on decreasing oxygen solubility. A 0.5 M Na_2SO_4 solution ($I = 1.5$) results in a 50% decrease in oxygen solubility. These results show that sodium salts exhibit a stronger effect that their acid counterparts. The data for H_2SO_4 was calculated based on the dissociation behavior of H_2SO_4 and HSO_4^- in sulfuric acid solutions (Young and Blatz 1949).

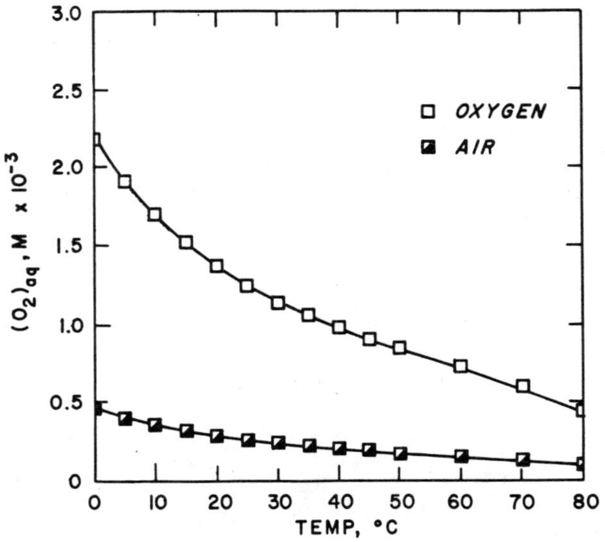

Fig. 2. Effect of temperature on the solubility of oxygen in water under one atmosphere pressure of oxygen and air.

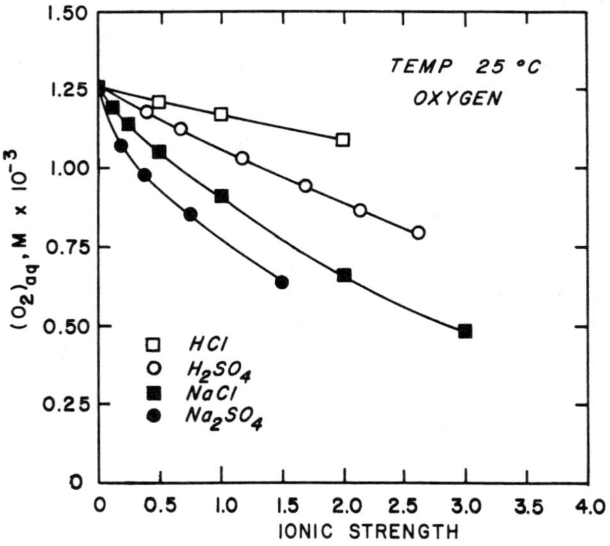

Fig. 3. Effect of ionic strength on the solubility of oxygen in aqueous solutions of acids and salts at 25°C and one atmosphere pressure of oxygen.

Dump leach solutions are typically loaded with dissolved salts. The chemical analyses, pH, and ionic strength of various copper leach liquors are presented in Table 1.

TABLE 1 Typical Copper Dump Leach Solution Composition

	Leach Dumps Sampled*				
	Concentration g/l				
Component	A	B	C	D	E**
Cu	0.69	0.46	0.63	0.38	0.5
Fe(total)	1.15	0.07	1.55	0.24	1.5
Al	3.22	0.69	5.62	0.85	5.0
Mg	3.66	1.99	2.31	2.63	7.0
Ca	0.68	0.50	0.59	0.50	0.6
SO_4	49.6	24.0	59.80	28.5	60.0
pH	2.3	2.9	2.1	2.7	2.5
Ionic *** Strength	1.9	0.6	2.3	0.8	2.3

* (Hiskey 1986)
** (Schlitt and Jackson 1981)
*** Based on major components

Ionic strength values for these solutions range from approximately 0.6 to 2.3. Based on these ionic strengths one would expect dissolved oxygen concentrations for air at 25°C to be of the order $1x10^{-4}$ to $2x10^{-4}$ mol/liter.

Copper Sulfide Minerals

There are a large number of copper sulfide minerals. A list of some of the more common copper sulfide minerals is presented in Table 2.

TABLE 2 Important Copper Sulfide Minerals

Mineral Name	Formula
Bornite	Cu_5FeS_4
Chalcocite	Cu_2S
Chalcopyrite	$CuFeS_2$
Covellite	CuS
Cubanite	$CuFe_2S_3$
Digenite	Cu_9S_5
Enargite	Cu_3AsS_4
Idaite	Cu_5FeS_6
Tetrahedrite	$Cu_{12}Sb_4S_{13}$

Of these, chalcocite (Cu_2S) and covellite (CuS) are important secondary minerals occurring in the supergene enrichment zone of mineralization. Chalcopyrite is the most important primary copper sulfide mineral and occurs mainly in the hypogene zone. In prophyry deposits, most low grade waste, from which copper is leached, is obtained from a pyritic zone outside the zone of ore-grade mineralization (Lowell and Guilbert 1970). This outer zone or halo contains primarily pyrite and chalcopyrite and is characterized by high pyrite-to-chalcopyrite ratios. Therefore, it is appropriate to consider chalcopyrite and pyrite as the most important sulfide minerals and that their oxidation controls the leaching of copper from waste dumps.

The analysis of pyrite oxidation will be considered first. Hiskey and Schlitt (1982) have reviewed the aqueous oxidation of pyrite. They discuss a model proposed by Singer and Strumm (1969) which incorporates an initiation step and a propagation cycle to explain the aqueous oxidation of pyrite. The mechanism can be expressed as follows:

Initiation reaction:

$$FeS_2 \xrightarrow{(+\ O_2)} Fe^{2+} + \text{S-compound} \qquad (3)$$

Propagation cycle:

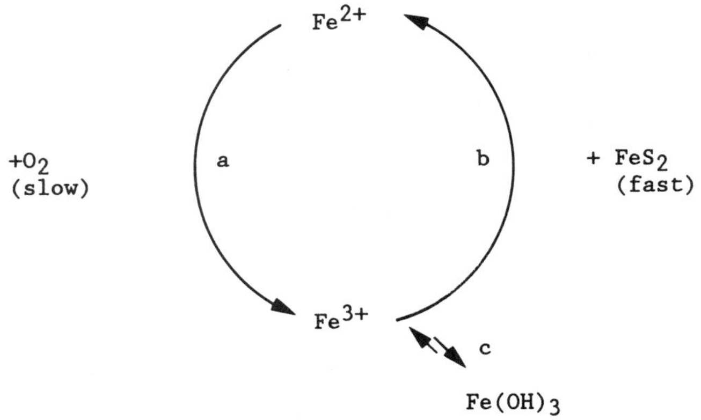

(4)

Reaction (3) initiates the overall dissolution process and provides a constant supply of ferrous iron to the system. In the absence of bacteria, ferrous ion is oxidized to ferric ion by oxygen reaction (4a) at a relatively slow rate under ambient or near ambient conditions. Reaction (4a) is formally represented by the following reaction:

$$2Fe^{2+} + 1/2O_2 + 2H^+ = 2Fe^{3+} + H_2O \qquad (4a)$$

Ferric ion reacts rapidly with FeS_2 in step (4b). At steady state oxygen is not directly involved in the oxidation of pyrite, but maintains the propagation cycle by regenerating ferric ion. In side reaction (4c), ferric iron is precipitated as $Fe(OH)_3$. This allows for removal of iron from the system and balances the iron derived from pyrite oxidation.

The dissolution of pyrite is controlled by an electrochemical mechanism involving both sulfate and elemental sulfur forming anodic reactions (Bailey and Peters 1976):

$$FeS_2 = Fe^{2+} + 2S^o + 2e \qquad (5)$$

$$FeS_2 + 8H_2O = Fe^{2+} + 2SO_4^{2-} + 16H^+ + 14e \qquad (6)$$

The partitioning or yield of sulfate and elemental sulfur is sensitive to the potential established on the pyrite electrode. The cathodic reaction that complements reactions (5) and (6) is

$$O_2 + 4H^+ + 4e = 2H_2O \quad (7)$$

Lower oxidation potentials favor a situation where both reactions (5) and (6) contribute to the dissolution of pyrite (i.e. both sulfate and elemental sulfur are produced). However, as the potential increases (application of overpressures of oxygen), the sulfate route, reaction (6), becomes dominant.

The elemental sulfur produced by the oxidation of pyrite is metastable under dump leaching conditions. The presence of sulfur oxidizing bacteria and the long leach cycles yield very little elemental sulfur. Therefore, the oxygen requirement for pyrite oxidation is 3.5 moles O_2 per mole FeS_2 dissolved.

The dissolution of copper sulfide minerals under dump leaching conditions is principally by the action of ferric ion. Ferric ion concentrations are typically two order of magnitude greater than those of dissolved oxygen. Copper sulfides yield elemental sulfur when leached with acidic ferric sulfate solutions. As indicated above, the influence of bacterial activity and the long leach times in dump leaching result in sulfate formation as illustrated by the following reactions

Chalcopyrite

$$CuFeS_2 + 16Fe^{3+} + 8H_2O = Cu^{2+} + 17Fe^{2+} + 2SO_4^{2-} + 16H^+ \quad (8)$$

Covellite

$$CuS + 8Fe^{3+} + 4H_2O = Cu^{2+} + 8Fe^{2+} + SO_4^{2-} + 8H^+ \quad (9)$$

Chalcocite

$$Cu_2S + 2Fe^{3+} = CuS + Cu^{2+} + 2Fe^{2+} \quad \text{stage 1} \quad (10)$$
$$CuS + 8Fe^{3+} + 4H_2O = Cu^{2+} + 8Fe^{2+} + SO_4^{2-} + 8H^+ \quad \text{stage 2} \quad (9)$$
$$Cu_2S + 1Fe^{3+} + 4H_2O = 2Cu^{2+} + 10Fe^{2+} + SO_4^{2-} + 8H^+ \quad \text{overall} \quad (11)$$

Chalcopyrite requires 16 moles of Fe^{3+} per mole Cu leached which is equivalent to 4 moles O_2 per mole Cu based on the oxidation of ferrous ion according to reaction (4a). Covellite consumes 2 moles O_2 per mole Cu, and the overall reaction for chalcocite dissolution, reaction (11), requires 2.5 moles O_2 per mole Cu leached.

If *FPY* represents the moles of pyrite oxidized per mole of sulfide copper, the oxygen requirement for chalcopyrite, covellite, and chalcocite is defined according to the relationships shown in Table 3. Typical copper bearing waste contains about 20 moles of pyrite for every mole of sulfide copper (Lowell and Guilbert 1970). Based on these proportions (*FPY* = 20), pyrite is certainly the most abundant sulfide and the most important consumer of oxygen. For a chalcopyrite waste, the air requirement would be about 75 liters of air/liter of leach solution. As pointed out by Cathles (1979), the only means by which this amount of oxidant can be supplied to the dump is by convection.

TABLE 3 Oxygen Consumption and Air Requirement of Copper Sulfides

Mineral	Theoretical Oxygen Consumption O_2/Cu leached (grams)	Air Requirement* liters air/ liter solution
$CuFeS_2$	2.00 + 1.75 *FPY*	4.00 + 3.5 *FPY*
CuS	1.00 + 1.75 *FPY*	2.00 + 3.5 *FPY*
Cu_2S	0.63 + 1.75 *FPY*	1.25 + 3.5 *FPY*

*0.25 grams O_2/liter of air and 0.5 gram Cu/liter of solution.

Role of Bacteria

The iron-oxidizing bacterium, Thiobacillus ferrooxidans, was initially identified in acid mine drainage and in the mine waters of the Rio Tinto leach dumps in Spain in 1947 (Colmer and Hinkle 1947). This microorganism was soon associated with the leaching of sulfide minerals along with the sulfur-oxidizing bacterium, Thiobacillus thiooxidans. Today, it is an accepted fact that a variety of microorganisms, such as mesophilic and thermophilic heterotrophs and autotrophs are involved in commercial heap and dump leaching operations.

As pointed out above, the presence of acid and ferric iron is all important in heap and dump leaching of sub-marginal sulfide copper ores. Also, that the major source of iron in the dumps is pyrite, present in most of the porphyry ores. It is not believed that the bacterial oxidation of pyrite in dump leaching is much more important than the chemical oxidation since the kinetics of pyrite oxidation under the influence of Thiobacillus ferroxidans increases the rate of oxidation by 20 to 1000-fold as compared to the purely chemical process (Lacey and Lawson 1970).

It has been suggested that the oxidation of pyrite in the presence of microorganisms proceeds as follows:

$$2FeS_2 + 7.5\ O_2 + H_2O \xrightarrow{bac} Fe_2(SO_4)_3 + H_2SO_4 \tag{12}$$

Thus, the end products of oxidation are Fe^{3+} and sulfuric acid it is also believed that the bacteria activity on copper sulfides can be direct or indirect. As an example, the indirect oxidation of covellite is given by:

$$CuS + 2Fe^{3+} \rightarrow Cu^{2+} + 2Fe^{2+} + S^o \tag{13}$$

followed by bacterial oxidation of Fe^{+2} and S^o to form Fe^{3+} and H_2SO_4. On the other hand, direct mechanism of bacterial action on covellite is given by:

$$CuS + 2O_2 \xrightarrow{bac} CuSO_4 \tag{14}$$

In a recent study by Berry et al (1978) the leaching of chalcopyrite by bacteria was described by electrochemical reaction as follows:

Anodic:

$$2CuFeS_2 + 16H_2O + H_2SO_4 \rightarrow 2Cu^{2+} + 2Fe^{3+} + 5SO_4{}^{2-} + 34H^+ + 34e \tag{15}$$

Cathodic:

$$8.5O_2 + 34H^+ + 34e + \rightarrow 17\ H_2O \quad (16)$$

Sum Reaction:

$$2CuFeS_2 + 8.5O_2 + H_2SO_4 \xrightarrow{bac} 2CuSO_4 + Fe_2(SO_4)_3 + H_2O \quad (17)$$

The bacterial oxidation of chalcocite may be expressed by the following equation:

$$Cu_2S + 1/2O_2 + H_2SO_4 \xrightarrow{bac} CuS + CuSO_4 + H_2O \quad (18)$$

The CuS formed is then further oxidized according to equations (13) and (14).

These microorganisms involved in dump leaching utilize carbon dioxide as their source of carbon and also requires a source of nitrogen as well as phosphates for chemosynthesis and growth. The energy available for the bacterial activity is in the form of electrons transferred during catalytic oxidation of ferrous iron, insoluble metal sulfides and elemental sulfur. The optimum conditions for bacteria activity are: pH 2 to 3; Eh < 500 mv; and temperature 30-35°C.

Because of the importance of bacteria in dump leaching, a comprehensive field study was carried out by Bhappu and co-workers (1969) to locate active zones of bacteria activity in the dumps. This study showed that in actively leaching dumps, the bacterial oxidation activity was confined to the top-most portion of the dumps and that there was hardly any activity below the first eight feet of the dump surface. This observation clearly indicates that the major contribution of the bacteria in dump leaching systems is the oxidation of ferrous iron to ferric in the leach solution along with the generation of acid with subsequent dissolution of copper sulfides by Fe^{3+}. The above observation also indicates that the construction of the dumps as practiced currently does not allow for maximum utilization of the bacterial oxidation activity and that the design of the dumps need to be modified in order to optimize bacterial activity throughout the dump.

Oxygen Distribution in Dumps

The transport of oxygen into waste dumps and the associated oxygen solution chemistry are key factors in determining the rate of copper extraction from low grade copper bearing waste material. A realistic picture of air circulation in dumps is one where air flows through the faces of the dump and convects horizontally before it ascends vertically. The driving force for air convection is the air density gradients within the piles of rock. Air density gradients relate directly to the temperature distribution within the dump. The exothermic sulfide oxidation reactions are the source of heat that drives the movement of air.

Cathles and Schlitt (1980) have developed a two dimensional model for air convection in low grade waste dumps. Fig. 4 shows the results of a typical solution of the model. Depicted is half of a dump cross-section, with the center of the dump at the left edge. It is assumed the dump is much longer than it is wide. In the figure, the air flow streamlines are shown by the solid lines with arrows, temperature contours by the dashed lines, and constant oxygen concentration contours by the dotted lines. The oxygen concentration is expressed as the fraction of normal air. Air enters from the free face at the left and convects along the streamlines with velocities the highest when the lines are closest together. In this example, the central portion of the dumps heats up to about 35°C which is about optimum for bacteria to sustain the oxidation of sulfides. At these conditions there is vigorous convection of air throughout the dump, thus maintaining the supply of oxygen.

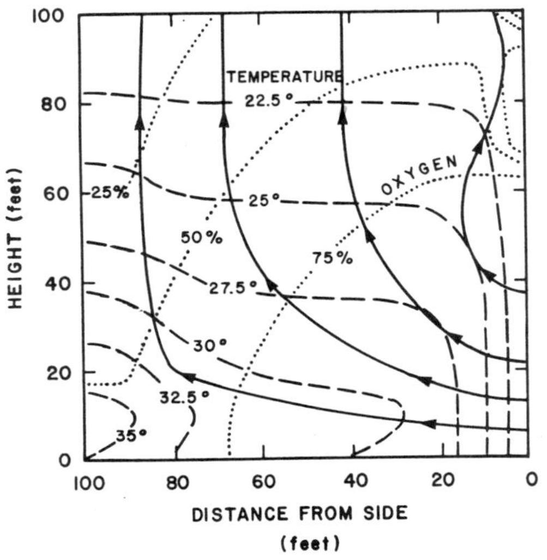

Fig. 4. Air convection through a cross-section of a leach dump under the following conditions: *FPY* = 20, permeability of 1000 Darcies, irrigation rate of 0.5 gal/ft^{2}hr, 15.6% copper solubilization, and 24 mo of leaching. (After Cathles and Schlitt 1980)

It is interesting to examine the effect of dump height, irrigation rate, and dump permeability on air convection behavior. First, the model clearly shows that the optimum dump height is of the order of 15 to 30 m. Dumps higher than this are characterized by air convection only near the dump face, and by large regions that are oxygen starved. Furthermore, the interior of large dumps tend to heat up to temperatures (45-50°C) which limit bacterial activity especially if the irrigation rate is low. Increasing the solution application rate has a quenching effect on large dumps that tend to heat up. Therefore, increasing the irrigation rate on large dumps improves the air convection characteristics. However, the opposite is true for smaller dumps where high solution flow rates can decrease temperatures within the dump, thereby limiting the driving force for air convection. Lower temperatures also decrease the rate of the chemical reactions involved. Finally, air convection in large dumps is very sensitive to dump permeability. This is shown by the copper extraction from a 60 m high dump as a function of permeability. Increasing the dump permeability form 500 Darcies to 2000 Darcies, increase the rate of copper extraction about 3 times. In small dumps (i.e. 15 m), permeability in the range of 500-2000 Darcies has little effect on copper extraction. However, the permeability must be sufficiently high enough to allow good percolation of leach solution through the dump.

Oxygen distribution in the Midas Test Dump (constructed by the Utah Copper Division of Kennecott Copper at Bingham Canyon, Utah and leached starting in April, 1969) has been previously reported (Cathles and Apps 1975). This dump is approximately 120 m long by 60 m wide with an average depth of 6 m. The dump has a maximum depth of 12 m at its front edge. The total amount of waste contained is 93,000 tons of 0.15 % and chalcopyrite is the principal sulfide mineral. Test holes were drilled into the interior of the dump to sample both oxygen concentration and temperature. Fig. 5 shows the oxygen concentration after about 5 months of operation. The contour indicate that air convection is sufficiently strong to oxygenate nearly the entire dump, only the extreme rear portion is depleted in oxygen. The maximum temperature measured in this dump was about 54°C and was located near test hole C7.

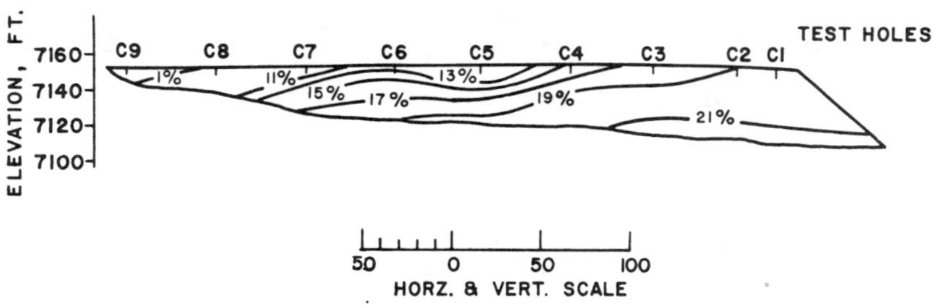

Fig. 5. Oxygen distribution in a section through a test dump at Bingham Canyon, utah. (After Cathles and Apps 1975)

ROLE OF IRON IN DUMP LEACHING

In the past, considerable thought has been given to the role of iron in heap and dump leaching by operators as well as researchers. Although several interesting and informative papers on the subject have appeared in the technical literature (Dutrizac and MacDonald 1974) neither the role of iron nor its importance in such systems is clearly understood. As far as dump leaching is concerned, the interest in iron is twofold. Iron, especially ferric iron, has been proved beneficial as an effective oxidant for dissolution of copper sulfides. On the other hand, iron in leach solutions may be harmful because of its tendency to precipitate under dump leaching conditions which may even extend to permanent sealing of the dump. Thus, in industrial practices, the operator is confronted with the dilemma of "to have or not to have" iron present in leach solutions. The answer to this quandary lies in the clear understanding of the mechanism involved in leaching of sulfide minerals and in the precipitation of ferric iron. As stated earlier the topic was thoroughly covered.

In heap and dump leaching, the oxide copper minerals such as malachite, azurite and chrysocolla are readily soluble in sulfuric acid. However, the corresponding sulfide minerals such as chalcocite, chalcopyrite and bornite require the presence of an oxidant to effect dissolution. With sulfuric acid leaching, the most common iron species are Fe^{2+}, Fe^{3+}, and $FeSO_4^+$ and $Fe(SO_4)_2^{2-}$. Of these, ferric iron is of considerable importance since it is known to be a potential oxidizing agent (Sullivan 1942) responsible for the dissolution of copper sulfide minerals.

In the leaching of copper sulfide minerals, the oxidizing ability of ferric iron can be calculated thermodynamically by using the Pourbaix (Eh-pH) diagrams (Pourbaix 1966) which show the function both of pH (acidity) and Eh (oxidation potential). The Eh of the leach system can be readily measured by using the electrical equivalent by means of instrument or by analyzing for Fe^{2+} and total iron. The oxidation and thus the solubility of the sulfide mineral in a leach solution with a given Fe^{3+} concentration can then be determined by comparing the mineral oxidation potential with ferric iron's reduction potential. Thus, according to Peters (1970), the oxidation of chalocite occurs by:

$$Cu_2S + 4H_2O \rightarrow 2Cu^{2+} + H_2SO_4 + 6H^+ + 10e \quad (19)$$

and the oxidation potential is given by:

$$E = 0.438 - 0.0355\ pH + 0.0059 \log (H_2SO_4)(Cu^{2+})^2 \quad (20)$$

Based on the above data, it can be shown that Fe^{3+} iron will oxidize chalcocite as long as the Fe^{3+}/Fe^{2+} ratio is greater than about 10^{-6}. Using this system, several diagrams of the stability regions of sulfide minerals in solution can be drawn and inspection of these diagrams will quickly indicate the possibility of reaction. Fig. 6 shows the Eh-pH diagram of the Fe-H_2O System by Pourbaix (1966) and Fig. 7 illustrates the stability regions for the copper sulfide minerals in the Cu-Fe-S-H_2O system by Peters (1970). These figures clearly show that if the Eh is maintained above 0.5 volts and below pH 4 copper minerals will leach. Moreover, comparison of these figures indicates that almost any amount of Fe^{3+} iron present will be helpful in leaching copper minerals.

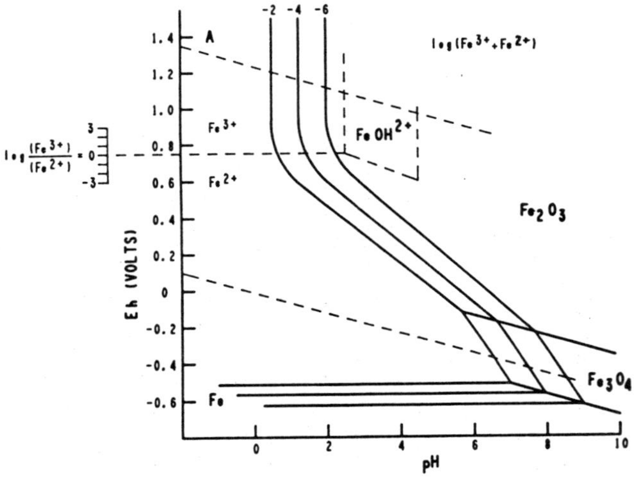

Fig. 6. Eh-pH diagram of the Fe-H_2O system according to Pourbaix (1966).

FIELD STUDIES AND OBSERVATIONS

Although a large number of laboratory and large-diameter column leach studies have been carried out to simulate and model heap and dump leaching, not many field studies have been conducted to demonstrate the leach chemistry, biological activities and the behavior of iron in the practical dump leaching operations. However, a detailed cooperative field study carried out by Kennecott Minerals and New Mexico Bureau of Mines during 1968-69 (1969) provides a comprehensive picture of what takes place in a typical dump leaching operation.

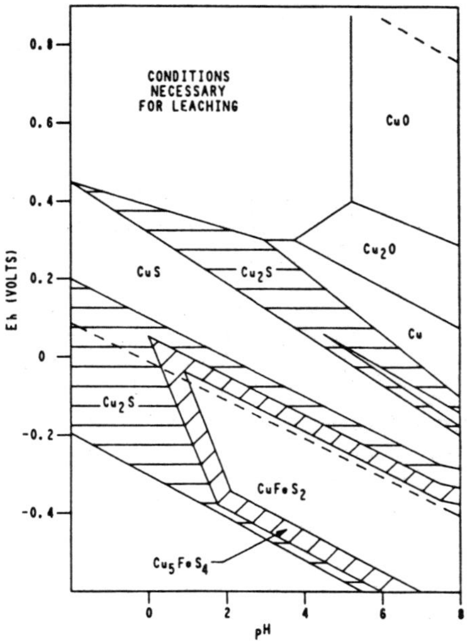

Fig. 7. The stability regions for the copper sulfide minerals in the Cu-Fe-S-H_2O system according to Peters (1970).

This field study was carried out at the dump leaching operation at Chino Mines Division of Kennecott Minerals at Hurley, N.M. A pneumatic (Becker) drill was used to obtain samples from active leaching dumps. Eight-feet interval samples were collected to obtain vertical profiles of moisture; soluble-salt content; the concentration of Fe^{3+}, Fe^{2+}, Cu^{2+}, and H^+ in the interstitial solutions; the iron, copper and sulfur content of solids (washed samples); and the bacteria activity analysis. Corresponding column leach studies in the laboratory were conducted on the samples to confirm and explain observations made on the field data.

The results of this field test are shown in Table 4. These results suggest the following conclusions:

Ore Solution Contact

1) In an active permeable dump, leach solutions are transported through the dump at variable, and rather slow flow rates as indicated by the moisture content which varied from 8 to 11 percent; 13 percent moisture content being observed in the laboratory column leach test.

2) Sufficient concentration of ferric iron, acid and copper were present in the interstitial solutions, revealing a transport of active leach solutions into and from different zones of dump.

TABLE 4 Results of Field Tests at Kennecott Chino Mines

By Bhappu, and others

Interval	50 ml Wash Solutions				Moisture	Soluble*	900 ml Rinse Residues			Bacterial Count
(ft.)	(Fe , Gpl)	(Fe , Gpl)	(Cu , Gpl)	pH	Wt %	Salts, %	% Cu	% Fe	% S	(per g dry wt)
0-8	1.242	4.900	1.16	2.59	7.43	54.7	0.080	4.69	1.99	1.64×10^5
8-16	0.565	5.230	1.92	2.73	7.84	54.0	0.053	3.10	1.12	1.08×10^3
16-24	0.028	0.480	0.078	2.47	10.74	33.4	0.075	3.67	0.78	1.08×10^2
24-32	-	-	-	-	10.74	33.4	-	-	-	1.08×10^2
32-40	0.056	0.125	0.460	3.07	11.00	18.8	0.065	2.54	0.85	-
40-48	0.056	0.153	0.209	2.91	11.00	23.6	0.047	3.67	1.05	1.08×10^2
48-56	0.028	0.113	0.245	2.64	8.17	23.2	0.025	2.82	0.59	-
56-64	0.028	0.000	0.064	2.80	10.20	17.0	0.084	2.54	0.34	1.07×10^1
64-72	0.028	0.170	0.151	2.81	8.93	9.7	0.039	5.08	0.73	-
72-80	0.028	0.017	0.655	3.66	8.07	31.0	0.082	1.13	0.31	1.07×0^1
80-88	0.028	0.057	1.09	2.95	5.67	22.4	0.063	3.39	0.10	-
88-96	0.028	0.113	2.26	2.85	5.24	32.5	0.051	2.82	0.33	1
0-32 Average	0.611	3.54	1.05	2.59	9.17	43.9	0.0692	3.62	1.299	1.08×10^3
32-80 Average	0.035	0.0935	0.265	2.96	8.55	22.3	0.057	2.96	0.538	1.07×10^1

* Percent soluble salts in intersticial solutions.
Solution going to and from dumps contain approximately 5.52% and 5.67% dissolved solids, respectively.

3) Accumulation of large amounts of soluble salts within leach dump proves that leach solutions do circulate in the dumps but that rates of the solution ingress and egress are rather slow and prevent thorough rinsing of soluble salts from the solids. Alternately, there is some adsorption

<u>Leaching Phase</u>

1) Nearly complete conversion (oxidation) of the onflowing ferrous iron solution to ferric iron occurs in the upper 10 m of the dump as indicated by the ratio of ferric/ferrous iron and the pH of the interstitial solutions. On the other hand, below a depth of 10 m, the content of ferrous and ferric, acid and iron varies considerably from one vertical zone to the next.

2) The presence of sufficient ferric sulfate-acid in the upper portion of the dump is responsible for effective dissolution of copper sulfide minerals in the upper as well as lower portions of the dump. The copper assay of the composite 0-30 m is 0.06 percent copper as compared to the head assay of the dump material reported to be about 0.15 percent copper.

3) The dissolution of copper sulfide minerals cannot be attributed to ferric iron solely since in the lower portions of the dump below 10 m, there is not sufficient ferric iron present to dissolve copper sulfides. Stoichiometrically 4.34 grams of Fe^{3+} is required to leach 1 gram of copper from chalcocite (Cu_2S) and to oxidize the sulfur that is formed. Certainly, such a concentration of Fe^{3+} does not exist under even the best of conditions. Additional mechanisms such as alternate wetting and drying cycles with leach solutions and air, electrolytic reactions between different sulfide minerals (such as chalcopyrite and pyrite), and dissolution by acid in combination with neutral salts such as NaCl may also be responsible for effective dissolution of copper sulfides.

<u>Iron Oxidation and Hydrolysis</u>

1) Iron oxidation and hydrolysis reactions occur in the upper zone of the actively leaching dumps and are primarily due to oxidation resulting from bacterial activity, air, and ground-water oxidation of a surface zone of retained iron salts.

2) Profiles of the concentration of various constituents (Fe^{3+}, Fe^{2+}, H^+, and Cu^{2+}) of the interstitial solutions show an alteration of ferrous iron to ferric iron and acid in the top zone of the dump. Likewise, iron and sulfur which are constituents of residual iron hydroxide product diminish in an exponential fashion with dump depth.

3) Bacteria activity seems to be confined to the first few inches of the dump and is responsible for increasing the rate of leach solution oxidation and hydrolysis at the surface of the dump. This limited activity appears to be related to the solution flow rate with increased activity associated with slower flow rates. Laboratory tests have shown that bacteria oxidation and hydrolysis of the ferrous sulfate solution was appreciable at very slow rates but at higher flow rates concomitant with field operations little or no bacterial oxidation was achieved.

4) The accumulation of soluble salts in the upper zone of the dump may also be a major factor promoting the oxidation and hydrolysis of ferrous sulfate solutions. The retained ferrous sulfate is then oxidized by bacteria and/or natural air oxidation during the leach solution addition cycle (generally 6 to 12 month periods). The semi-dry condition in this zone and the presence of catalytic agents such as Cu^{2+}, MnO_2 and Co^{2+} may assist significantly in the air oxidation of ferrous sulfate salts.

IMPORTANCE OF ALTERNATE WETTING AND DRYING

1) Regardless of the mechanisms involved in leaching, it is obvious that the repeated wetting and drying cycles practiced in the dump leaching operations are essential for effective dissolution of copper sulfide minerals, especially where larger ore pieces are concerned.

2) Alternate wetting and drying is also effective in removal and crystallization of soluble salts at the surface of the ore particles through evaporation and capillary action. These salts are formed during the drying cycle and are washed out during the wetting cycle.

3) Diffusion also plays an important role in dump leaching since on permeation of the leach solution into the dry ore mass through capillary action followed by dissolution of copper minerals a diffusion pressure is created that allows the soluble salts to migrate to the less concentrated solution near and at the surface of the ore pieces and between the mineral particles. It is understandable that a single drying cycle will not bring all the soluble salts to the surface and a large number of alternate wetting and drying cycles will be required to recover a reasonable amount of copper in the dump leaching operations. With chalcocite ores, which are easily oxidized, the recoveries may amount to 30 to 50 percent in 4 to 5 years under dump leaching conditions and about 50 to 80 percent under heap leaching environment. On the other hand, for slow leaching chalcopyrite ores, a recovery of 6 to 15 percent in 4 to 5 years is the best one could expect. The only reason why the dump leaching practices are economically successful is that a large tonnage of low-grade ores are subject to dump leaching

DUMP LEACH OPTIMIZATION

The above discussions on the role of iron in heap and dump leaching dictates that the heaps and dumps should be designed and constructed to provide optimum conditions for aeration and bacteria activity. Since most dumps are deposited near the mine on the existing topography, the locations are selected to provide an impermeable surface and natural drainage for positive collection of pregnant solution. In several instances dumps have been built on specially prepared areas using asphalt or plastic liners to prevent seepage. Moreover, leach dumps are constructed to facilitate access of leach solution, bacterial nutrients and oxygen to the surfaces of sulfide minerals. If dump permeability is adequate, natural updrafts through the dump, resulting from heat generated by oxidation of sulfide minerals, should ensure sufficient oxygen supply. Sometimes, a network of piping or tunnels may be installed within the dumps to provide effective aeration. These pipes may be connected to a blower to force air into the dump as was practiced at Anaconda's Butte operation in Montana (Malouf 1986).

The dimensions of the dumps should be such that the surface to volume ratio allows for efficient conservation of heat generated by oxidation of sulfides and ensures that leach operation is not adversely affected by low external temperatures. The required information should be collected through realistic laboratory and pilot plant tests prior to dump construction. Cathles and Schlitt (1980) have concluded that, in general, the optimum dump height should be about 15 m and the ideal width should be roughly twice the height or about 30 m. Also, in order to provide effective aeration from all sides, Ballard (1966) has proposed the construction of dumps spread out in form of fingers (Finger Dumps).

In actual practice, however, the dumps are raised in lifts of 15 to 30 m with dump widths of several hundred feet in order to minimize haulage and dumping costs. The ore is generally hauled from the open-pit mines in trucks or trains and bulldozers are used to level the surfaces. This results in unnecessary compaction of the dump and reduces its porosity thus reducing the solution percolation rates. In order to increase percolation rates some operators resort to ripping of the dump surface prior to leaching. In recent years Mountain States Engineers (Almond and Schwalm 1982) has advocated the use of in-pit crushing, belt conveying and ore stacking to reduce haulage costs and to avoid compaction of the dumps. Such a technique also necessitates primary crushing of the ore which results in improved solid-liquid contact and thus the overall recovery of copper.

OXYGEN INJECTION

Since oxygen plays a very important role in dump leaching of low-grade copper ores, it would be beneficial to utilize forced aeration or air/oxygen injection in such operations. The primary reason for the lack of oxygen in the dumps, especially high dumps built through truck haulage, is the lack of convective air flow through the dumps due to compaction of the ore. This results in lower copper recoveries due to poor oxidation of the sulfide minerals. It is believed that injection of low pressure air into the dumps through bore holes should stimulate oxidation of sulfide minerals and thus result in improved copper recoveries.

During the late sixties, Anaconda attempted to induce aeration of the Butte dumps by blowing low pressure air as well as oxygen at the bottom of the dumps through large air-fans located in tunnels (Malouf 1986). Such a technique resulted in improved aeration of the dumps and increased copper recoveries, especially during the winter months. It is unfortunate that this procedure was not introduced as a permanent dump leaching practice. Kennecott Copper, in recent years, has also attempted to obtain increased copper recoveries through forced aeration of leach dumps by injecting air into the Bingham dumps through bore holes (Jackson 1983).

In another study by Lewis et al (1974) involving nuclear solution mining of copper, it was also shown that rapid dissolution of the sulfide copper minerals is achieved in a rubblized ore column by operating at elevated temperatures and at relatively high oxygen concentrations. In this case oxygen gas is introduced into the bottom of the chimney through bore holes. Part of the oxygen is dissolved in the leach solution under high hydrostatic pressures while the undissolved portion rises through the chimney, providing a lifting force that induces sufficient circulation so as to carry dissolved oxygen to all parts of the chimney. It is understandable that in this case the broken ore is inundated by the leach solution and it is possible to disperse dissolved oxygen throughout the broken mass. This is not feasible under dump leaching conditions and therefore alternate means must be devised for injecting oxygen/air into the dumps and for dispersing it throughout the dump.

Based on the above information, it appears that in the dump leaching operation, the broken ore should be subjected to oxidation over a sufficient period of time (several months depending on the total volume of the ore and the height of the dump) by injecting oxygen and/or air and through the bore holes (with the bottom 50% perforated). After aeration, the leaching should be initiated and continued until the copper concentration in the leach solution drops below the economic level. The aeration and the leaching cycles should be repeated as long as sufficient extra copper is recovered to meet the operating costs. An alternate method of oxidation and leaching would involve enriching the leach solution with dissolved oxygen or addition of an oxidant. In this case, oxygen would be injected into the leach solution at the bottom of the pipe. The maximum achievable oxygen concentration would be governed by the temperature and hydrostatic pressure at the point in the dump. This technique may be very useful for high dumps varying from 180 to 360 m.

It should be noted that the effective oxygen/air injection technique for increased copper recovery is still in its infancy and much more bench scale and filed studies are needed to optimize the process for industrial use. Also, since such a process can be modeled readily, it would be feasible to come up with an effective, computerized operational and economic model.

SUMMARY

Oxygen, along with moisture, is a basic ingredient for copper sulfide dump leaching. In porphyry copper environments, pyrite is the most abundant sulfide mineral and the aqueous oxidation of pyrite with oxygen initiates the entire leaching process. Pyrite oxidation supplies the acid necessary for leaching, yields dissolved iron which in the form of ferric ion is responsible for the oxidation of copper sulfide minerals, and generates heat which is the driving force for air convection through the dump. An analysis of the mineral chemistry indicates that for a chalcopyrite waste, the air requirement is about 75 liters of air/liter of leach solution. The only way this amount of oxidant can be supplied is by air convection through the dump. Dump height, solution application rate, and dump permeability are the most important factors controlling air convection and the rate of copper leaching. Furthermore, without oxygen, bacterial activity in the dumps would be severely retarded resulting in poor copper extraction.

Dumps should be designed to provide opitmum conditions for aeration and bacterial activity. The optimum dump height should be approximately 15 to 30 m (50-100 ft) and the ideal width should be roughly twice the height. If the situation dictates that high dump must be built, then care must be taken to insure as great a permeability as possible and to operate with a sufficient enough irrigation rate to moderate internal dump temperatures. Large dumps that are normally oxygen starved can be reactivated by induced or forced aeration through bore holes.

Solution management and the application of alternate wetting and drying (leach/rest cycles) are also important factors in the overall extraction process. With regards to air movement in the dump, the rest cycle permits drainage of void spaces normally flooded with solution and the ingress of air. During the rest cycle, the avialability of oxygen allows dissolution of copper sulfides. The next leaching cycle washes the solubilized values from the dump.

Dump leaching is in general terms a very simple process, but is a process that involves a large number of critical parameters. Oxygen is one of the most important. Given the tremendous resource of copper available in waste dumps constructed as part of open-pit copper mining operations in the western United States, it is vital that we continue both basic and applied research efforts to elucidate the role of oxygen in dump leaching.

REFERENCES

Almond, M. R., and R. J. Schwalm (1982). Paper presented at the 1982 American Mining Congress, Las Vegas, Nevada.

Bailey, L. K., and E. Peters (1976). Can. Met. Quart., 15(4), 333-344.

Ballard, J. K. (1966). Paper presented at the Annual Meeting of AIME, New York.

Berry, V. K., L. E. Murr, and J. B. Hiskey (1978). Hydrometallurgy, 3, 309-326.

Bhappu, R. B. (1986). Iron Control in Hydrometallurgy, CIM-Ellis Horword Ltd., Chichester, England, 183-201.

Bhappu, R. B., et al. (1969). Trans. Soc. Min. Engrs., 244, 307-315.

Brooke, J. N. (1977). Min. Congr. J., 63(8), 38-41.

Chase, C. K., and G. M. Potter (1982). Paper presented at the Annual Meeting of AIME, Denver, Colo.

Cathles, L. M. (1979). Mathematical Geology, 11(2), 175-191.

Cathles, L. M., and J. A. Apps (1975). Met. Trans. B, 6B, 617-624.

Cathles, L. M., and W. J. Schlitt (1980). Leaching and Recovering Copper from As-Mined Materials, SME-AIME, New York.

Colmer, A. R., and M. E. Hinkle (1947). Science, 106, 253-255.

Dutrizac, J. E., and R. J. C. MacDonald (1974). Minerals Sci. Engng., 6(2), 59-100.

Hiskey, J. B. (1986). Paper presented at the Workshop on Biotechnology Applied to the Mining and Mineral Processing Industry, Idaho Falls, Idaho.

Hiskey, J. B., and W. J. Schlitt (1982). Interfacing Technologies in Solution Mining, SME-AIME, New York.

Jackson, J. S. (1983). Private communications.

Lacey, D. T., and F. Lawson (1970). Biotechnol. Bioengng, 12, 696-703.

Linke, W. F. (1965). Solubilities - Inorganic and Metal Organic Compounds, Vol II, Am. Chem. Soc., Washington, D. C.

Lowell, J. D., and J. M. Guilbert (1970). Econ. Geol., 65(4), 373-408.

Malouf, E. E. (1986). Private communications.

Peters, E. (1970). Thermodynamics of Kinetic Factors in the Leaching of Sulfide Minerals from Ore Deposits and Dumps, SME Short Course - Bioext45active Mining, 46-75.

Pourbaix, M. J. N. (1966). Atlas of Electrochemical Equilibria in Aqueous Solutions, Pergamon Press, New York.

Schlitt, W. J., and J. S. Jackson (1981). In Situ, 5(2), 103-131.

Singer, P. C., and W. Stumm (1969). Amer. Chem. Soc., Div. Fuel Chem. Prepr., 13(2), 80-87.

Sullivan, J. D. (1942). Trans. AIME. 106, 515-546.
Taylor, J. H., and P. F. Whelan (1942). Inst. Min. Met. Bulletin, 457, 1-36.
Young, T. F., and L. A. Blatz (1949). Chem. Reviews. 44, 98-105.

OXYGEN USE AT NORANDA'S HORNE SMELTER

S. El-Barnachawy, G. Kachaniwsky*, H. Persson and D. Poggi**

* Noranda Minerals Inc., Horne Division, Noranda, Quebec

** Noranda Research Centre, Pointe Claire, Quebec

ABSTRACT

Over the last decade, Noranda's Horne smelter has gained considerable expertise in operating with tonnage oxygen. The oxygen production capacity is currently 550 tonnes/day, distributed between the Noranda Process Reactor (370 tonnes/day) and the wet charge reverberatory furnace (180 tonnes/day). This paper describes current operation at the Horne and the impact of oxygen on productivity and fuel consumption. The ability to optimize smelter performance will be highlighted.

KEYWORDS:

Copper smelting; Noranda Horne smelter; Oxygen; Noranda Process; Reverberatory furnace; Smelter strategy.

INTRODUCTION

The Noranda smelter was commissioned in 1927 to treat direct smelting copper-gold ore from the Horne Mine. Initially, two calcine-charged reverberatory furnaces smelted 900 tonnes/day of ore. In the following years, the smelter treated an increasing tonnage of custom concentrates from the surrounding area. In 1957 a wet-charge reverberatory furnace was built to treat additional custom concentrate[1].

Following development work in the sixties and seventies, a decision was made in 1971 to build a full-scale Noranda Process reactor. This plant, with a nominal capacity of 720 tonnes/day was started up in 1973[2]. By the time the Horne Mine was shut down in 1976, the smelter was operating three reverberatory furnaces and one Noranda Process reactor. Smelting capacity exceeded 1,000,000 tonnes/year of which approximately 85% was custom concentrates.

Since 1975, the use of tonnage oxygen has contributed significantly to increased productivity and reduced operating costs. Today the smelter uses one reverberatory furnace and one

Noranda Process reactor; both employ oxygen technology. There are five Peirce Smith converters, with three normally on line, and three rotary anode furnaces. The smelter capacity is in excess of 850,000 tonnes/year of copper-bearing feed.

OXYGEN PLANTS

Oxygen is produced in two plants, one rated at 85 tonnes/day and the other at 465 tonnes/day, for a combined capacity of 550 tonnes/day.

The smaller unit was commissioned at the Horne smelter in 1975. It was originally built in 1965 by American Cryogenics Inc. for an Imperial Oil Ltd. project in Nova Scotia. Designed as a low pressure plant, it is capable of producing oxygen at a purity of up to 99.5%. The plant now produces oxygen at a purity of 97%. The main components of the unit are: a Cooper-Bessemer air compressor, reversing heat exchangers, blower-loaded expansion turbines, a column cold box and two Sulzer oxygen compressors.

In 1981-82, a smelter modernization program was implemented which included: (i) conversion of No. 2 reverberatory furnace to oxy-fuel firing, (ii) closure of No. 3 reverberatory furnace and (iii) increased reactor smelting rate brought about by the use of tonnage oxygen[3].

The new 465 tonnes/day oxygen plant, supplied by Air Products & Chemicals Inc., came on line in October 1982. The plant is of the reversing exchanger type with the following major components: a Hitachi air compressor (4 stages), Siemens motor, reversing heat exchangers, two expansion turbines (one motor/generator and the other blower loaded), a column cold box and a Lotema oxygen compressor. This plant has given excellent service; at full capacity it exceeds the rated output by almost 5%, producing oxygen at a purity of 95.7% and at a power consumption of only 60% of the small plant.

If required, the large plant can be turned down to 280 tonnes/day to save power. The small plant cannot be turned down. The only possible power saving procedure is to shut down the oxygen compressors and vent the oxygen. Both plants can be brought back to full capacity within half an hour.

Routine maintenance is carried out on both plants. The smaller unit is de-rimed and serviced every year, while the larger plant is on a two year de-rime schedule. Competent operating personnel with a good understanding of the process are available to maintain a safe and stable operation.

NORANDA PROCESS

The Noranda Process[2,4,5] uses a single 5.2-m diameter by 21.3-m long cylindrical reactor as shown in Figure 1. Copper-bearing feed and fluxes are fed into the vessel using a high speed slinger. The bath of matte and slag is maintained in a highly turbulent state by oxygen-enriched air injected through a series of tuyeres. Exothermic oxidation of iron and sulfur provides most of the heat for smelting; additional heat is supplied by the addition of coal to the feed and by a small gas burner.

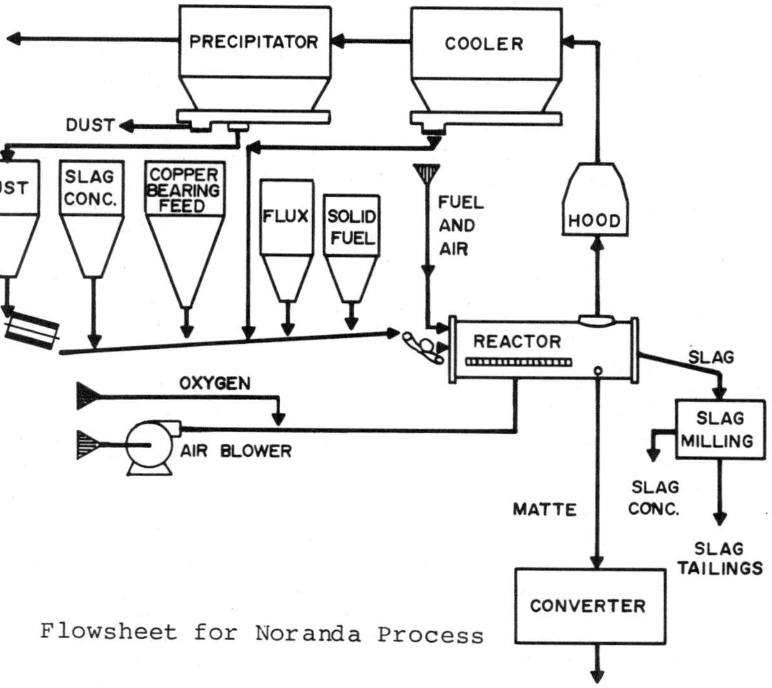

Flowsheet for Noranda Process

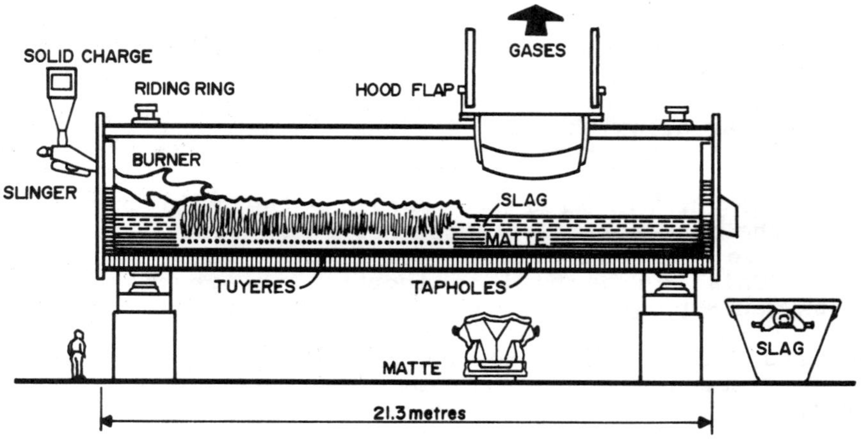

Longitudinal section through Noranda Process reactor

Figure 1. Noranda Process

Copper matte, containing 55-75% Cu, is periodically tapped into ladles from a tap hole in the barrel of the reactor. Slag is skimmed from a tap hole in the end plate and slow-cooled in ladles. It is then crushed and treated in the concentrator for copper recovery by flotation. The slag concentrate is recycled to the reactor while the tailings, containing 0.3% Cu, are discarded.

Off-gases leave the reactor through a water-cooled hood to an evaporative spray cooler and then to an electrostatic precipitator. An acid plant having a nominal capacity of 1950 tonnes/day of sulfuric acid is under construction and is scheduled to start up in 1989.

The use of oxygen to increase concentrate smelting rates, to reduce fuel consumption and to increase the strength of SO_2 in the off-gases was recognized at an early stage in the development of the Noranda Process. Tests with purchased liquid oxygen were carried out on the 100 tonnes/day pilot reactor in 1972 and on the present reactor in 1973. The results of these tests have been reported elsewhere[6,7].

All tonnage oxygen used in the Noranda Process, in the range 320 to 370 tonnes/day, is introduced through the tuyeres. The enrichment level is usually 32 to 37% O_2, but levels as high as 40% O_2 and oxygen addition rates up to 465 tonnes/day have been tested for shorter periods. Oxygen enrichment to these levels has had no adverse effect on the tuyere line life. It has been found that a refractory slag coating effectively protects the tuyere line during operation. Figure 2 shows that while the oxygen enrichment of the reactor has increased over the last 10 years, the unit refractory consumption has remained virtually constant. Concurrently, reactor campaign life has increased substantially.

Methods of increasing the enrichment levels above 40-42% O_2 in the reactor tuyeres are actively being sought in R&D programs. Pilot and plant-scale test work has indicated that higher pressures may allow the use of higher levels of oxygen enrichment without causing a deterioration of the tuyere line[8,9].

Tonnage oxygen can also be added to the main burner air and through the feed port. Test work has shown that any excess oxygen added above the bath reacts efficiently with concentrate and splash[7]. This can be used to increase the reactor throughput when oxygen supply is limited by the maximum tuyere blowing rate.

Figures 3 and 4 illustrate the metallurgical performance of the Noranda Process using tonnage oxygen. They are based on monthly operating data for the reactor recorded since start-up in 1973. The use of oxygen has allowed the reactor concentrate throughput to increase by more than 100% and the fuel consumption to be decreased by up to 60%. The SO_2 strength in the off-gas has more than doubled.

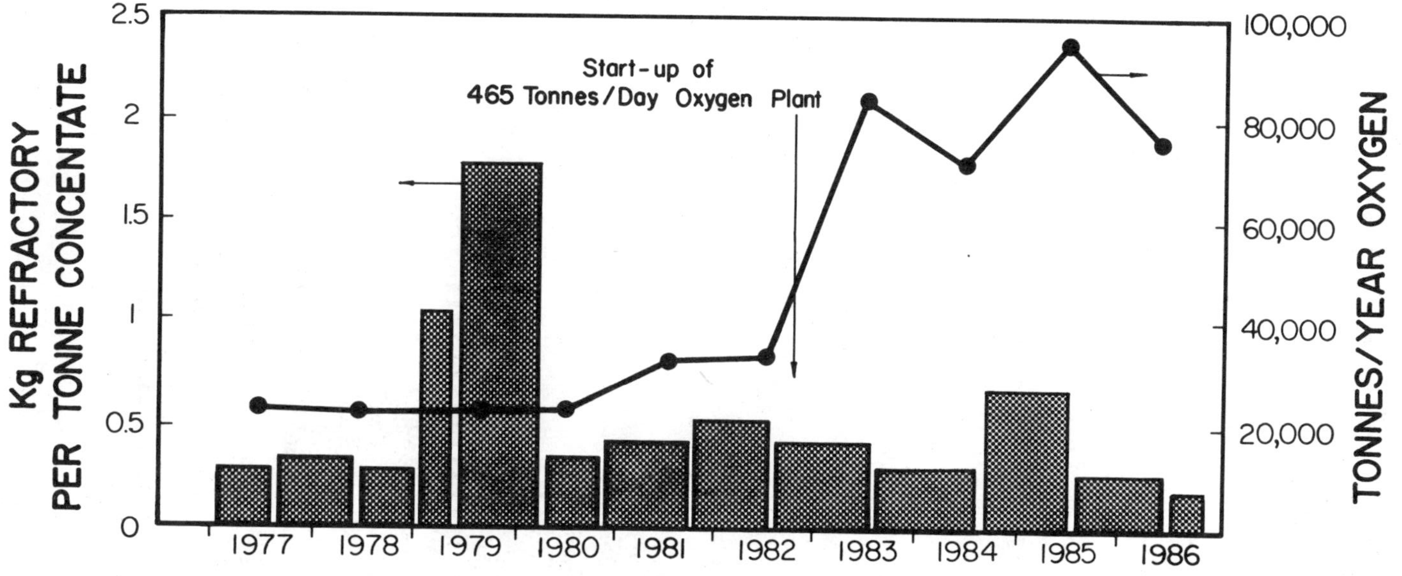

Figure 2. Refractory consumption in Noranda Process

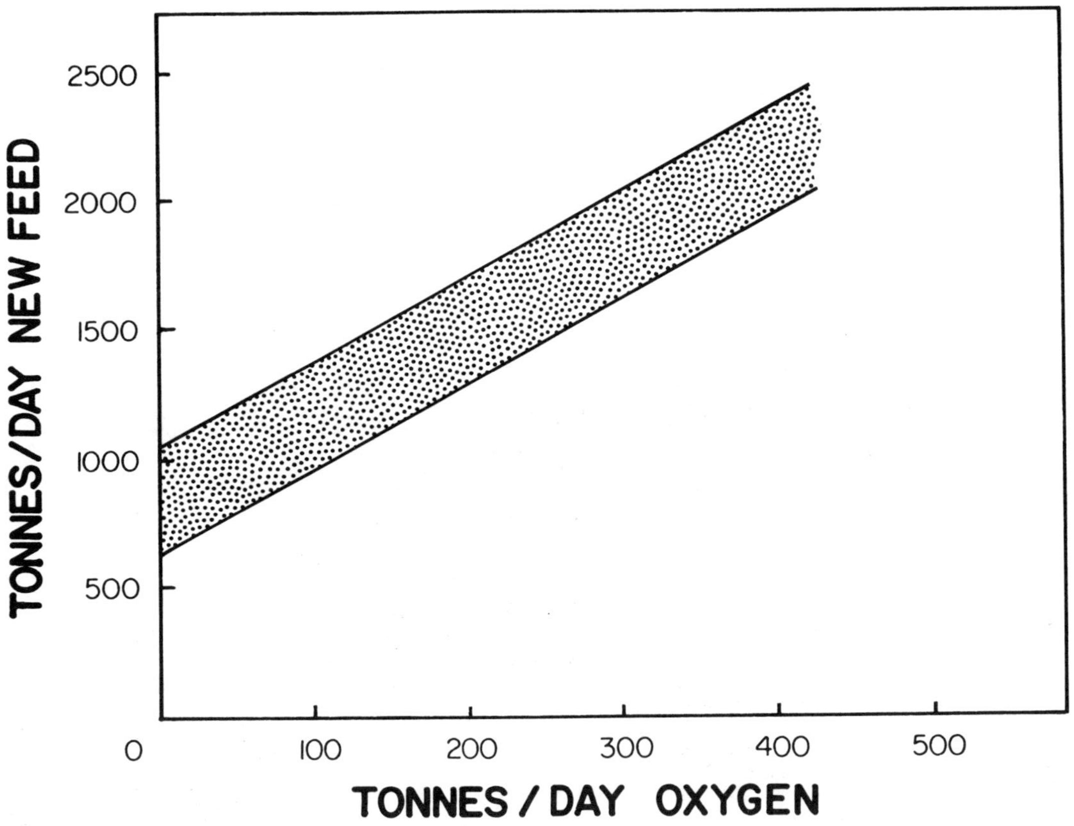

Figure 3. Noranda Process - effect of tonnage oxygen on smelting rate

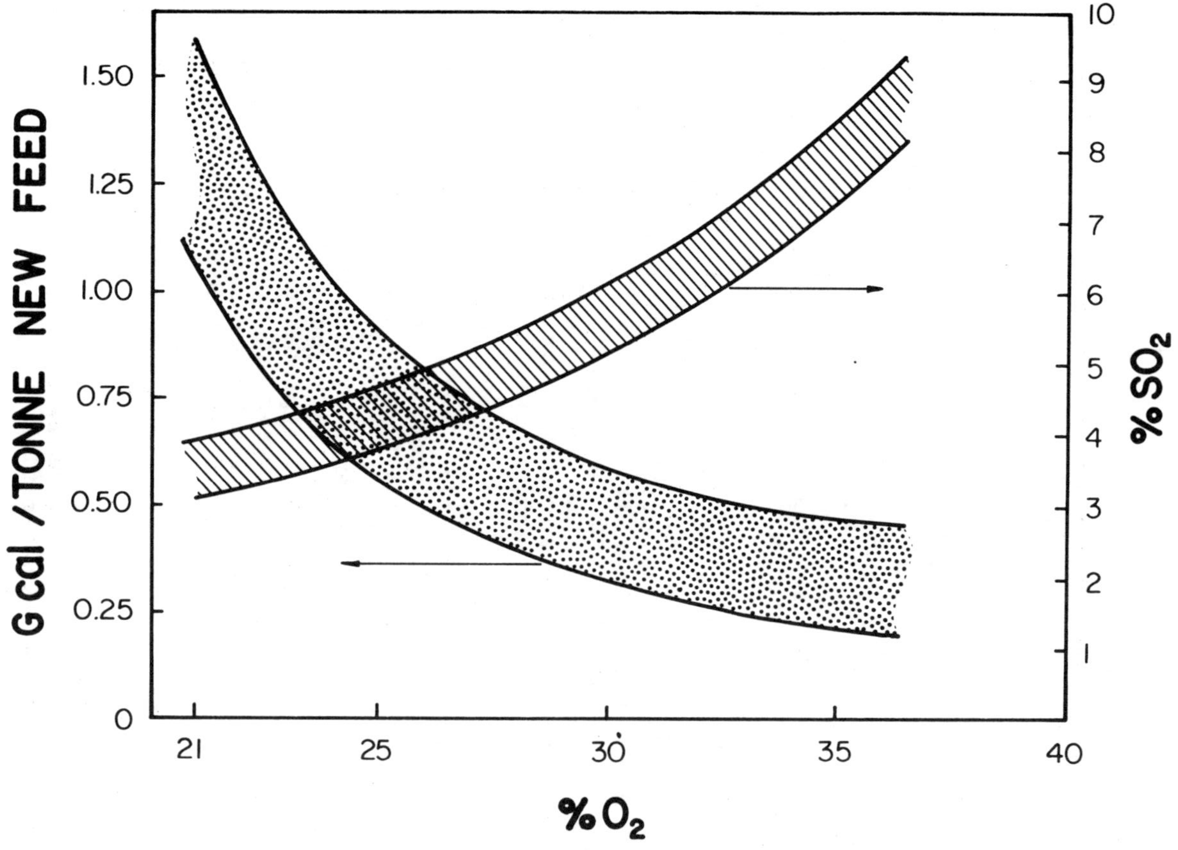

Figure 4. Noranda Process - effect of tuyere oxygen enrichment on fuel consumption and SO_2 concentration in off-gas.

REVERBERATORY FURNACE

No. 2 reverberatory furnace was commissioned when the smelter started in 1927. It is of conventional design (32-m long by 10.5-m wide) with a suspended arch. Originally operated as a calcine-charged furnace, it was converted to wet-charging when it was rebuilt in 1976[3]. Two shuttle conveyors feed the concentrate through 56 drop chutes four times per shift. The furnace is fitted with two conventional burners for oil or natural gas firing, however these burners are normally blanked off. A natural gas fired preheater can supply 425°C combustion air to this system.

Oxygen use in the reverberatory furnaces started in late 1975, following the installation of the small oxygen plant. When the reactor was not operating, oxygen was supplied to the reverberatory furnaces through two 15-cm lances alongside the two conventional burners. The oxygen addition rate was limited to 40 tonnes/day per furnace to prevent burning of the roof and slagging of the uptake. The combustion air was enriched to 22 to 24% O_2, which decreased fuel consumption by an average 3 to 4%. This practice was terminated with the start-up of the oxy-fuel burners in late 1982.

For the higher tonnages of oxygen available with the large oxygen plant, it was decided to use several roof mounted oxy-fuel burners. This technology was originally developed by Codelco[10] and was licenced by Noranda from INCO. Eight burners were installed, four on each side of the furnace, so that the flames are directed to the base of the charge banks (Figure 5). The burners were designed for a maximum oxygen consumption of 40 tonnes/day per burner, corresponding to a firing rate of 5 Gcal/hour.

A gradual start-up of the oxy-fuel burners was carried out to allow the operators to become familiar with the new burner systems and controls. The original plan called for an operation with half the fuel supplied by six oxy-fuel burners and the balance by the conventional burners. By the summer of 1983 it became clear that under this so called "hybrid firing mode", the furnace performance did not meet expectations. There was an uneven high rate of refractory wear mainly on the furnace shoulders and roof as well as thermal corrosion of the bottom part of the feed chutes. The furnace draft had been increased from 1.8 to 3.0-mm H_2O to increase air infiltration and protect the refractory. The fuel consumption was 20% above that planned.

A multi-disciplinary task force, including personnel from operations and Noranda Research Centre was appointed to study the operating problems. The task force recommendations resulted in the following changes and modifications:

(a) The conventional burners were shut-down by late 1983 and the furnace was fired using the oxy-fuel burners alone.

(b) The furnace draft was decreased to 1.9-mm H_2O to decrease air infiltration.

(c) Two additional oxy-fuel burners were installed near the front end of the furnace, to keep the recycled converter slag hot, bringing the total number of burners to ten.

(d) A natural gas supply was installed to allow the smelter to fire these burners with either oil or gas to achieve the lowest fuel cost.

(e) Stainless steel replaced the mild steel for the bottom part of the feed chutes.

(f) Burnt chrome-magnesite brick replaced the chemically bonded chrome-magnesite brick on the roof panels.

These modifications have resulted in significantly improved performance of the reverberatory operation, as described below.

Figure 6 shows that firing with only oxy-fuel burners has reduced the fuel ratio by approximately 30% (down to 1.0 Gcal/tonne concentrate), compared to the hybrid firing and by 50% compared to conventional firing. The oxy-fuel burners now supply about 90% of the stoichiometric oxygen for fuel combustion. The concentrate throughput, shown in Figure 7, has increased by more than 40% as a result of oxy-fuel firing.

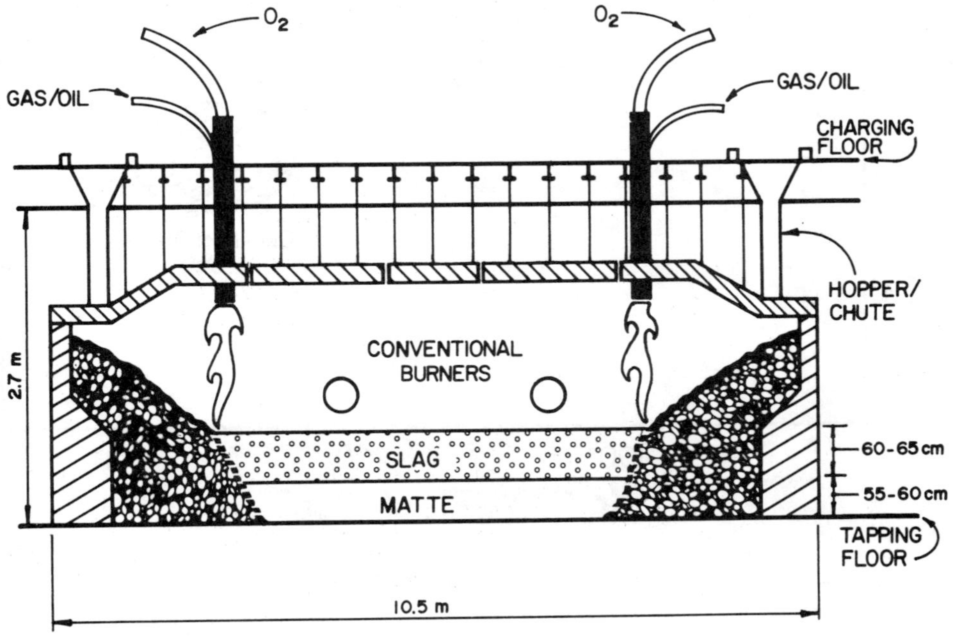

Figure 5. Cross section through oxy-fuel fired reverberatory furnace.

Table I shows the refractory consumption in No. 2 reverberatory furnace since 1981. The introduction of oxy-fuel burners has increased the unit consumption by about 0.5 kg/tonne concentrate to about 3 kg/tonne concentrate. The higher unit consumption in 1985 and 1986 is not entirely due to oxy-fuel firing. Commercial considerations made it necessary to operate periodically with stand-by fires that slowly eroded the arch lining.

There has been no oxidation of the charge banks and the matte grade has remained stable at 32 to 35% Cu. Copper losses in slag have increased from 0.4% Cu to 0.5% Cu following the conversion to oxy-fuel firing. The increase is attributed to the fact that

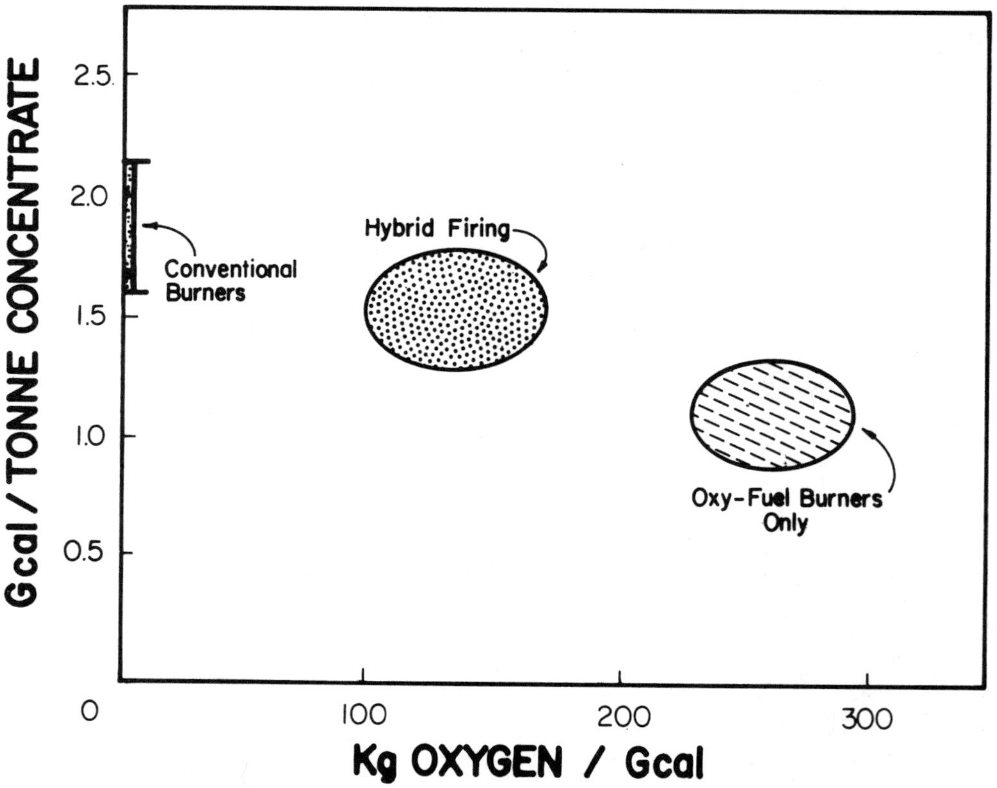

Figure 6. Reverberatory furnace - effect of tonnage oxygen on fuel consumption.

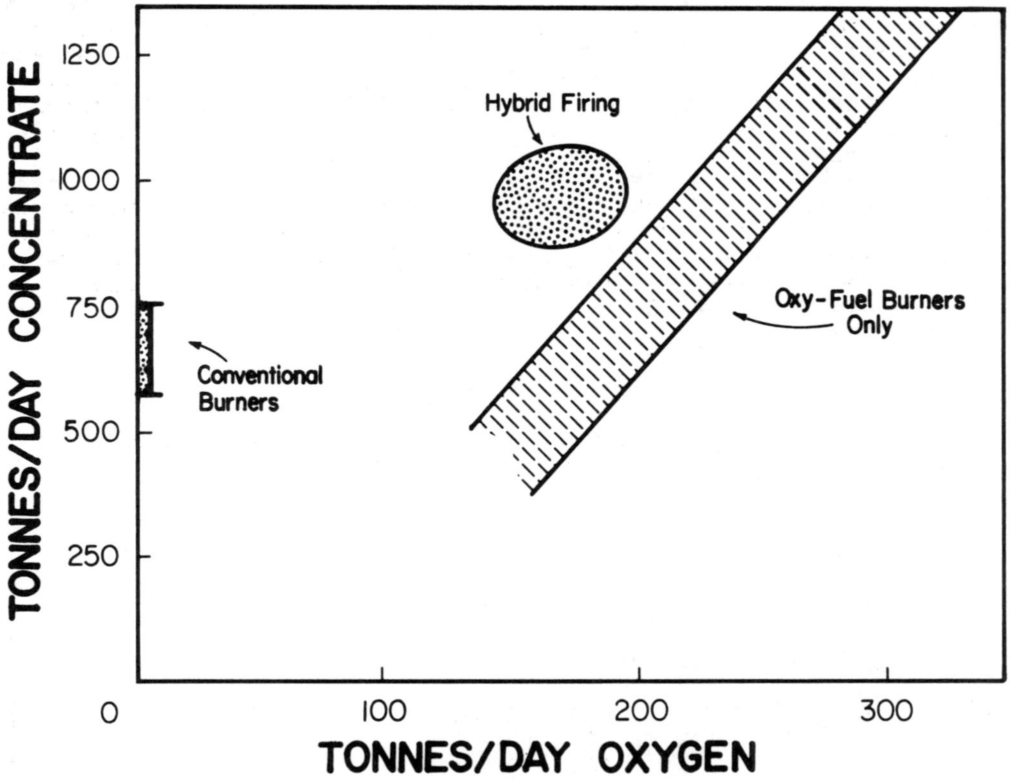

Figure 7. Reverberatory furnace - effect of tonnage oxygen on concentrate throughput.

whereas converter slag was historically recycled to two or three reverberatory furnaces, all converter slag is now being recycled to one remaining furnace.

Table I

Refractory Consumption - No. 2 Reverberatory

	1981	1982	1983	1984	1985	1986
kg refractory/tonne concentrate						
Before oxy-fuel firing	2.1	1.6				
After oxy-fuel firing		2.5	2.6	2.3	3.2	3.6
Total O_2 consumed, tonnes/year			55,000	76,000	63,000	50,000

SMELTER FLEXIBILITY AND STRATEGY

As a custom smelter, Noranda must remain competitive for both locally and internationally produced concentrates. These concentrates can often be metallurgically complex, refractory or difficult to smelt in some other way. The versatility necessary to treat this mix of materials profitably is supplied in part by the use of tonnage oxygen and the use of two distinct smelting processes.

The feed materials are distributed between the two smelting processes to optimize metal recoveries and minimize smelter costs, while maintaining an acceptable anode quality. In general, the reactor treats:

(a) High magnetite and refractory feed.

(b) Complex concentrates benefiting from enhanced impurity elimination at high matte grades.

(c) Feed high in precious metals due to the reactor excellent recoveries.

(d) Shredded scrap and wet materials which cannot be charged to the reverberatory furnace.

In general, the reverberatory furnace treats:

(a) Complex concentrates benefiting from enhanced impurity elimination at low matte grades.

(b) Copper concentrates which tend to be dry to reduce overall fuel requirements.

Due to the low operating cost and the excellent metal recovery of the Noranda reactor, the current smelter strategy calls for reactor operation at maximum capacity. On average the reactor treats about 70% of the concentrate tonnage using 70% of the tonnage oxygen, with the balance handled by the reverberatory furnace.

Progress in shifting more concentrate and tonnage oxygen from the reverberatory furnace to the reactor is continuing and as yet the upper limit on reactor capacity does not appear to have been reached.

There is however a lower practical limit on reverberatory furnace operation. When using oxy-fuel burners alone, the reverberatory furnace has a lower firing limit of about 24 Gcal/hour corresponding to about 150 tonnes/day of oxygen. Below this rate the furnace concentrate smelting rate drops off rapidly and it becomes increasingly difficult to maintain slag, cottrell and stack temperatures.

The oxygen distribution between the two smelting processes is decided at weekly production meetings based on projected feed tonnage and composition. This decision is only a guideline since day-to-day changes may be necessary to accommodate temporary changes in the operating performance of the plant. A common interruption is the need to curtail production to meet ground level SO_2 concentration targets.

Several levels of SO_2 control are available starting with the curtailment of converters then the reactor and finally the reverberatory furnace. If the reactor is required to reduce production, the operator will first lower the blowing rate while maintaining the oxygen rate. If further curtailment is needed, oxygen tonnage may be lowered to the point that the reactor is turned to stand-by.

During periods of reactor production curtailment it is possible to switch additional oxygen to the reverberatory furnace to boost its throughput. This is decided between the converter scheduler and the reverberatory furnace operator based on the predicted length of the control period, the status of the converters and the matte levels in the smelting furnaces.

The smelter can operate with the reverberatory furnace alone on higher tonnage oxygen during periods when the reactor is shut-down for maintenance. The smelting rate is then limited only by the maximum tonnage capacity of the oxygen line. During one of these periods the reverberatory furnace used 350 tonnes/day of oxygen, bringing the smelting rate to 1420 tonnes/day of concentrate.

In the event of concentrate shortages, the reverberatory furnace fires can be completely shut down. The intense heat of the oxy-fuel burners allows the furnace to come back on stream within three days of a cold start-up and to be at full production capacity after two weeks. However, bottom build-ups may take up to 4-6 weeks to disappear completely.

FUTURE OPERATION

With the start-up of the new acid plant fixing reactor gases it will be important to have a continuous high-strength SO_2 gas stream. Tonnage oxygen will play a major role in achieving this objective.

If reduced future concentrate supply warrants the shut down of the reverberatory furnace, the reactor will become the single smelting vessel. Under this scenario the converters may be operated with some oxygen to improve scrap and revert smelting capacity. The converter slag would then be treated by milling and flotation for copper recovery.

ACKNOWLEDGEMENT

The authors wish to thank the management of Noranda Minerals Inc., Horne Division, for permission to publish this paper.

REFERENCES

1. Anderson, J.N., Reverberatory Furnace Practice at the Noranda and Gaspé Smelters, in Extractive Metallurgy of Copper, Nickel and Cobalt, ed. by P. Queneau, AIME, New York, 1961, pp. 133-158.

2. Mills, L.A., Hallett, G.D., and Newman, C.J., Design and Operation of the Noranda Continuous Smelting Process, in Extractive Metallurgy of Copper, ed. by J.C. Yannopoulos and J.C. Agarwal, The Metallurgical Society of AIME, New York, 1976, pp. 458-487.

3. Mackey, P.J., Bailey, J.B.W., and Hallett, G.D., The Noranda Smelter - 1965 to 1983, in Advances in Sulfide Smelting, Ed. H.Y. Sohn, D.B. George and A.D. Zunkel, AIME, New York, N.Y., 1983.

4. Mackey, P.J., Bailey, J.B.W., and Hallett, G.D., The Noranda Process - An Update, in Copper Smelting - An Update, ed. by D.B. George and J.C. Taylor, AIME, New York, 1981, pp. 213-236.

5. Persson, H., Iwanic, M., El-Barnachawy, S., and Mackey, P.J., The Noranda Process at Different Matte Grades, Journal of Metals, Vol. 38, No. 9, 1986, pp. 34-37.

6. McKerrow, G.C., Hallett, G.D., and Tarassoff, P., Oxygen in the Noranda Process, Paper presented at the Latin American Congress on Mining and Extractive Metallurgy, Santiago, Chile, August 27-31, 1973.

7. Bailey, J.B.W., Beck, R.R., Hallett, G.D., Washburn, C. and Weddick, A.J., Oxygen Smelting in the Noranda Process, Paper presented at the 104th AIME Annual Meeting, New York City, N.Y., February 16-20, 1975.

8. Bustos, A.A., Brimacombe, J.K., Richards, G.G., Vahed, A. and Pelletier, A., Development of Punchless Operation of Peirce-Smith Converters, Paper to be presented at Copper 87, Vina de Mar, Chile, 1987.

10. Achurra, J., Espinoza, R. and Torres, L., Improvements in Full Use of Oxygen in Reverb Furnaces at Caletones Smelter, Paper presented at the 106th AIME Annual Meeting, Georgia, March, 1977.

THE BENEFITS OF OPTIMIZING AIR SEPARATION PLANT PERFORMANCE

D. C. King, R. L. Hutchison, K. J. Murphy, A. P. Odorski

Allentown, PA U.S.A.

ABSTRACT

Cryogenic air separation technology is used to supply nearly all of the oxygen consumed by industry today. The process separates the constituents of air by cryogenic distillation. Because cryogenic air separation is a mature technology, only small improvements in the cycle efficiency are likely in the future. Once a plant is installed, improvements can be achieved through advances in plant operation. The focus of this paper will be to discuss several methods for optimizing air separation plant performance. The methods include: 1) computer control and optimization, 2) improved maintenance, and 3) equipment upgrades. A computer control and optimization system is used to insure efficient operation while matching production to demand. Compressor maintenance techniques restore performance to prior levels, while turboexpander upgrades improve plant performance beyond prior levels. These methods have been selectively applied to many of the several hundred air separation plants owned and operated by Air Products with excellent results. This technology is also available for application to plants manufactured, owned, or operated by others.

KEYWORDS

Oxygen; cryogenics; air separation; computer control; optimization; compressor performance; turboexpanders.

INTRODUCTION

A flowsheet for a typical cryogenic air separation plant is shown in Figure 1. Air is compressed in a centrifugal compressor and cooled to near its dew point in the main heat exchangers. Water and carbon dioxide can be removed in front end mole sieve adsorbers or in reversing passages in the main exchangers. The dry stream is fed to the high pressure distillation column, where a nitrogen-rich overheads product is produced. This overhead stream is fed to a turboexpander, which produces refrigeration for the plant. Several reflux streams are taken from the high pressure column to the low pressure column where the final separation into high purity products takes place. These products are returned to the main heat exchanger to cool the incoming air stream.

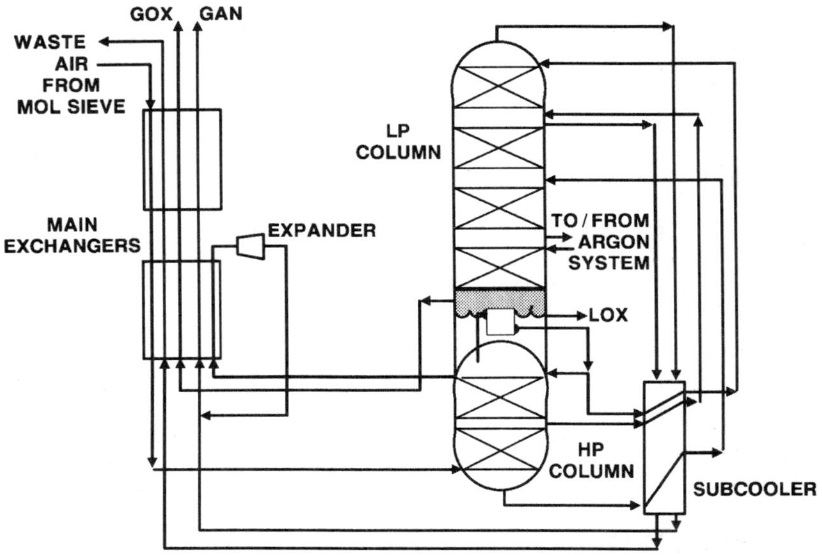

Fig. 1. Low pressure cryogenic cycle.

The methods used to optimize the performance of an air separation plant can focus on plant operation, maintenance, and process equipment upgrading. An audit of the day-to-day operation of the plant is the main tool for identifying operational inefficiencies and recommending courses of action. The plant may be running fairly efficiently, however it may be producing more or less than the required production at any given time. If the plant can always be run most efficiently at the current demand, then the performance of the plant is optimized and power costs will be minimized. This is what Air Products terms the "best operator" philosophy. A supervisory host computer using Air Products proprietary optimizing software will operate a plant on a "best operator basis" continuously.

The centrifugal air compressor is the plant's primary power consumer. This compressor uses atmospheric air as the process stream, which is subject to local contaminants. These contaminants can affect the efficiency of the compressor by deposition on the critical rotating parts. If these deposits are periodically removed, then the compressor efficiency can be restored to near the original efficiency. The turboexpander extracts work from the process to provide refrigeration. If the work can be recovered, then the overall plant efficiency will increase. If a more efficient expander is used, then less expander flow will be needed to produce the same amount of refrigeration and more reflux can be taken to the low pressure column to increase the product recovery. This increased recovery results in less air feed and power consumption for a given production rate.

COMPUTER CONTROL AND OPTIMIZATION

Application

In 1975, Air Products embarked on a program of using process control computers to improve the energy efficiency of its industrial gas processing facilities. Today over 100 plants that Air Products owns and operates have computer systems. Additionally, this technology has been applied to 15 plants that are owned and operated by others. The focus of the computerization effort has been cryogenic air separation plants that produce oxygen, nitrogen, and argon products in either liquid or gaseous form, but plants that produce other industrial gases such as hydrogen, helium, carbon monoxide, carbon dioxide, and ammonia have also been computerized.

Cryogenic air separation plants are very suitable for the use of process control computers, since the major production cost is electric power for driving compression equipment. Relatively small power consumption savings will offset the capital investment necessary to add a computer system, even when performed on a retrofit basis where pneumatic instrumentation must be replaced with computer-compatible instrumentation. The cryogenic distillation section of an air separation plant involves multiple components in up to three interconnected high purity distillation columns. This process exhibits nonlinear behavior to which a computer control system can more effectively respond than an operator. Most air separation plants are operating to supply an on-line customer who has a variable product demand. A computer system can monitor this demand and quickly adjust the operation of the plant to match production requirements.

Hardware and Software

The two parts of a computer control system consist of: a base line control system and an optimizing host computer. The base line control system provides basic monitoring and regulatory control. This system can be panel-based and consist of either pneumatic or electronic instrumentation. Plants with pneumatic instrumentation must be retrofitted with electronic transmitters and controllers when a computer control system is added (Fig. 2). Interface equipment is required to convert the electronic field inputs into a digital format for host computer processing, and vice versa. Alternately, the base line control system could be a digital control system (DCS) which uses microprocessor-based controllers and CRT operator interface units (Fig. 3). Linking a host computer to a DCS system consists of simply adding an additional node onto the plant loop. A DCS base line control system only has limited optimizing capabilities so adding a host computer to either base line system has the same benefits.

The second part of the computer control system is the host computer which provides advanced process monitoring, control, and optimization. The host computer consists of a processing unit and associated peripherals. The optimization applications programs reside in the host computer. These programs are written in FORTRAN or some other high level process control language. The basis of the monitoring package is a resident data base of information required for optimization which is frequently updated. Efficiency calculations such as specific powers and recoveries are performed on-line. The operator can access this information through a CRT and compare the current plant performance against standards that are stored in computer memory. An integration and logging package builds upon the resident data base and efficiency calculations. Plant logs are automatically generated which contain integrated plant production, consumption, performance, and efficiency information.

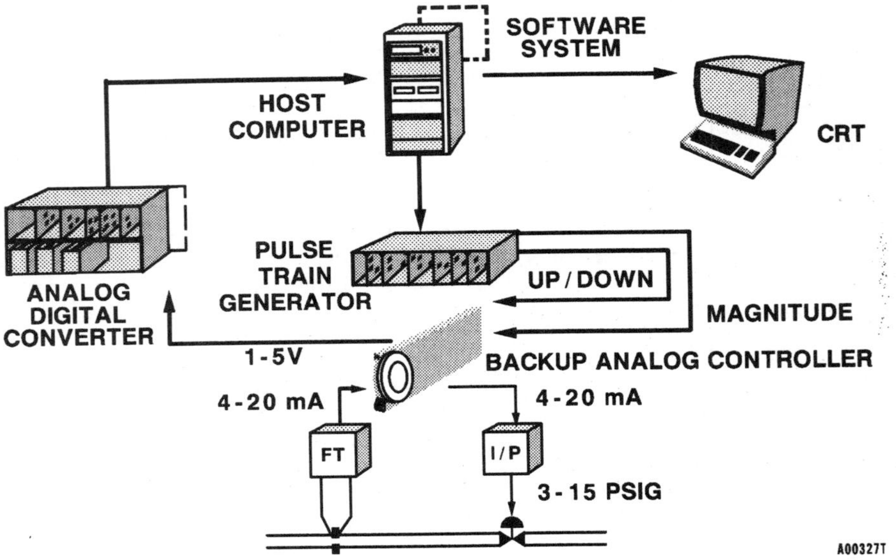

Fig. 2. Direct digital control.

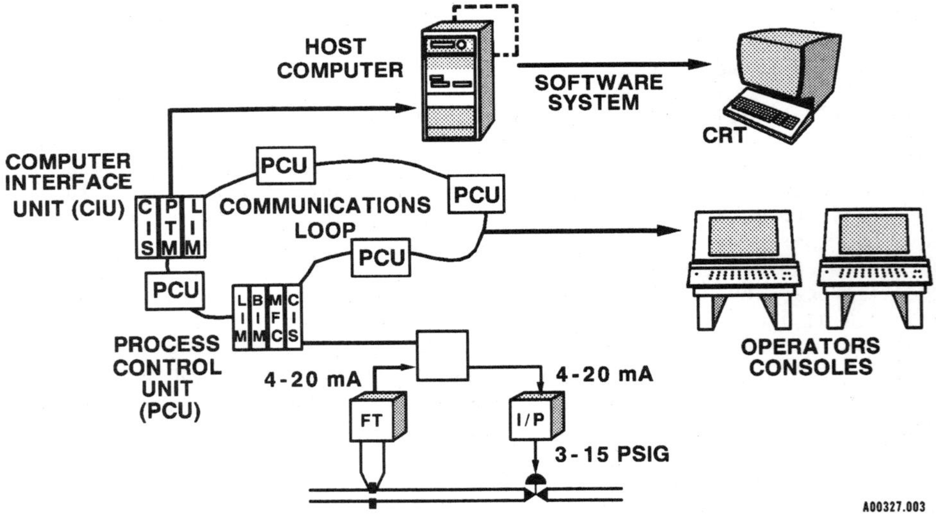

Fig. 3. Supervisory control with digital control system.

Host computer control, in its minimum scope, is a means of transmitting set point changes that are generated by the optimization program to the base line control system. In a panel-based, base line system, the critical loops are actually controlled by the host computer in direct digital control (DDC). The computer contains the control algorithm and parameters, calculates a control valve movement, and sends this change through the electronic controller to the valve.

This type of control is superior to conventional analog control due to the advanced capability of the computer. Techniques such as rate-of-change limiters, adaptive tuning, decoupling algorithms, logarithmic signal conditioning, and multiple cascade control can be implemented. The computer provides flexibility for trying new control strategies and control algorithms. In a DCS base line system, DDC is used for difficult loops, but most of the loops are run in supervisory mode where the host computer calculates a new set point for each DCS controller.

Steady-state and Transient Optimization

The primary justification for a host computer is process optimization. Optimization can take several forms. Steady-state optimization is used to calculate the controller set points that will minimize the production cost for a given production rate, subject to process and machinery constraints. Load-following optimization is used to change the plant production rate in response to varying customer demand. Dynamic ON/OFF control is used to assist in the start-up or shutdown of a plant.

The objective of steady-state optimization is to simultaneously solve the heat and material balances to produce a given production at minimum cost. Extensive plant testing aided by process simulation is used to establish correlations to predict recovery as a function of several parameters. These correlations are continuously updated via purity feedback to create a dynamic correlation. Air Products has consistently demonstrated a better than 5 percent reduction in power consumption for a given production rate or 5 percent increase in production rate for the same power consumption when steady-state optimization is implemented as compared to manual operation. If argon is being produced, recovery can be increased by 15 to 20 percent. These savings result from mass flow control, cascade purity control, and operating closer to the constraints. The conservatism that is associated with manual operation is eliminated.

As an example, Air Products installed an optimizing computer control system on the Passaic Valley Sewage Commission oxygen generators in 1982. Savings due to steady-state optimization between 4 and 14 percent have been documented (Patrylak, 1985). Fig. 4 presents these savings as a function of oxygen production rate. Higher savings are evident at turndown due to a greater turndown capability with the computer system. A system was installed in 1985 on an Air Products oxygen plant that supplies a copper smelter in Arizona. This plant had a DCS base line control system and 8 percent steady-state savings were documented above the DCS control system performance. Fig. 5 is a plot of specific power, defined as power consumption divided by oxygen production versus production for both the manually controlled and computer controlled cases at the Arizona smelter.

The objective of load-following optimization programs is to automatically match the plant production to the gas demand, subject to process and equipment limitations. The program can be based upon a feedback system, where a process parameter such as pipeline pressure is used to determine the required production rate. This is sufficient for systems where the demand changes slowly, such as the need for oxygen in a wastewater treatment plant. Fig. 6 shows the diurnal demand curve for Passaic Valley with the corresponding manual and computer generated responses in plant production. The distance between the computer and manual curves represents the savings attributed to the load-following optimization program. For systems that have much more severe demand changes, a feed-forward mechanism is needed. The oxygen needs of a copper smelter are more variable due to furnace burners being taken in and out of service and due to converters rolling in and out. The flows at the oxygen consumption points were

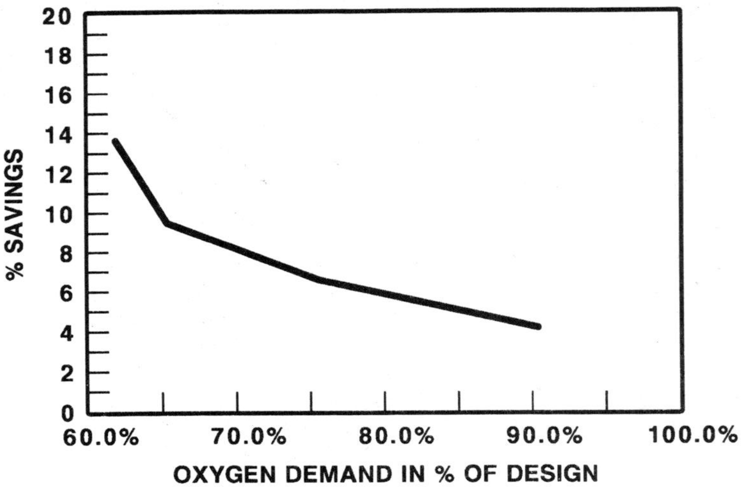

Fig. 4. Computer optimization savings.

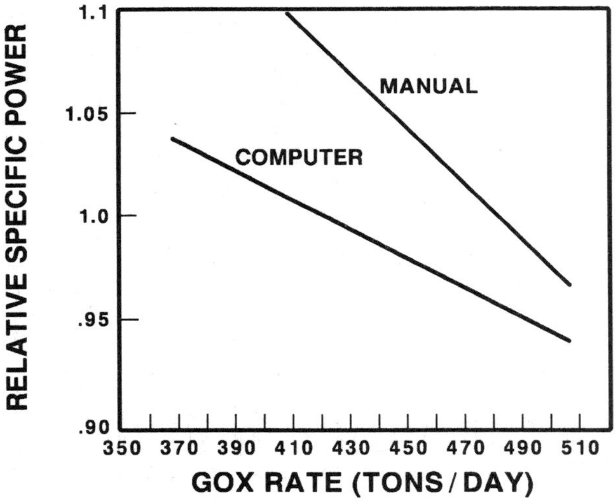

Fig. 5. Optimization benefits (manual vs. computer).

measured on the Arizona smelter so that a feed-forward algorithm could be used to determine the oxygen production rate. The performance of a load-following optimization system is dependent upon the maximum ramping rate of the air separation plant, the severity of demand changes, and the capacitance of the system. The objective is to minimize overproduction that is vented while avoiding supplemental liquid vaporization that makes up for production deficiencies.

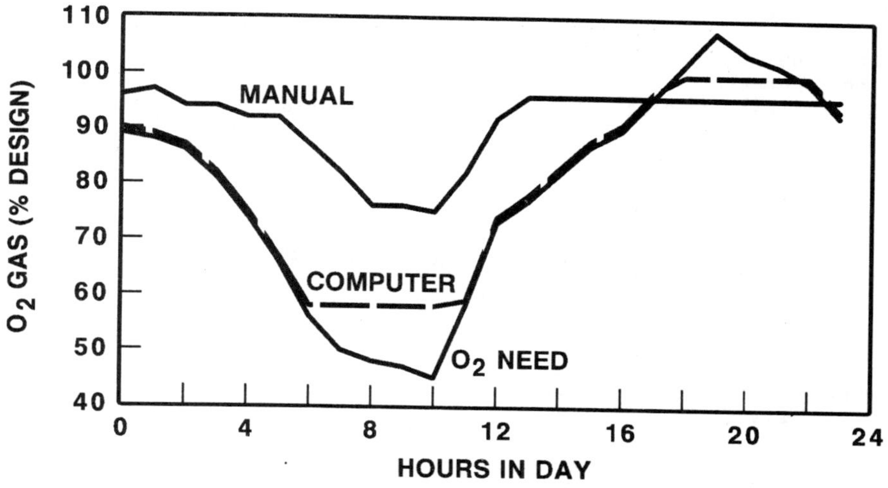

Fig. 6. Gaseous oxygen load following (manual vs. computer).

Computer-assisted plant start-up and shutdown is an example of an application of dynamic ON/OFF control. A system to accomplish this was implemented by Air Products at an oxygen plant owned and operated by the City of Houston (Russek, 1985). Due to electric utility economics, it was advantageous to shut down the plant daily to avoid on-peak power costs and to be an interruptable electric power consumer. Additional valves were automated, and a software package to assist the operator in starting up and shutting down the plant was written. The operator uses the CRT to monitor the start-up and to enter information into the computer after certain tasks are completed. Using a computer in this application has several advantages. It allows one operator to start-up or shutdown the plant expediently, which minimizes the unproductive power associated with a plant start-up. Also, the increased risk of process equipment failure associated with frequent start-up and shutdowns is minimized.

Central Plant Monitoring

Once a network of optimized plants is established, the benefits of linking the plant computers to a central location is evident. In 1981 Air Products initiated a program to link all of its plant computers to a central computer in Allentown, PA where the central engineering and operating departments are located (Chatterjee, 1981). The central computer initiates a daily phone call to each of the sites to retrieve integrated and instantaneous production, consumption, and efficiency information. This data is stored for on-line and ad-hoc inquiries and for generating exception reports. Trends in data are observed for each plant and degradation in equipment performance is quickly brought to the attention of line management and the plant efficiency engineer. Corrective action can be taken to minimize the cost impact of a problem. This same system is available to customers who purchase optimization systems for their own plants. Plant performance can be monitored in a remote location such as the customer's own corporate or central engineering office.

Benefit Summary

Implementation of computer monitoring, control, and optimization improves operating plant efficiently between 5 and 15 percent, typically resulting in a less than two year payback. Techniques such as linear programming and dynamic optimization will produce additional energy efficiency benefits while artificial intelligence and similar packages have potential to reduce the programming effort and further improve the profitability of computer projects.

MAINTAINING EQUIPMENT PERFORMANCE AND UPGRADING EXISTING FACILITIES

The Plant Audit

Air Products routinely evaluates the status of owned and operated facilities and offers this service to other owners of cryogenic plants. Following a comprehensive audit, maintenance services or equipment replacement can be conducted by Air Products to maximize equipment life and optimize plant performance. There are three distinct types of plant audit that can be performed: Energy Efficiency, Operation and Safety. Different teams of specialists are brought together for each type of audit. Following data collection and evaluation, a detailed report is issued outlining findings and recommendations.

An energy audit should be considered for plants that have been in operation for five or more years. Energy costs are usually the largest share of the cost of producing oxygen and represent the area of greatest opportunity to reduce costs. The audit addresses all items that impact plant efficiency including: distillation column performance, machinery efficiency, argon and liquid production rates if applicable, turndown and maximum production limits, plant operating modes and equipment reliability. The customer receives a detailed account of the areas that can be improved along with specific recommendations for maintenance and plant equipment modifications.

The operations audit can be performed on both new and old facilities. It is led by an operations engineer who reviews equipment operation, plant logs and operating philosophy and practices. Different specialists are also available to assist in inspection of specific plant equipment. The results of this study can improve maintenance procedures as well as operating efficiency.

Safety audits consist of two main elements: a physical inspection of equipment and the safety procedures being followed, and a presentation, using video cassettes and slides, to operating and maintenance staff personnel on various aspects of safety. Two areas where our audits have uncovered inefficiencies are the expander and the main air compressor.

Maintaining Air Compressor Performance

Our past operations audits have shown that air compressor pinion shaft vibration levels tend to increase with time due to the gradual or sometimes rapid deposit of foreign material on the impellers. Nonuniform depositing causes a change in mass distribution which leads to imbalance and resultant increase in synchronous vibration amplitude. Quite often the deposit build up is fairly uniform and the unbalance occurs when deposited material "flies" off the wheel. Deposits in the compressor also cause a gradual deterioration of performance characteristics. These deposits may generally be categorized as atmospheric dirt, chemical scale, or rust.

One method to remove accumulated deposits is to water wash the internals of the compressor. Clean, cool water is sprayed into the air piping directly upstream of the impeller. The presence of water in the air effectively breaks up and washes off the deposits which have built-up on the impeller. The water is not intended to impinge as a high velocity jet on the impeller, but rather to blanket it with water spray. The impeller, rotating at high speed, strikes the water droplets which results in a washing action. Water droplets carried through the machine provide for additional cleansing of the diffusors, interstage piping, and the intercooler air passages.

This method of water washing is desirable since the decrease of rotor vibration and increase in compressor efficiency is quickly noticeable. After 30 to 60 minutes of washing each stage (one at a time), the impellers are restored to a condition of balance and vibration levels return to normal. Restoring vibration levels to normal results in longer life of the machine and reduces maintenance and operating costs. Bearings, seals, and gears will run longer between replacements and major overhauls are conducted less frequently. The typical time between overhauls without water washing is 18 to 24 months; it can be increased significantly (in some cases to 6 years or more) by instituting water wash procedures.

The water wash system has been successfully installed on centrifugal compressors at over 140 Air Products facilities and typically yields plant energy savings of 3 percent. Additionally, this technology has been applied to 4 plants that are owned and operated by others. The water wash system is installed using proprietary spray nozzles developed by Air Products. The system is permanent once installed and can be operated by plant personnel with a minimum amount of training by Air Products. The compressor can be water washed at desired intervals, while keeping the plant on-line. The system can be installed during overhaul of the compressor.

Turboexpander Retrofits

In the 1970's when power costs soared, the need for high efficiency equipment was met with a high efficiency turboexpander designed and manufactured by Air Products. Turboexpanders for many requirements have been provided including: horsepower up to 2500 HP; temperatures from -450°F to +500°F; pressures up to 1000 PSIG and speeds up to 100,000 rpm. The on-stream time of Air Products' turboexpanders exceeds 99 percent.

Most of the turboexpanders for oxygen plants commissioned prior to 1978 were compressor loaded expanders and had isentropic efficiencies in the low 70 percent range. Today, Air Products' turboexpanders are approaching 90 percent isentropic efficiency. The current typical turboexpander is generator loaded to allow power recovery into the existing power grid. The net result of the improved expander efficiency plus generated power is a reduction of 5 to 10 percent of the total oxygen plant power requirement. The higher efficiency expander reduces air flow requirements which result in power savings. The adaptation of the generator allows approximately 95 percent recovery of the available expander gas horsepower.

Tonnage oxygen plants built before 1978, or newer plants with poorly performing expanders, can be retrofitted with the high efficiency generator loaded turboexpanders. The results easily justify the capital cost requirements. In many cases, the existing accessory system and controls can be used to minimize capital expenditures. One added benefit in an expander retrofit is the increased production capability of the oxygen plant. With the higher efficiency expander, the air flow required for a fixed oxygen production is lower; alternatively for a fixed air flow the oxygen production can be higher with the higher.

SUMMARY

Since cryogenic air separation is a mature technology, major improvements in plant efficiency will come from advances in plant operation. An audit of the day-to-day operation of the plant is the main tool for identifying operational inefficiencies. Three methods for optimizing performance have been discussed. They include computer control and optimization, improved maintenance, and equipment upgrades. These methods have been developed and refined on Air Products owned and operated plants and are also available for application to plants manufactured, owned, or operated by others.

REFERENCES

Patrylak, A. J., W. E. Garber, and B. G. Bryson (1985). Computer Control and Optimization in an Oxygen Plant Supplying a wastewater Treatment Facility. 4th IAWPRC Workshop on Instrumentation and Control of Water and Wastewater Treatment and Transport Systems, Houston, TX.

Russek, S. L., T. M. Beckowski, and D. R. Vinson (1985). Computer Automated Start-up and Shutdown of an Air Separation Plant. 1985 Industrial Energy Technology Conference and Exhibit, Houston, TX.

Chatterjee, N., D. J. Hersh, and J. R. Couch (1981). Total Energy Management with Mini-computer Systems. Eighth Energy Technology Conference Proceedings, Washington, D.C.

ELECTROLYTIC OXYGEN FOR INDUSTRY

INTEGRATED APPLICATION OF HYDROGEN AND OXYGEN

S. Benzimra, R.L. LeRoy and A.K. Stuart

Electrolyser Inc., 122 The West Mall, Toronto, Ontario, Canada M9C 1B9

ABSTRACT

This paper reviews the process design features of water electrolysis and examines the economic conditions which make electrolytic oxygen competitive in metallurgical applications.

KEYWORDS

Water electrolysis; electrolytic hydrogen and oxygen; cryogenic oxygen.

INTRODUCTION

Cryogenic air separation plants supply most of the pure oxygen which is used in industry. This process generates nitrogen as a by-product of limited value.

In recent years, water electrolysis has been gaining increased acceptance in industrial applications due to a significant reduction in capital cost and improved energy efficiency. The economics of electrolytic oxygen must be viewed in conjunction with the co-generation of hydrogen as a highly valued co-product. Since the investment cost per unit of electrolytic oxygen production is not particularly sensitive to economies of scale - in contrast with cryogenic oxygen - it follows that the electrolytic route is most economic for relatively small oxygen demands, although site specific conditions may considerably enlarge the scope of economic applications.

ELECTROLYSIS: GENERAL REVIEW

Extensive literature has been published over the past ten years concerning developments in the technology of water electrolysis. References 1 to 4 are of particular interest. The most important recent developments achieved in this field lead to a significant reduction in capital cost and increasingly higher energy efficiencies. This section summarizes the current status of water electrolysis processes.

Basic Theoretical Considerations

A direct electric current passing between two electrodes immersed in a water solution containing a suitable electrolyte (typically potassium hydroxide of 25-30% w/w concentration) results in water decomposition according to the following equations:

(a) at the cathode, the negative electrode: (1)

$$2H_2O + 2e^- \longrightarrow H_2 + 2OH^-$$

(b) at the anode, the positive electrode: (2)

$$2OH^- \longrightarrow 1/2\ O_2 + H_2O + 2e^-$$

(c) Overall: (3)

$$H_2O \longrightarrow \underset{\text{(cathode)}}{H_2} + \underset{\text{(anode)}}{1/2\ O_2}$$

The electrical requirements necessary to achieve Reaction (3) are defined in terms of current (Amperes) and voltage (Volts), the product of which represents electric power (Watts).

192,970 Coulombs (Ampere-seconds) of direct current (D.C.) are required to split one gram-mole of water (18 grams) to produce one mole of hydrogen (22.4 litres) and half a mole of oxygen (11.2 litres).

The voltage of an electrolytic cell producing hydrogen and oxygen at economic currents must be sufficient to compensate for:

- the reversible potential for water decomposition (1.19 Volts at 70°C)
- the ohmic resistance of the solution (0.10 - 0.30 V)
- the overvoltage at the cathode, due to kinetic limitations on the rate of the hydrogen production reaction (0.10 - 0.35 V); and
- the overvoltage at the anode (0.25 - 0.40 V)

Overall cell voltages typically fall in the range of 1.7 to 2.1 Volts.

The current efficiency in hydrogen production by water electrolysis ranges from 95 - 98% in bipolar equipment to 99.8% in unipolar cells. Thus, the quantity of electricity required to produce a given quantity of hydrogen and oxygen is approximately constant for all water electrolysis processes. Where significant progress has been made in recent years is reduction of the overall cell voltage at economic current densities, and thus in increase of energy efficiency.

The minimum theoretical electrical energy required to produce electrolytic hydrogen and oxygen at 70°C (a typical operating temperature in commercial application) is

(1.19 x 192970)/(3600 x 22.4) = 2.85 kWh per normal cubic metre of hydrogen.

In actual practice, energy consumption is 50 to 60% higher than this theoretical value because of losses due to cell resistance and the overvoltages at the electrodes. In addition, there are losses in the power conditioning equipment used to produce direct current, amounting to 2 to 5%.

The reversible potential depends primarily on cell temperature. Ohmic resistance of the solution is affected by the temperature of the bath, the geometry of the electrolytic cell, the concentration of the electrolyte and the selection of the material used to separate the anodic from the cathodic compartments. Overvoltages at the electrodes can be substantially reduced by using activated electrodes consisting of specially coated surfaces, typically having nickel-based substrates (cf. Reference 3).

Cell geometry, temperature, pressure, separator arrangement and electrolyte concentration are the main features which differentiate the various technologies currently available in the water electrolysis industry.

Water Electrolysis Technologies

Water electrolysers in commercial use fall into two major classifications: unipolar and bipolar.

In the unipolar "tank-type" design each electrode has the same polarity on both surfaces and carries out a single electrode process, i.e. oxygen or hydrogen evolution. A single cell contains a number of electrodes, with all electrodes of the same polarity being connected electrically in parallel. The result is that the overall cell voltage is equivalent to that of one anode/cathode pair, or 1.7-2.1 Volts. A sufficient number of cells are connected electrically in series by copper bus bars to form a cell battery of the desired gas output.

Electrolysers of the bipolar design may consist of a single massive assembly of a relatively large number of electrodes, each of which is cathodic on one side and anodic on the other. Each electrode is insulated from and electrically in series with its neighbour; and each pair of electrodes, with separating diaphgrams, forms an individual cell unit. A bipolar electrolyser may thus contain from thirty to several hundred individual cells in series at 1.7-2.1 Volts each, so that the corresponding applied voltage ranges from 50 to 600 Volts DC, depending on the required capacity.

While both electrolyser types exhibit roughly the same energy requirements, in the range of 8.6 - 9.2 kWh/Nm³O_2 (4.3 - 4.6 kWh/Nm³ H_2) and comparable floor space requirements, unipolar systems offer significant advantages in the simplicity of their design.

ELECTROLYSER INC. has been particularly successful in the development of a high capacity unipolar cell design ("EI-250") rated at 100,000 Amperes, with two major installations currently in place.

Main Design Feature of the "EI-250" Cell

TABLE I: Process Design Characteristics

Cell current:	100,000 Amperes
Current density:	2.50 kiloAmps/m^2 of electrode
Cell voltage:	1.85 Volts
Oxygen output:	21 Nm^3/hour per cell (*)
Hydrogen output:	42 Nm^3/hour per cell (*)
Energy consumption:	9.0 kWh AC/Nm^3 oxygen (*)
Energy efficiency:	Above 80%
Operating temperature:	70°C
Electrolyte:	KOH solution at 25% conc. (w/w)
Cell dimensions:	2.1 m x 1.1 m x 1.8 m (L) (W) (H)
Specific plant area:	0.40 m^2/Nm^3/h oxygen (*)
Feedstock and Utilities	
Deionized water:	2 litres/Nm^3 oxygen (*)
Cooling Water:	120 litres/Nm^3 oxygen (*)
Nitrogen and Instrumentation Air:	minor

(*) Normal conditions: 0°C, 101.3 kPa

The usual battery limit conditions of an oxygen/hydrogen plant based on "EI-250" cells are as follows:

Deionized water:	atmospheric pressure, ambient temperature
Cooling water:	30°C maximum inlet temperature at 350 kPa
AC power:	at any available voltage. Prices quoted in Table 2 would apply within 0.6 to 25 kV range
Oxygen gas:	saturated at 40°C, 120 mm WC, 99.7% purity
Hydrogen gas:	saturated at 40°C, 120 mm WC, 99.9% purity

A particularly interesting feature of electrolytic processes is the very wide range of throughput; since gas generation is in direct proportion to current intensity, the plant can be turned down to 10% of design capacity in less than one minute. At this reduced current density, ohmic resistance voltage drops significantly, thus increasing energy efficiency. Conversely, the plant can easily tolerate a peak throughput of 125% under sustained conditions while incurring a relatively small drop in efficiency.

EVALUATION OF CAPITAL AND OPERATING COSTS

Since most industrial requirements of pure oxygen are currently supplied by cryogenic air separation plants, it is natural to assess the costs of the electrolytic route against the cryogenic process. As precise price criteria can be greatly affected by site specific conditions, the figures presented here are order of magnitude estimates and should be used only comparatively as a preliminary indication of a potentially attractive alternative.

Capital cost refers to installed plant cost. All costs are expressed in U.S. dollars, for a typical North American site. Plant capacities are expressed in metric tonnes of oxygen per day (T/d). In the simplified approach used, all the non-power operating costs - labour, maintenance, utilities, etc. - have been omitted for comparison purposes since the costs involved are similar for both the electrolytic and cryogenic routes.

Attention is focused on relatively small scale plants, under 250 T/d, for which the basic costs are listed on Table 2.

TABLE 2: Basic Cost Data for Cryogenic and Electrolytic Oxygen Plants

	CRYOGENIC	ELECTROLYTIC EI-250
Capital costs:		
210 T/d:	$ 6.5 million	$ 15.7 million
25 T/d:	$ 1.46 million	$ 2.31 million
Approximate cost exponent:	0.7	0.9
Energy consumption:	330 kWh/T O_2	6300 kWh/T O_2
Hydrogen production: (higher heating value)		17.9 GJ/T O_2(*)

* I Gigajoule (GJ) corresponds to 78 normal cubic metres of hydrogen

COMPARATIVE ECONOMIC EVALUATION

Figures listed in Table 2 lead to a preliminary comparative estimate of the cost of one tonne of oxygen, excluding the relatively minor non-power operating costs. Three parameters are used:

- the unit cost of power, "E", expressed in U.S. cents per kilowatt hour
- the value of hydrogen co-product, "H", expressed in energy terms as dollars per gigajoule ($GJ); and
- a capital recovery factor, "CRF", defined as the fraction of the investment which must be borne by the annual production of oxygen calculated on the basis of 350 on-stream operating days for both technologies.

Case 1 - Captive Oxygen Production:

For a reference plant capacity of 25 T/d, the costs in U.S. \$/T of oxygen produced in an air separation plant ("AOX") and of electrolytic oxygen ("EOX") may be expressed as follows:

$$AOX = 3.3\ E + 1460000\ CRF/(25 \times 350) = 3.3\ E + 167\ CRF \quad (4)$$

$$EOX = 63\ E + 2310000\ CRF/8750 - 17.9\ H = 63\ E + 264\ CRF - 17.9\ H \quad (5)$$

As an example, using a power cost of 2 cents/kWh and 10% CRF, the cost of AOX is \$23.30/T. To match this cost by the electrolytic route, the hydrogen co-produced must be valued at:

$$[(63 \times 2) + 26.4 - 23.3]/17.9 = \$7.21\ GJ$$

Equations (4) and (5) can be generalized for various plant capacities ("K") up to about 250 T/d using the approximate plant cost exponents shown in Table 2:

$$AOX = 3.3\ E + 439\ CRF/K^{0.3} \quad (6)$$

$$EOX = 63\ E + 364\ CRF/K^{0.1} - 17.9\ H \quad (7)$$

The equalization of oxygen costs (AOX = EOX) determines the breakeven value H of hydrogen co-produced above which the electrolytic route can be considered as a viable alternative to the conventional cryogenic process,

$$H = 3.3\ E + (20.3/K^{0.1} - 24.5/K^{0.3})CRF \quad (8)$$

Equation (9) shows that the breakeven value of the hydrogen co-product is affected, in decreasing order of importance, by:

- the power rate "E"
- the capital recovery factor "CRF"
- the plant capacity "K"

Power rates are essentially site specific and are treated here as an input parameter. Capital recovery factors were introduced to compare processes requiring different investment costs. If the oxygen product is to be used captively and the oxygen plant is not considered as a profit centre, the value of CRF may well be around 10%. There is a relationship between the value of CRF and the more commonly used discounted cashflow return on investment (DCFROI): for an investment requiring a one year construction period and having a 20 year life time, a 10% CRF corresponds to 7.8% DCFROI. Since all these percentages are in real terms, 7.8% is in reasonable agreement with prevailing borrowing rates. Thus, a 10% CRF is considered reasonable for comparison purposes.

As for plant capacity, we examined oxygen generating plants ranging from 10 to 200 T/day, although larger units could well be contemplated since there is no limit on the capacity of modular electrolytic installations. Equation (8) shows, however, that for very large capacities the breakeven hydrogen cost may increase somewhat.

Equation (8) is graphically represented in Figure 1 for CRF = .10, "E" ranging from 1 to 3 cents/kWh and "K" within the 10 to 200 T/day range.

The right-hand ordinate of Figure 1 shows what the equivalent price of natural gas should be (in \$/GJ, HHV basis) if hydrogen were to be produced by steam methane reforming (SMR) rather than by electrolysis. In should be noted that SMR units could vary in design and cost depending on the export steam requirements. The values indicated in Figure 1 represent approximate prices of natural gas based on average steam production, as per Equation (9):

$$G = 0.7H - 1.4 \qquad (9)$$

where: "G" = natural gas price in \$/GJ, HHV basis
"H" = hydrogen value in \$GJ, HHV basis

<u>Case 2 - Comparison with Purchased Oxygen</u>

In some instances, the alternative to electrolytic production of hydrogen and oxygen is purchase of oxygen on an over-the-fence basis, for example from an adjacent air separation plant or from delivered merchant liquid oxygen. Typical purchase costs in 1987 U.S. dollars range from \$30 to \$70 per tonne.

To determine the breakeven value of co-product hydrogen in this case, equation (7) must be expanded to include an allowance for non-power operating costs, covering labour, maintenance, feed water, cooling water, make-up caustic, etc. A global value of these costs is estimated at \$8.5/T. The resulting plant gate cost of electrolytic oxygen is shown in Equation (10):

$$EOX = 63\ E + 364\ CRF/K^{0.1} + 8.5 - 17.9\ H \qquad (10)$$

If comparison is being made with a purchased oxygen cost of "P" U.S.\$/T, the breakeven hydrogen value becomes:

$$H = 3.5\ E + 20.3\ CRF/K^{0.1} + 0.47 - 0.056\ P \qquad (11)$$

Breakeven hydrogen values for purchased oxygen prices of \$30, \$50 and \$70 per tonne are indicated by the three bands in Figure 2. As in Figure 1, equivalent natural gas prices for production of the co-product hydrogen are also shown. The breakeven hydrogen value decreases substantially as the price of purchased oxygen increases.

CONCLUSIONS

Water electrolysis, as a source of high purity <u>oxygen</u>, should be seriously considered as long as <u>hydrogen</u> can be used.

Hydrogen gas has broad applications as a building block in the chemical industry and as a reducing agent in a number of processes. The actual potential value of hydrogen depends on site specific conditions. Possible integration with adjacent operations can be, in some instances, of great mutual advantage.

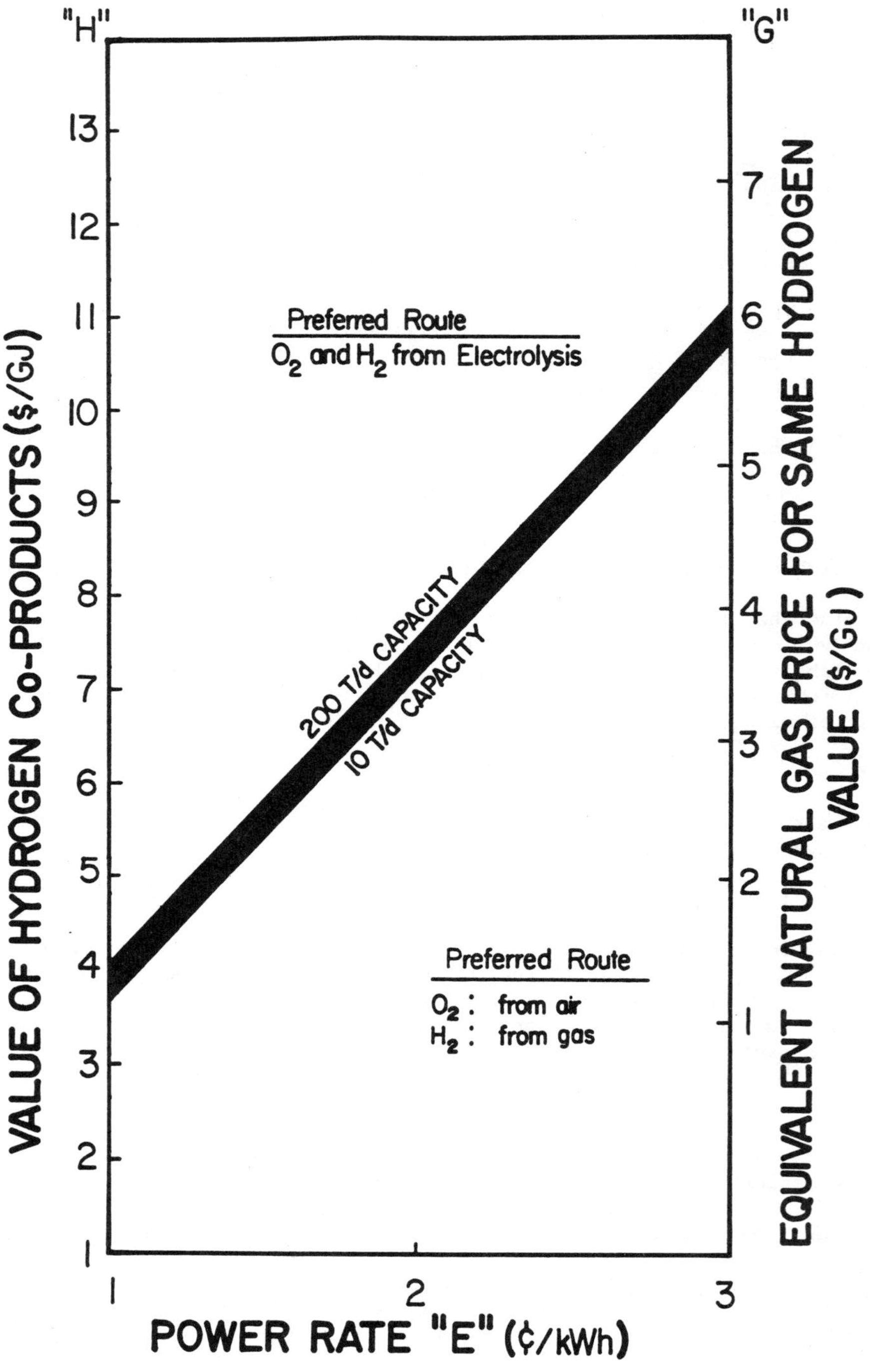

Fig. 1. Values of co-product hydrogen at which electrolytic production of oxygen and hydrogen becomes preferred over captive production of oxygen from air and hydrogen by steam methane reforming.

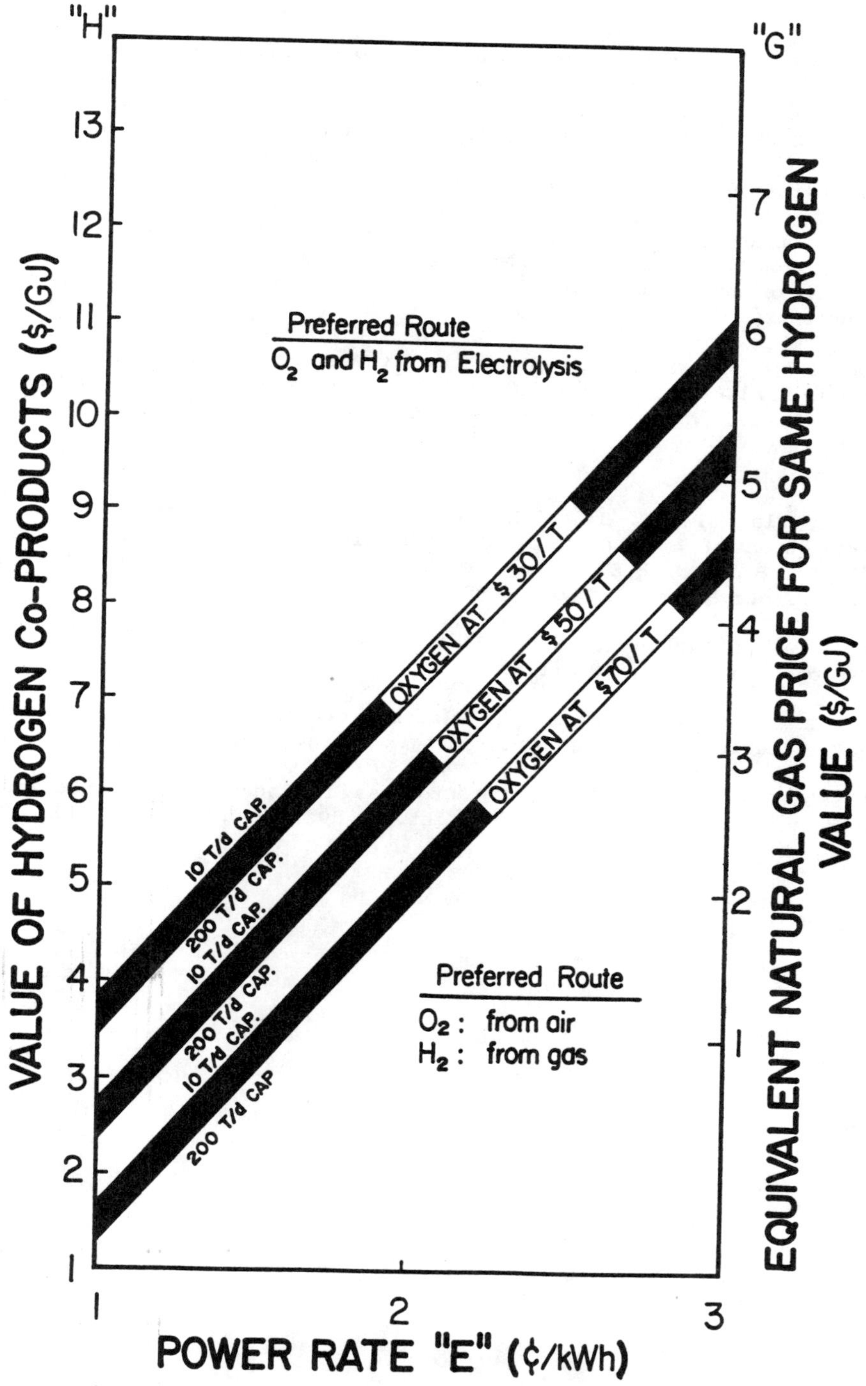

Fig. 2. Values of co-product hydrogen at which electrolytic production of oxygen and hydrogen becomes preferred over purchase of merchant oxygen and hydrogen production by steam methane reforming.

Small to medium oxygen requirements and relatively low power rates enhance the attractiveness of the electrolytic route. This is particularly true when the oxygen demand is sufficiently low that the investment in an air separation plant is not warranted and oxygen must be purchased at market prices.

From an operational point of view, electrolytic plants offer the advantage of a wide range of throughputs - from 10% to 125% of design capacity - with an almost instantaneous modulation capability, whereas the turndown ratio of an air separation plant is generally limited to 60%. The modular nature of electrolytic plants also allows for a smooth gradual build-up in capacity, to match any increasing oxygen demand.

The number of factors involved in the assessment of the economics of an electrolytic plant is such that each situation must be evaluated on a case-by-case basis before any firm conclusion can be established.

REFERENCES

1. P.V. Tilak, P.W.T. Lu, J.E. Colman and S. Srinivasan, "Electrolytic Production of Hydrogen", in Comprehensive Treatise of Electrochemistry (J.O'M. Bockris, B.E. Conway, E. Yeager and R.E. White, eds.), Vol. 2, p.1, Plenum Press, New York (1981).

2. R.L. LeRoy, "Industrial Water Electrolysis: Present and Future", Int. J. Hydrogen Energy, 8, 401 (1983).

3. M.B.I. Janjua and R.L. LeRoy, "Electrocatalyst Performance in Industrial Water Electrolysers", Int. J. Hydrogen Energy, 10, 11 (1985).

4. R.L. LeRoy and A.F. Hufnagl, "Progress in Industrial Demonstration of Advanced Unipolar Electrolysis", Int. J. Hydrogen Energy, 8, 581 (1983).

OXYGEN PRODUCTION TECHNOLOGIES FOR NON-FERROUS SMELTING APPLICATIONS

K. J. Murphy, A. P. Odorski, A. R. Smith and T. J. Ward

Allentown, PA, U.S.A.

ABSTRACT

Oxygen production in 1985 totaled 379 billion cubic feet, making it the fifth largest product in the U.S. chemical industry. Although cryogenic air separation processes have been typically used to supply most industrial requirements, newer technologies are already available and more are being developed to supply oxygen more efficiently for the non-ferrous smelting industry. This paper reviews the technical features and the state of development of adsorption, chemical, and membrane processes and describes the latest improvements in computer control of cryogenic plants. The paper also discusses selected non-ferrous oxygen applications and the compatibility of these new oxygen production technologies.

KEYWORDS

Oxygen; non-ferrous smelting; cryogenics; adsorption; membranes; air separation; chemical air separation; computer control.

INTRODUCTION

In 1985 oxygen was the fifth leading product manufactured by the U.S. chemical industry (Anon., 1986). Over 379 billion cubic feet (10.7 billion cubic meters), equivalent to 31.4 billion pounds (14.2 billion kilograms) were produced. The majority of production came from cryogenic air separation plants, the preferred method for most industrial applications. Over the past few years, the range of industrial uses and consumption patterns have expanded to not only include large, continuous duties; but smaller and more cyclical requirements as well. Applications for purities below the traditional industrial standard of 99.5% have also been increasing. As a result, a need for new technologies to handle this broader range of applications has developed. Adsorption technologies have been employed in lower purity applications (Schaedel, 1985). Membrane based systems may prove to be a better fit when the usage is small and intermittent, and when only "enriched air" (<50% O_2) is required. Chemical based systems are also being developed that will offer reduced energy costs for larger requirements. Cryogenic plants are continually being improved through efficiency enhancements such as computer control. All of these technologies trade off capital cost and efficiency to find their market niche.

This paper will explore the current state of development of the four technologies: Adsorption, Membranes, Chemical, and Cryogenics; examine their purity limits and operating characteristics, and give examples where these technologies mesh with non-ferrous processes. Figure 1 shows the oxygen recovery from the feed air stream and the specific power to produce a contained ton of oxygen as a function of oxygen purity for the four technologies. Power consumption is a major portion of the cost to produce oxygen in larger plants, but decreases in importance as plant size decreases. Recovery is a relative measure of efficiency and also controls equipment size and cost to a certain extent. Due to the tradeoffs between power and capital costs, Adsorption and Membrane Systems compete with LOX for smaller production rates. As product rate increases and power cost begins to control the cost of oxygen, on-site cryogenic plants become the economic choice. If high purities (>95% O_2) are required, LOX or on-site cryogenic plants are the only currently available alternative.

OXYGEN PRODUCTION TECHNOLOGIES

Adsorption

Systems employing adsorption principles to separate the components of air were first used over twenty years ago. These processes are based on the ability of some natural and synthetic materials to preferentially adsorb nitrogen or oxygen molecules. In the case of zeolites (aluminosilicates), non-uniform electric fields exist in the void spaces of the material, causing polar molecules to be adsorbed more strongly than non-polar molecules. Thus in air separation, nitrogen molecules are more strongly adsorbed than oxygen or argon molecules. As air is passed through a bed of zeolitic material, nitrogen is retained and an oxygen-rich stream exits the bed. Carbon molecular sieves can also function as air separation media. The carbon molecular sieves have pore sizes on the same order of magnitude as the size of air molecules. Since oxygen molecules are slightly smaller than nitrogen molecules, they diffuse more quickly into the cavities of the adsorbent. Thus, carbon molecular sieves are selective for oxygen and zeolites are selective for nitrogen.

Zeolites are presently used in essentially all adsorption based processes for oxygen production. A typical flowsheet is shown in Fig. 2. Pressurized air enters a vessel containing the zeolite adsorbent. Nitrogen is adsorbed and an oxygen-rich effluent stream is produced until the bed has been saturated with nitrogen. At this point, the feed air is switched to a fresh vessel and regeneration of the first bed can begin. Regeneration can be accomplished by heating the bed or by reducing the pressure in the bed, which reduces the equilibrium nitrogen holding capacity of the adsorbent. Heat addition is commonly referred to as Temperature Swing Adsorption (TSA), and pressure reduction as Pressure or Vacuum Swing Adsorption (PSA or VSA). The faster cycle time and simplified operation associated with pressure reduction makes it the process of choice for air separation.

Variations in the process that effect operating efficiency include separate pretreatment of the air to remove water and/or carbon dioxide, multiple beds to permit pressure energy recovery during bed switching, and vacuum operation during depressurization. Optimization of the system is based on product flowrate, purity and pressure; energy cost and expected operating life.

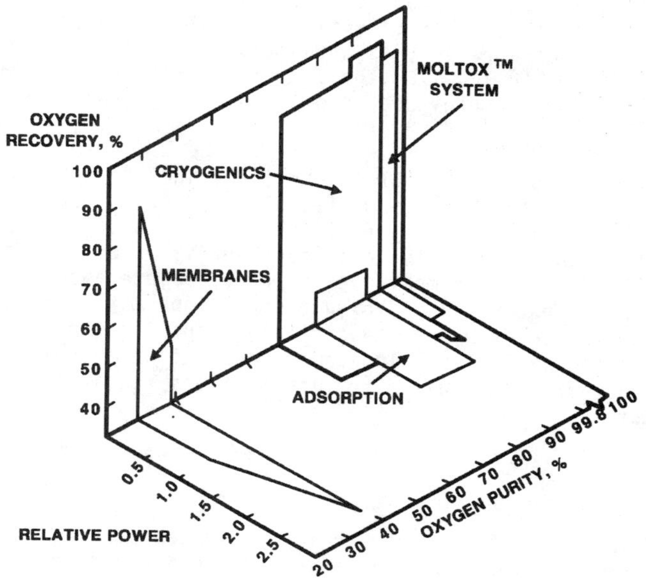

Fig. 1. Technology comparison.

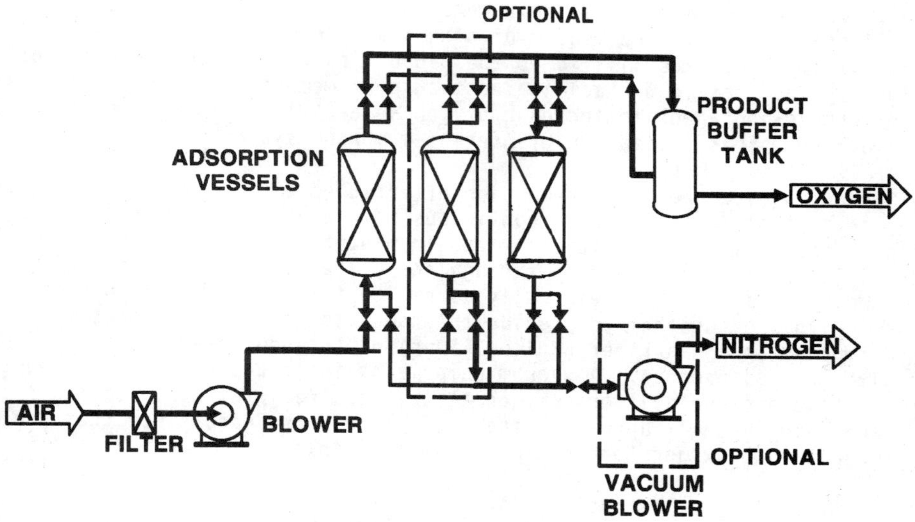

Fig. 2. Air Products/Seitetsu adsorption process.

However oxygen purity is limited to a maximum of 93-95%. Due to the cyclic nature of the adsorption process, bed size is the controlling factor in capital cost. Since production is directly proportional to bed volume, capital costs increase more rapidly as a function of production rate compared to cryogenic plants. Up to 100 T/D, the capital cost is relatively low and compensates for the higher energy usage compared to cryogenics (Paffenbarger, 1986).

Other factors that effect analysis of this process include operability and product storage. Since the system operates at near ambient conditions, start-up and shutdown are relatively fast operations. A unit idled over night or during a weekend can be restarted and reach purity in as little as several minutes. If product storage to meet peak demands, maintenance or other outages is required, a backup liquid oxygen (LOX) tank may be required. LOX would typically be hauled in by an industrial gas supplier, but can also be generated by an add-on liquefier. In summary adsorption systems can deliver 85 to 95% oxygen and are best suited for applications requiring less than 100 T/D of product.

Membranes

Recently, semipermeable membranes have been introduced for separating oxygen and nitrogen from air. The process is based on the difference in rates of diffusion of oxygen and nitrogen through a membrane which separates high pressure and low pressure process streams. Flux and selectivity are the two properties that determine the economics of membrane systems, and both are functions of the specific membrane material. Flux determines the membrane surface area, and is a function of the pressure difference divided by the membrane thickness. A constant of proportionality that varies with the type of membrane is called the permeability. Selectivity is the ratio of the permeabilities of the gases to be separated. Due to the smaller size of the oxygen molecule, most membrane materials are more permeable to oxygen than to nitrogen. However, commonly available membrane materials do not have a high enough selectivity to produce medium or high purity oxygen, so that available membrane systems today are limited to the production of "enriched air"; i.e., 25 to 50% oxygen. Active or facilitated transport membranes, which incorporate an oxygen-complexing agent to increase oxygen selectivity, are a potential means to overcome this limitation of present membrane systems.

Figure 3 shows a flowsheet for production of enriched air. A major benefit of membrane separation is the simple, continuous nature of the process and operation at near ambient conditions. An air blower supplies enough head pressure to overcome pressure drop through the filters, membrane tubes and piping. Membrane materials are usually assembled into cylindrical modules that are manifolded together to provide the required production capacity. Oxygen permeates though a fiber (Hollow Fiber) or through sheets (Spiral-Wound) and is withdrawn as product. A vacuum pump maintains the pressure difference across the membrane and delivers oxygen at the required pressure. Carbon dioxide and water usually appear in the enriched air product, since they are more permeable than oxygen for most membrane materials.

As with adsorption systems, capital is essentially a linear function of production rate and product backup is not available without a separate liquid oxygen storage facility. Membrane systems will probably fit applications up to 20 T/D. This technology is relatively new, and continued improvements could make membranes attractive for small oxygen-users with discontinuous usage patterns.

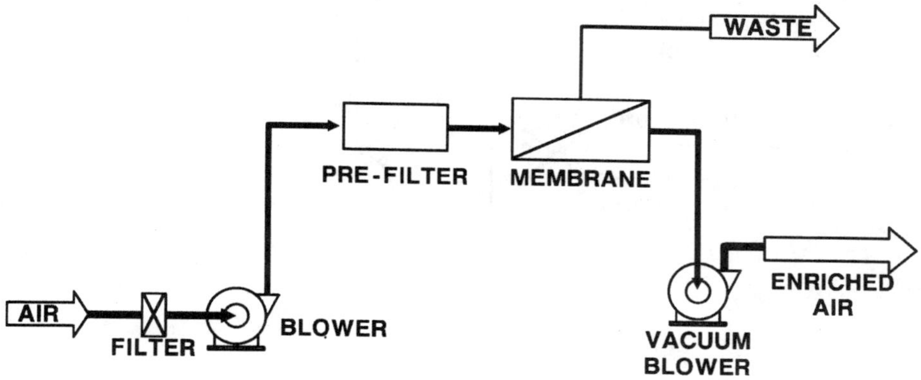

Fig. 3. Membrane process.

Chemical

Air Products and Chemicals is currently developing a continuous, liquid-phase absorption/desorption chemical air separation process that significantly reduces electric energy consumption compared to cryogenic cycles. The MOLTOX™ air separation technology was invented by D. C. Erickson, Energy Concepts Co., Annapolis, MD. Air Products has acquired the exclusive worldwide rights in an agreement with Energy Concepts Co. and began operation of a 1/4 T/D pilot plant, sponsored in part by the U. S. Department of Energy, in March 1986.

The two basic MOLTOX™ oxygen systems, PSA and TSA, are shown in Fig. 4. Actual design and operation can be a combination of these processes. Dry, carbon dioxide free air enters the absorber at a temperature of 900 to 1200°F and a pressure of 20 to 186 psia where it contacts molten liquid salt. Oxygen in the air reacts chemically with the salt and is removed with the liquid salt leaving the bottom of the absorber. The oxygen bearing salt flows to the desorber vessel where it is heated and/or reduced in pressure, generating gaseous oxygen. Lean salt is cooled before returning to the absorber to complete the loop.

The MOLTOX™ oxygen systems' major advantage is that the chemical reaction does not consume the compression energy of nitrogen, which is the major portion of the air stream. In the PSA mode, nitrogen leaving the absorber can be expanded to recover its energy; in the TSA mode air/nitrogen is compressed only high enough to overcome the pressure drop through heat exchange and absorption equipment. The TSA process is especially applicable for oxygen-users who have waste heat sources, low value fuel streams or boilers that can be integrated with the MOLTOX™ system.

1. PSA - (PRESSURE SWING ABSORPTION)

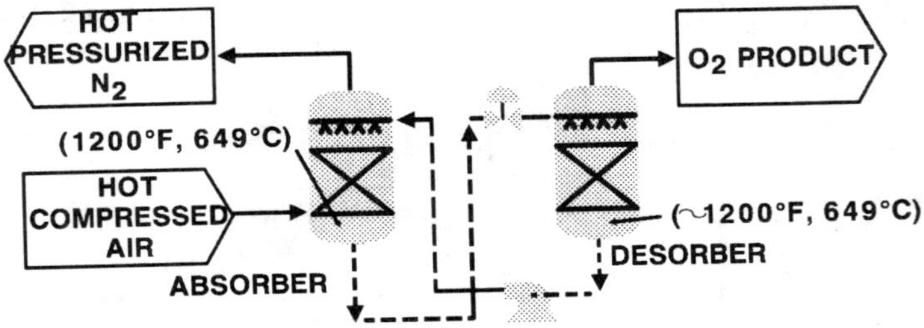

2. TSA - (TEMPERATURE SWING ABSORPTION)

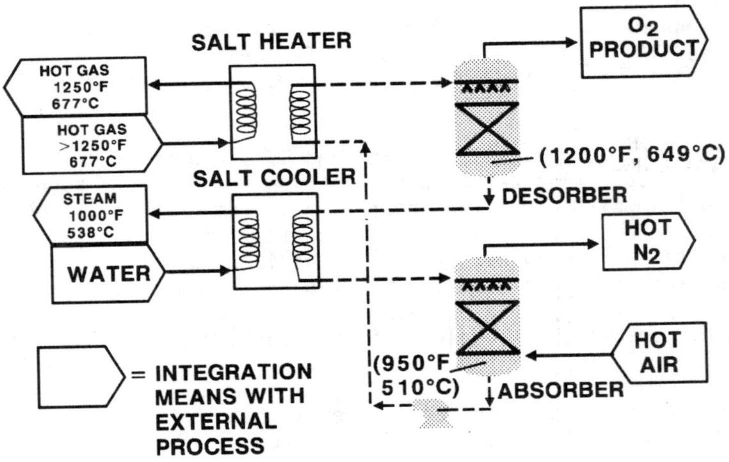

Fig. 4. Integrated MOLTOX™ system salt loops.

Pilot plant operation has demonstrated the closed loop, continuous process of absorbing oxygen from air and producing a very high purity oxygen product (99.9%). The next phase of development consists of further laboratory and pilot plant testing to support design and construction of a 5 to 50 T/D semiworks unit to be located at an oxygen-user's facility. Full sized commercial plants are planned for the early 1990's, and will be best suited for large oxygen requirements such as 500 T/D and larger where the operating efficiency offsets the extra capital associated with the equipment.

Cryogenics

Cryogenic air separation technology is used to supply nearly all the oxygen consumed by industry today. The process separates the constituents of air by cryogenic distillation. Refrigeration to overcome heat leak and to supply reflux to the distillation columns is obtained by expanding a portion of the compressed gases. Many articles have been published on the principles of cryogenic air separation; Fig. 5 is the flowsheet for the dual distillation column, low pressure cycle. This basic process, with variations, has been used extensively to produce oxygen in the 95 to 99.8% purity range. Reversing heat exchangers, which combine impurity removal with primary heat exchange, or molecular sieve adsorption systems can be used to pretreat the air before it enters the distillation section of the plant (Wilson, 1984). Below 95% purity, the specific power (kWh/ton of contained oxygen) remains nearly constant for the low pressure cycle. Therefore, other process cycles have been developed to optimize the cost of producing oxygen at purities from 70 to 95% (Bernstein, 1984).

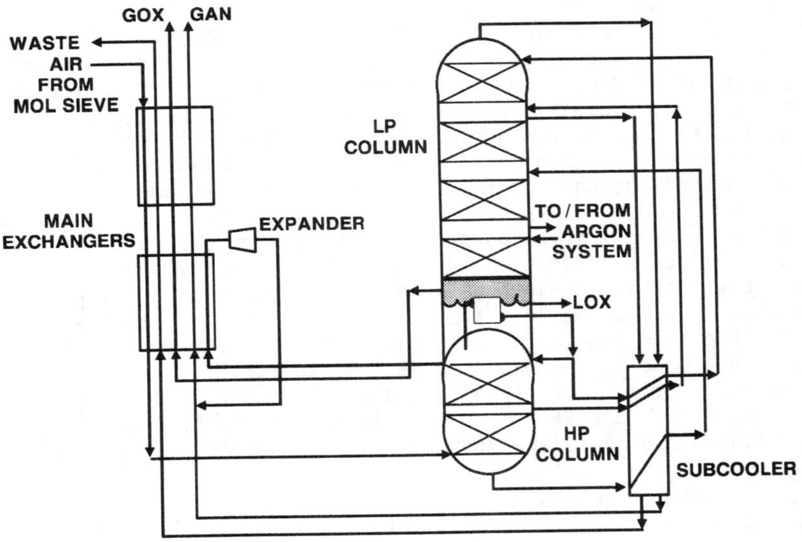

Fig. 5. Low pressure cryogenic cycle.

An advantage of cryogenic systems is their ability to produce nitrogen and argon by-products as well as liquefy a portion of the total output. The liquid is stored in vacuum insulated tanks where it is readily available for use during peak flow requirements, to serve as backup during outages, or for merchant sales. Operation of a cryogenic plant is not as flexible as an adsorption unit for oxygen-users with wide fluctuations in flow requirements. Depending on the length of time a cryogenic plant has been shutdown it can take several minutes to several hours to restart, cool down and reach purity. If production schedules are well known, restart times may not adversely effect operations and LOX is available as backup; however, the unproductive power cost to restart the unit must also be considered in determining the actual cost of supplying oxygen.

Because cryogenics is a mature technology, drastic improvements in the cycle efficiencies are not likely in the future. The major improvements over the past few years and in the near future will come from advances in the operation of the plants such as the application of computer control. Air Products has pioneered this technology and today, over 100 plants that Air Products owns and operates have computer systems. Additionally, this technology has been applied to 15 plants that are owned and operated by others.

Cryogenic air separation plants are good candidates for process control computers, since the major production cost is electric power for driving compression equipment. A small savings in power consumption will offset the capital investment necessary to add a computer system, even when performed on a retrofit basis where pneumatic instrumentation must be replaced with computer-compatible instrumentation. Most air separation plants are operating to supply an online customer who has a variable product demand. A computer system can monitor this demand and quickly adjust production accordingly.

A computer control system consists of two parts, a baseline control system and an optimizing host computer. The baseline control system provides basic monitoring and regulatory control. This system can take one of two forms: 1) A panel-based system consisting of either pneumatic or electronic instrumentation; or 2) a digital control system (DCS) which uses microprocessor-based controllers and CRT operator interface units (Fig. 6, 7). The host computer consists of a processing unit and associated peripherals. The optimization applications programs reside in the host computer.

The host computer provides advanced process monitoring, control, and optimization. The basis of the monitoring package is a resident database of information. Efficiency calculations such as specific powers and recoveries are performed on-line. Plant logs are automatically generated which contain integrated plant production, consumption, performance, and efficiency information. In a panel-based baseline system, the critical loops are actually controlled by the host computer in direct digital control (DDC). In a DCS baseline system, DDC is used for difficult loops, but most of the loops are run in supervisory mode where the host computer calculates a new setpoint for each DCS controller.

The primary justification for a host computer is process optimization. Optimization can take several forms. Steady-state optimization is used to calculate the controller setpoints that will minimize the production cost for a given production rate, subject to process and machinery constraints. Load-following optimization is used to change the plant production rate in response to varying customer demand. Dynamic on/off control is used to assist in the start-up or shutdown of a plant.

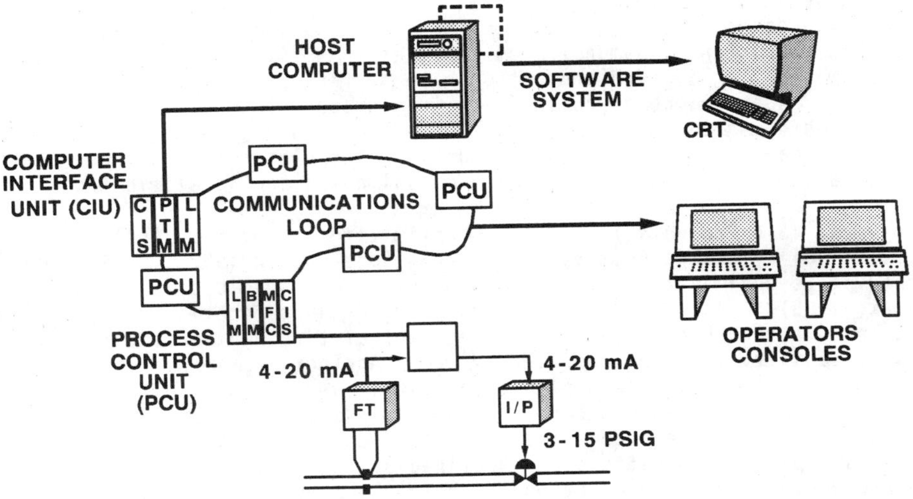

Fig. 6. Supervisory control with digital control system.

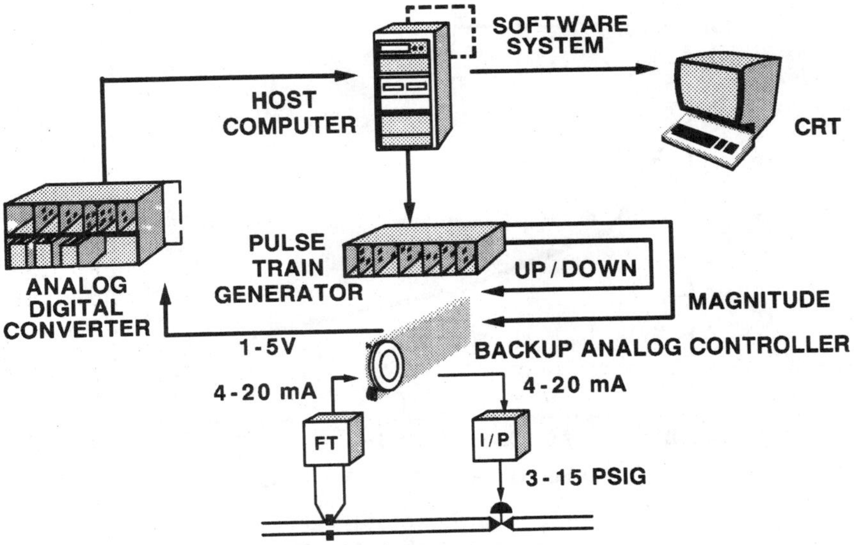

Fig. 7. Direct digital control.

The objective of steady-state optimization is to simultaneously solve the heat and material balances to yield a given production at minimum cost. Extensive plant testing aided by process simulation is used to establish correlations to predict recovery as a function of several parameters. These correlations are continuously updated via purity feedback to create a dynamic correlation. Air Products has consistantly demonstrated a better than five percent reduction in power consumption for a given production rate or five percent increase in production for the same power consumption when steady-state optimization is implemented as compared to manual operation. These savings result from mass flow control, cascade purity control, and operating closer to the constraints. The conservatism that is associated with manual operation is eliminated.

As an example, Air Products installed an optimizing computer control system on the Passaic Valley Sewage Commission oxygen generators in 1982. Savings due to steady-state optimization of up to 14 percent have been documented (Patrylak, 1985). Figure 8 presents these savings as a function of oxygen production rate, and higher savings are evident at turndown due to greater turndown capability with the computer system. Another system was installed in 1985 on an Air Products' oxygen plant that supplies a copper smelter in Arizona. This plant had a DCS baseline control system and 8 percent steady-state savings were documented above the DCS control system performance. Figure 9 is a plot of specific power, defined as power consumption divided by oxygen production, versus production for both the manually controlled and computer controlled cases at the Arizona plant.

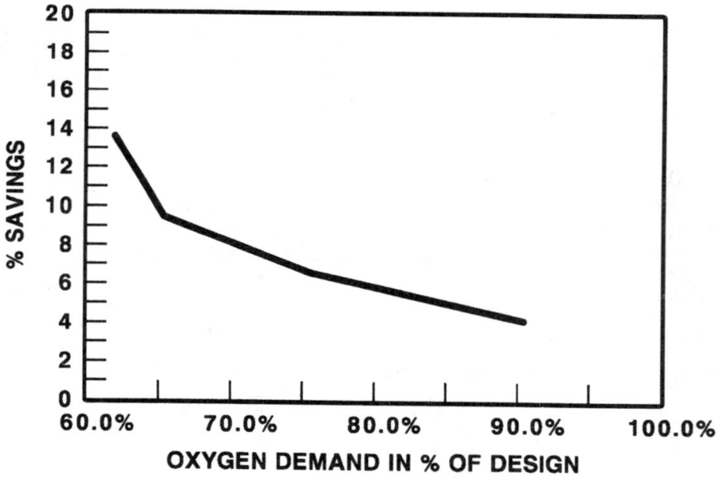

Fig. 8. Computer optimization savings.

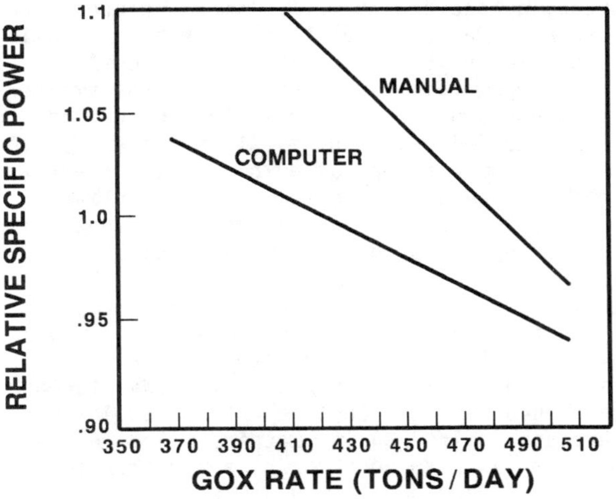

Fig. 9. Optimization benefits (manual vs. computer).

The objective of load-following optimization programs is to automatically match the plant production to the gas demand, subject to process and equipment limitations. The program can be based upon a feedback system, where a process parameter such as pipeline pressure is used to determine the required production rate. This is sufficient for systems where the demand changes slowly, such as the need for oxygen in a wastewater treatment plant. Figure 10 shows the diurnal demand curve for Passaic Valley, with the corresponding manual and computer generated responses in plant production. The distance

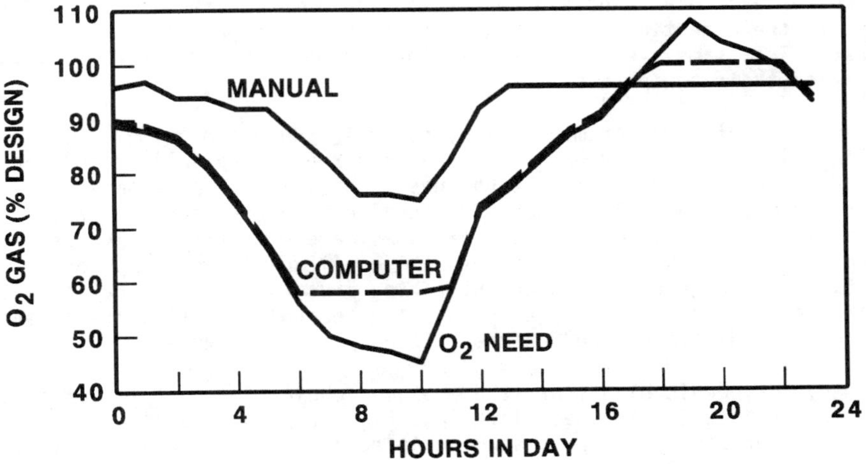

Fig. 10. Gaseous oxygen load following (manual vs. computer).

between the computer and manual curves represents the savings attributed to the load-following optimization program. For systems that have much more severe demand changes, a feedforward mechanism is needed. The oxygen needs of a copper smelter are more variable due to furnace burners being taken in and out of service and due to converters rolling in and out. The flows at the oxygen consumption points were measured on the Arizona smelter so that a feedforward algorithm could be used to determine the oxygen production rate. The performance of a load-following optimization system is dependent upon the maximum ramping rate of the air separation plant, the severity of demand changes, and the capacitance of the system. The objective is to minimize over-production that is vented, while avoiding supplemental liquid vaporization that makes up for production deficiencies.

Computer-assisted plant start-up and shutdown is an example of a application of dynamic on/off control. A system to accomplish this was implemented by Air Products at an oxygen plant owned and operated by the City of Houston (Russek, 1985). Due to electric utility economics, it was advantageous to shutdown the plant daily to avoid on-peak power costs and to be an interruptable electric power consumer. A propane-fired boiler was installed in the vaporization system to allow maximum power to be shed while interrupted. Additional valves were automated, and a software package to assist the operator in starting-up and shutting down the plant was written. The operator uses the CRT to monitor the start-up and to enter information into the computer after certain tasks are completed. Using a computer in this application has several advantages. It allows one operator to start-up or shutdown the plant expediently, which minimizes the unproductive power associated with a plant start-up. The increased risk of process equipment failure associated with frequent start-up and shutdowns is minimized.

Once a network of optimized plants is established, the benefits of linking the plant computers to a central location is evident. In 1981 Air Products initiated a program to link all of its plant computers to a central computer in Allentown, PA where the central engineering and operating departments are located (Chatterjee, 1981). The central computer initiates a daily phone call to each of the sites to retrieve integrated and instantaneous production, consumption, and efficiency information. This data is stored for on-line ad hoc inquiries and for generating exception reports. Trends in data are observed for each plant and degradation in equipment performance is quickly brought to the attention of line management and the plant efficiency engineer. This same system is offered to customers who purchase optimization systems for their own plants.

Implementation of computer monitoring, control, and optimization improves operating plant efficiency between 5 and 15 percent, typically resulting in less than a two year payback. Techniques such as linear programming and dynamic optimization will produce additional energy efficiency benefits, while artificial intelligence and similar packages have potential to reduce the programming effort and further improve the profitability of computer projects. Knowledge gained in computerizing cryogenic air separation plants can also be applied to new oxygen production technologies to increase the efficiency and ultimately reduce the cost of producing oxygen.

In summary, cryogenic plants are the economic choice for most medium to large scale requirements due to their lower power consumption and economies of scale. LOX for backup and peak shaving is readily produced on-site in the same equipment used to produce gaseous oxygen, eliminating the need for add-on liquefiers or supply contracts. Although cryogenic technology is mature, there is potential for future improvement in design, control and operation.

Technology Summary

Each of the four technologies have merits that make them amenable to certain applications. Although membrane systems are not as mature as adsorption and cryogenic technology, further developments could result in increased use for certain applications. Chemical air separation processes are expected to compete with cryogenic technology in the 1990's. Figure 11 shows the relative power to produce a contained ton of oxygen, normalized for an oxygen supply pressure of 0 psig. Blending with air has been shown to allow comparison of the technologies throughout the purity range. For smaller scale oxygen-users, power cost does not necessarily control product cost, and other factors need to be assessed. Figure 12 shows the capacity range where each technology is economically practical. Membrane information is based on relatively few current applications. On/off production patterns, peaking requirements and backup will all effect the actual cost and relative comparison of the technologies. Table 1 summarizes the relative advantages of the various technologies in terms of cost and operating characteristics.

Liquid oxygen (LOX) has not been addressed separately since it is a form of cryogenic production, but has been included for comparison purposes. The transportablility and the existing network to supply LOX to industrial customers brings the economies of scale of large plants to small users. Its flexibility in terms of variable flowrate, pressure and instantaneous supply are major advantages.

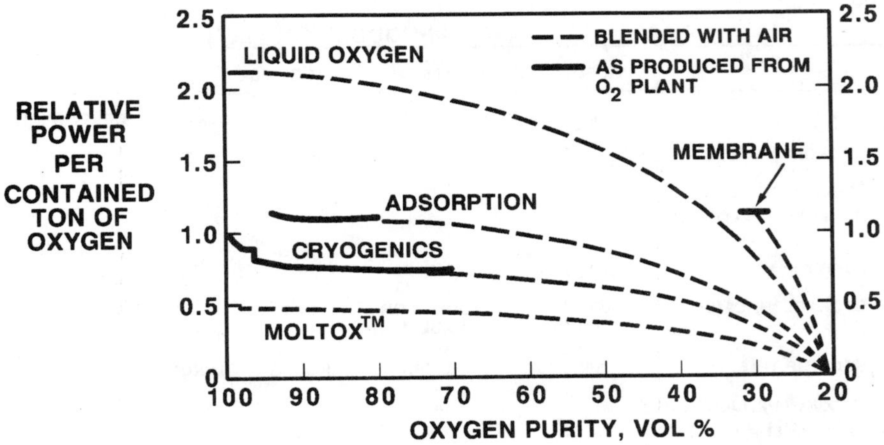

Fig. 11. Oxygen technology comparison at 0 psig.

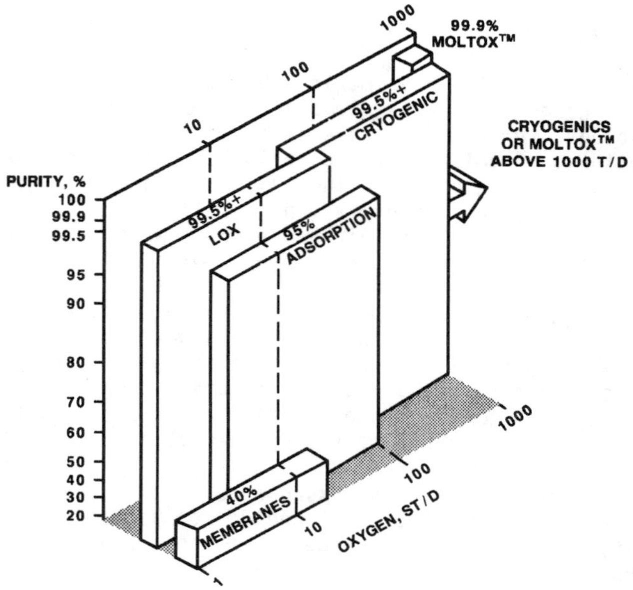

Fig. 12. Typical application ranges.

TABLE 1 Characteristics of Air Separation Technologies

	OXYGEN PRODUCTION TECHNOLOGY				
	ADSORPTION	CHEMICAL	CRYOGENIC	LOX	MEMBRANES
STATUS	SEMI-MATURE	PILOT PLT	MATURE	MATURE	DEVELOPING
ECONOMIC RANGE	<100T/D	>500T/D	>20T/D	<50T/D	<20T/D
SPECIFIC POWER RANKING (1 = LOWEST)	3	1	2	5	4
PURITY LIMIT	95%	99.9%+	99.8%	99.5%+	ca. 40%
STORAGE/BACKUP	ADD-ON OR HAUL IN	ADD-ON OR HAUL IN	STD	STD	ADD-ON OR HAUL IN
STARTUP TIME	MINUTES	HOURS	HOURS	MINUTES	MINUTES
FLEXIBLE PRODUCTION RANKING (1 = HIGHEST)	3	5	4	1	2

OXYGEN APPLICATIONS

Historically, oxygen plant requests have taken the general form of specifying production based on average daily volume and purities of 95% or 99.5% O_2. Current oxygen-using technologies have widely varying requirements for oxygen, and it is useful to know how these requirements will effect the cost of oxygen.

As described in the previous sections, oxygen plants are sensitive to a few key parameters which substantially influence the total cost of the product. The key parameters are:

- Plant Size
- Co-Products (nitrogen and/or argon in addition to oxygen)
- Purity
- Pressure
- Pattern of Use
- Storage Size and Liquid Making Requirements

It is important for the oxygen user to understand the cost impact of variations in each parameter as well as the allowable limits for each oxygen production technology. Slight variations in the parameters can often translate into large operating cost changes. For example, pressure requirements should be fully understood before issuing product specifications. High pressure product availability for convenience can be very expensive due to higher capital investment and higher power (compression) costs. Specifying a higher purity than is required also increases both capital and power costs. For cryogenic plants there is a breakpoint around 96% purity. Below 96% oxygen, a specialized process cycle can be used to substantially reduce the power requirements compared to the standard low pressure cycle. LOX and GOX storage can be beneficial in insuring continuous product supply and meeting transient requirements. When overspecified, however, capital and operating costs can increase significantly.

A useful comparison is obtained when applications are grouped based on oxygen utilization. Four such utilizations are illustrated below:

- Concentrate Combustion - Sulfides
  - Burners
  - Submerged Injection
- Pressure Oxidation
- Oxy-Fuel Burners
- Enrichment

Many of the current non-ferrous smelting technologies fit into one of the above categories with each having different requirements for the key oxygen parameters. Table 2 lists each process category, examples of associated smelting technologies, corresponding oxygen parameters, and oxygen supply modes, which fit the criteria. In addition to the conventional supply modes, future MOLTOX system applications can be evaluated by checking the availability of excess process heat. Continuity is an indication of the variation in flowrate requirements. "High" continuity indicates basically steady-state operation with few shutdowns and only minor peaks and valleys in the oxygen requirement. "Low" continuity processes exhibit the opposite trends and are typical of supply to processes that are cyclical or diurnal in nature. Pressure specifications vary from "Low" for burner enrichment processes (<100 psig) requiring a blower or simple reciprocating compressor,

TABLE 2 Oxygen Applications

O_2 USE	SMELTING PROCESS	EXCESS ENERGY AVAILABLE	CONTINUITY	PRESSURE	PURTIY	PROCESS
CONCENTRATE COMBUSTION						
- BURNERS	FLASH SMELTING - INCO, OUTOKUNPU, FCR	YES	MEDIUM	MEDIUM	70-100%	CRYOGENIC, ADSORPTION, OR CHEMICAL
- SUBMERGED INJECTION	NORANDA CONTINUOUS, QSL	YES	HIGH	HIGH	70-100%	CRYOGENIC, ADSORPTION, OR CHEMICAL
	MITSUBUSHI			MEDIUM		
PRESSURE OXIDATION	ZINC PRESSURE LEACH, GOLD ORE PRESSURE OX.	NO	HIGH	HIGH	HIGH	CRYOGENIC
OXY-FUEL COMBUSTION	CONVENTIONAL REVERB, TBRC	NO	LOW	MEDIUM	90-100%	CRYOGENIC, OR ADSORPTION
ENRICHMENT	CONVENTIONAL REVERB, P-S CONVERTERS, ROASTERS - Zn, Au, BLAST FURNACE - Pb	YES/NO	MEDIUM/ LOW	LOW	30-100%	CHEMICAL ADSORPTION, OR MEMBRANE

to "High" pressure oxidation or combustion processes requiring several stages of centrifugal and/or reciprocating compression up to 1200 psig. Purity requirement is the basic parameter that determines oxygen process selection.

As noted earlier, if the requirement exceeds about 95% purity only LOX or on-site cryogenics can be used. For purities higher than 40% ("enriched air") but below 95%, either adsorption or cryogenic processes can be applied. For enriched air applications membrane based systems can be added to the list of competing technologies. The oxygen production technology decision path is shown in Table 3 based on the following hierarchy: 1) Nitrogen and Argon co-product requirement, 2) purity, 3) size, 4) continuity, 5) storage/backup and 6) product cost - capital + power.

Tables 2 and 3 are useful to guide the oxygen-user in initial screening studies. Since the choice of oxygen parameters impacts both the oxygen-supply and oxygen-consuming processes, a total system cost optimization should be performed. This is best accomplished when the oxygen supplier and customer work closely to examine the exact nature of the process and the interrelationship between the processes.

TABLE 3 Oxygen Technology Screening

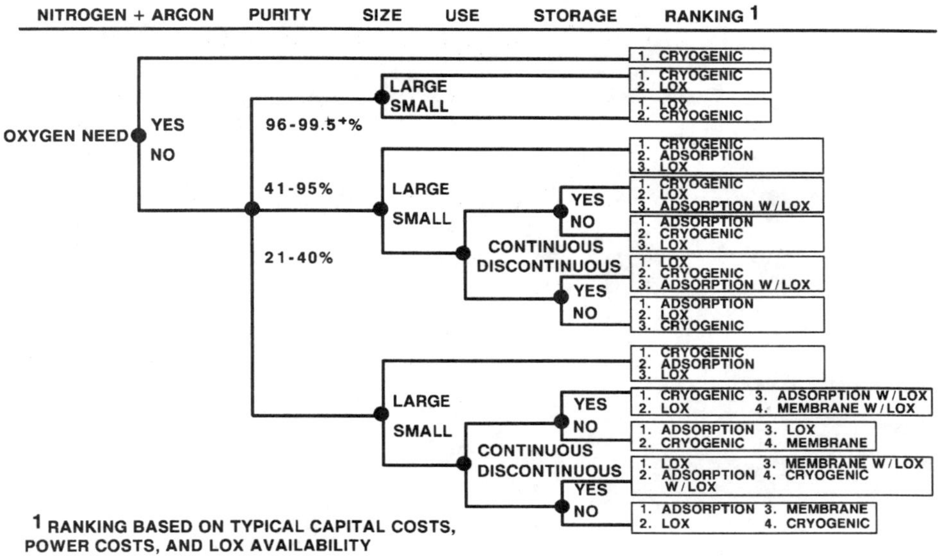

REFERENCES

Anon. (1986). Top 50 chemical production. Chemical and Engineering News, 13.

Bernstein, J. T. (1984). Low-purity oxygen production study. EPRI Report AP-3499.

Chatterjee, N., D. J. Hersh, and J. R. Couch (1981). Total energy management with mini-computer systems. Eighth Energy Technology Conference Proceedings, Washington, DC.

Paffenbarger, J. (1986). Review of oxygen production technologies for IGCC power plants. EPRI Report: RP8000-5.

Patrylak, A. J., W. E. Garber, and B. G. Bryson (1985). Computer control and optimization in an oxygen plant supplying a wastewater treatment facility. Fourth IAWPRC Workshop on Instrumentation and Control of Water and Wastewater Treatment and Transport Systems Proceedings, Houston, TX.

Russek, S. L., T. M. Beckowski, and D. R. Vinson (1985). Computer automated start-up and shutdown of an air separation plant. 1985 Industrial Energy Technology Conference Proceedings, Houston TX.

Schaedel, S. V. (1985). Low-cost oxygen breathes life into industrial furnace. GRID, 8, 10-17, 42, 43.

Wilson, K. B., A. Theobald, and A. R. Smith (1984). Air purification for cryogenic air separation units. IOMA Broadcaster, January - February.

UTILIZATION OF OXYGEN AT THE LA OROYA SMELTER

Hipolito Zevallos
Project Co-ordinator
Empresa Minera del Centro del Peru, S. A.
(CENTROMIN PERU S. A.)

ABSTRACT

Since 1975 Centromin Peru has been planning the modernization of its copper smelter. Several schemes were proposed by foreign consulting engineering firms in the past. Because of the copper price, and the high cost of financing, this project was postponed several times. The recent decision to go ahead with the implementation of oxygen utilization at the copper smelter with a very low investment, will permit a considerable cost reduction and improvement in productivity of the copper circuit. Oil, at the La Oroya smelter, is at present one of the most expensive items (approximately US$ 250/mt) due to the tax imposed on this consumable in the mining and metallurgical industry. Centromin Peru expects, according to preliminary estimates, a reduction of about 40% to 50% of the present oil consumption and improved productivity of the reverberatory and converting operations of 50% to 80%. Therefore, the present production of about 57,000 metric tonnes of blister copper should be attained with only one reverberatory furnace and two equivelent 13' x 30' Pierce Smith Converters.

INTRODUCTION

La Oroya smelter is located in the province of Yauli at an altitide of 3,755 meters above sea level. The copper and lead refineries are located some kilometers from the smelter on the road to Lima, Huancayo and Cerro de Pasco. The annual rainfall average is approximately 500 mm and the temperature varies from 0° to 20° C; the atmospheric pressure is 495 mm Hg. The railway from La Oroya to Cerro de Pasco was completed in 1904. Smelting operations began at La Oroya in 1922 and the first blister copper cake was cast in November of the same year. The smelter was constructed to treat copper ores and concentrates from the surrounding mines. Cobriza Mine did not exist. It was not until 1942 that a pilot plant was constructed for the electrolytic refining of copper. A copper refinery with a capacity of 40,000 tonnes/year was completed in 1948, and was expanded to 57,000 tonnes/year in 1976. In 1957, the width of the copper reverberatory furnace was increased. The installation and commissioning of the ASARCO furnace at the copper refinery took place in 1965. New Cobriza Mine started operation in 1965 at 900 tonnes/day, providing copper concentrates to the smelter. Two years later, in 1967, a copper rod plant was started up. No. 1 reverberatory furnace was fitted with a suspended basic roof in 1971. In 1966, a new waste-heat boiler and air preheater were installed for use in conjunction with the No. 2 copper reverberatory furnace. The original Cottrell system was completed in 1941 and modernization was initiated in 1967 and completed in 1969. Expansion of the copper refinery was put into operation in 1976. On January 1st. 1974, Cerro de Pasco Corporation was nationalized and became Emprosa Minera del Centro del Peru, also known as Centromin Peru. In 1982, the new Cobriza

TABLE 1 La Oroya Copper Smelter Production Data - 1985

	Tonnes	Copper tonnes	Silver ounces	Gold ounces	Bismuth tonnes	Antimony tonnes	Lead tonnes	Arsenic tonnes
1-Centromin ores and concentrates	197,428	46,224	4,256,806	6124				
Purchased ores and concentrates	52,175	11,404	5,463,087	29,751				
Total	249,602	57,628	9,719,893	35,875				
2-Fluxes and pyrites	223,762	239	1,984,497	-				
Miscellaneous	7,294	453	98,197	-				
3-Transfers	11,696	4,327	1,010,608	15				
4-Circulating load	77,841	16,346	3,365,890	1,557				
Total charge treated	570,196	79,084	16,178,995	45,082	744	2,377	14,085	6,570
Analysis		14%	28.4oz/t	0.08oz/t	0.13%	0.4%	2.5%	1.2%
5-Matte produced	210,386	67,324						
Slag produced	321,964							
Blister copper	60,504	59,306	12,425,878	34,248				

1- Includes 123,100 tonnes of clean Cobriza concentrate
2- Includes silica flux, lime and pyrite
3- Materials from anode residue treatment plant
4- De-arsenized copper calcine, dust, clean-up
5- Average matte grade 32%

concentrator came into operation treating 9,100 tonnes/day of ore. The smelting/refining complex at La Oroya treats concentrates and ores produced at the following locations; Cerro de Pasco, Casapalca, Morococha, Yauricocha and Cobriza. Purchased copper concentrates, which used to form about 60% of the feed to the copper smelter, gradually diminished in the last four years to about 20% due to the expanded production from the Cobriza concentrator.

REVERBERATORY FURNACE CHARACTERISTICS

The copper smelter at La Oroya contains two reverberatory furnaces, the approximate dimensions of which are currently 30.5 m long by 8.7 m wide. In 1962 the sprung silica arch on No. 2 reverberatory furnace was replaced with a suspended basic roof. A similar roof was installed on No. 1 furnace in 1971. No. 1 reverb furnace operates with two Stirling waste-heat boilers and No. 2 reverb operates with one Babcock and Wilcox bi-drum waste-heat boiler. A basic roof permits an increase in the intensity of firing which increases furnace throughput due to the increased smelting temperatures which are obtained. The installation of a suspended basic roof on calcine charged furnaces in several smelters has been a significant factor in improving fuel ratios, increasing furnace throughput and, by using hot patching methods for repair, increasing the length of furnace campaigns. In addition, a basic roof, because it does not undergo the phase transformations which occur in a silica arch, can be heated and cooled more rapidly and is less susceptible to spalling caused by thermal shock during interruption of furnace operation. A further advantage with a suspended construction is that it makes more practical a roof contour designed for efficient combustion of fuel and utilization of heat in the smelting zone.

For a reverberatory furnace the relationship between the height and width above the slag line, particularly in the smelting zone, is important to give a balanced distribution of heat between the charge banks and the molten bath. These dimensions determine the cross-sectional area of the furnace available for combustion gases which for a given firing rate, determines the gas velocity. The gas velocity in the furnace was recommended not to exceed about 7 meter/second to avoid excessive dust and heat carryover. Once the combustion volume required for a given firing rate is determined, the ratio of furnace height to width above the slag line must be such that the furnace is wide enough to provide a clear channel in the centre of the furnace for free flow of matte and slag to the settling zone. This ratio is a function of the angle of repose of the charge banks and the physical characteristics of the charge. A height to width ratio which is too large will result in the furnace becoming over-charged and "bridging" of the charge-banks will occur so that converter slag will occasionally flow back into the burners, a condition which occurs once in a while at La Oroya. On the contrary, too small a height to width ratio relative to the furnace throughput will reduce melting capacity and give rise to a molten bath which is too shallow resulting in an inadequate heat sink, poor matte-slag separation and insufficient depth of the matte and slag in the settling zone to permit a satisfactory tapping and skimming operation. The installation of the suspended basic roofs on both reverberatory furnaces at La Oroya was accompanied by raising the roof elevation to increase combustion volume and furnace throughput.

The reverberatory furnaces at La Oroya are fired with heavy fuel oil (bunker C) having a gross calorific value of generally better than 10,000 KCal/kg permitting high flame temperatures. Oil burners are located in a horizontal line at the burner wall opposite the slag tapping end of the furnace. In theory certain prerequisites are needed for high flame temperature: good atomization of the fuel oil and thorough mixing of the oil with combustion air to ensure rapid ignition, good flame propagation, intensive combustion and a short hot flame. Oxy-fuel burners will contribute to fulfill these requirements.

FURNACE CHARGE

The charge to the furnace consists principally of prefluxed hot calcine from copper and arsenic wedge roasters which is delivered in calcine cars to the charge hoppers along the sides of each of the two furnaces at a temperature of 400°C. Miscellaneous new material (high arsenic concentrate) is also charged to the furnace after roasting together with a circulating load which is mainly high copper (24.7%) and lead (5%) slag. Material fed resulting from inter-plant transfer is mainly antimony slag from the anode residue treatment furnace containing 1.3% copper and 99 oz/tonne silver. Recently, the availability of clean copper concentrate has allowed the addition of high arsenic copper concentrate to the smelter to be reduced. This has resulted in a gradual increase in copper grade giving a decrease in S:Cu ratio from almost 2.0 (which was the ratio 5 or 8 years ago) to the present 1.85. This ratio could have gone down further were it not for the silver containing pyrite ore that is fed to the reverberatory furnace after roasting. The iron/insol ratio has remained almost the same at 1.9. Other impurities such as lead, zinc, antimony and arsenic have a general tendancy to decrease.

Concentrates treated by Centromin Peru in 1985 consist of 80% of their own concentrates, including 123,000 tonnes of clean copper concentrate from their Cobriza mine, and 20% purchased concentrates. Some felt that the clean Cobriza concentrate at 25% copper, 32% sulphur and 30% iron, should be fed directly to the reverberatory furnace as a green feed avoiding unnecessary roasting and saving energy. Lime, silica flux and pyrite are also fed. The pyrite, besides adding sulphur to the system, added 1,600,000 oz of silver. Most of this silver - 1,100,000 oz - came from Pyrite San Expedito. The additional iron in this pyrite requires more silica which is the reason why tonnages of flux used and discard slag produced are high.

Centromin's metallurgical staff is confident that the use of oxy-fuel technology will help in implementing green feed smelting especially when clean concentrates are available as is reported by other smelters in the world with subsequent cost savings. For the present, however, the first step will be to use calcine feed alone.

NEW TECHNOLOGY

Important advances to replace conventional methods of copper smelting of 40 years ago have been achieved through changes in equipment and operating procedures. In the past few years Centromin Peru has been reviewing the possibility of either expanding and/or retrofitting its copper smelter with the latest technology to improve its profitability and justify the investment. Unfortunately, due to low copper prices, the results were marginal for a large investment. Present thoughts are now to limit modifications to give operating cost savings and increased productivity. The most effective oxygen utilization system for the traditional reverberatory furnace operation is the oxy-fuel technology. In converting operations, the use of oxygen enrichment has become the best practical method of increasing the smelting of cold materials, also resulting in time saving in the blowing operation. This all has a cost benefit and in general gives a saving in energy consumption.

Oxy-fuel Technology Description

Today, in the pyrometallurgy industry of copper, there are a few smelters such as Onahama in Japan, Inco in Sudbury, Canada, and Caletones in Chile, that operate very successfully with this technology. Centromin Peru is confident that the contribution of oxygen utilization at 4000 meters above sea level is definitely advantageous. According to this technique, industrial oxygen at 85%-95% and fuel oil are used for combustion purposes. However, the oxy-fuel system can be used with other fuels - solids or gases - or oxygen can be used to enrich air in air-fuel burners. The oxy-fuel reverberatory technology allows for partial or total replacement of standard air-fuel long flame burners located at the front wall of the furnaces by higher temperature short flame burners on their roofs. This improvement to the conventional smelting technology results in a better use of consumed fuel. Along the length and over the furnace roof, the oxy-fuel burners are adequately located and set apart. As a result of their number and location, generated heat, higher in comparison to conventional air-fuel systems, is more evenly distributed along the reverberatory furnace combustion chamber with subsequently better fuel utilization by the furnace charge. The use of oxy-fuel burners is a well proven technology and would involve less capital expenditure. There also exists the possibility that in view of the present smelter feed, where 80% of the concentrate treated is a clean concentrate, that the roasting operation could be by-passed in the future.

Two important features of the use of oxy-fuel burners over oxygen enrichment of fuel combustion air are firstly, a major reduction in waste gas volume which is known to be a limiting factor to increasing production rates with the same reverb boilers, and secondly, a small increase in the grade of matte produced to 35%-38% which helps the converting cycle. The reduction in waste gas volume would give significantly improved furnace draft conditions, even with an increase in furnace throughput, and the improvement in matte grade would have a beneficial effect on the capacity of the existing Peirce-Smith converters.

Since the matte grade would be increased by combustion of some sulphur in the furnace charge this would change the furnace atmosphere from that experienced with the conventional firing system. It would be likely that the amount and composition of flue dust carried out of the furnaces would change and the dusts would have a higher lead oxide and zinc oxide level than at present. Bottom build-up is one of the limiting factors to extending the life of the reverberatory furnace at La Oroya under the present operation. We trust that the oxy-fuel burner flames hitting on the molten bath will provide good matte and slag temperatures and formation of magnetite will be controlled to avoid bottom build-up. Oxy-fuel off-gas volume is about 60% lower than in a conventional furnace, and the heat to the waste-heat boilers is consequently lower due to the smaller volume of combustion gases per unit of fuel. The residence time of these gases in the furnace increases, resulting in a more complete heat transfer to the solid charge to be smelted.

When a sufficient number of oxy-fuel burners are installed in the furnace roof, the conventional horizontal burners at the burner wall of the reverb may be completely eliminated. The oxy-fuel burners are arranged in such a way, that the flame impinges directly on the charge. This gives a better thermal transfer because the flame acts by convection and radiation. Furthermore, the temperature difference between the flame and the furnace charge is greater than with a conventional burner operating with ambient air or oxygen enriched air. These two factors, the extra heat transfer by convection, and the higher thermal driving force, result in an increase in the capacity of an existing reverb as well as an efficient use of oxygen and fuel.

CONCLUSIONS

We consider oxy-fuel a very attractive technology to retrofit the reverberatory furnace into a most energy efficient and productive unit. The advantages can be summarized as follows:

- Low capital investment is required, since no major furnace modifications will be needed. The major item is the cost of the oxygen plant.

- Easy and fast implementation of equipment replacement or conversion.

- Increasing productivity of the reverberatory furnace and converters.

- Additional substantial operating cost savings will be possible due to reduction in oil consumption and extended brickwork life.

ACKNOWLEDGEMENT

The author wishes to express his gratitude to the management of Centromin Peru authorizing the publication of this paper.

RECENT ADVANCES IN OXYGEN COMBUSTION TECHNOLOGY

By

Richard A. Browning*, Robert De Wilde** and Louis S. Silver*

*Union Carbide Corporation, Linde Division
Tarrytown, New York

**Union Carbide Benelux, N.V.,
Antwerp, Belgium

ABSTRACT

Productivity increases, improvements in energy efficiency, reductions in flue emissions, and improved process control are important considerations for today's industrial furnace applications. The use of oxygen can play an important role in obtaining these process improvements. Traditional methods of using oxygen for combustion and the results obtained are discussed. A new patented combustion technique, the Linde "A" Burner System, is described which allows the use of up to 100% oxygen for combustion without creating a localized high temperature flame. Both an aluminum remelt furnace and a copper anode furnace have been successfully converted to 100% oxygen combustion using the "A" Burner System.

KEY WORDS

Oxygen; oxygen enrichment; oxygen/fuel burners; oxygen/fuel combustion; fuel savings; production increase; Linde "A" Burner System; aluminum melting, copper anode furnace.

INTRODUCTION

Oxygen has historically been used in industrial combustion processes to increase production and to reduce fuel consumption. When oxygen is used for combustion in place of air in high temperature process furnaces, a greater percentage of energy provided to the furnace is available for useful heating due to a reduction in the volume of combustion products leaving the furnace. In a typical metallurgical heating or melting furnace fired with natural gas and cold air, only about 30% of the heat of combustion is available to provide useful heat to the furnace. The remainder of the heat leaves the furnace with the products of combustion through the flue. Although a portion of this flue loss can be recovered by using a recuperator to preheat combustion air, the net available energy remains at about 50% of the fuel input even with a combustion air preheat temperature of 540°C (1000°F).

By replacing air with oxygen for combustion, the corresponding reduction of nitrogen in the flue gas lowers the sensible heat loss dramatically. As a result, 70% or more of the heat of combustion of the fuel is available energy for the furnace. This higher energy efficiency made possible by using oxygen for combustion results in the higher productivity and fuel savings sought in many metallurgical furnace applications.

Conventional techniques involve replacing only a portion of the combustion air with oxygen. Such enrichment techniques are often limited to 30% oxygen in the combustion oxidant or less. Factors contributing to this limitation include high flame temperatures and lowered furnace gas circulations which contribute to temperature uniformity problems and high NO_x emissions.

Despite these limitations, careful application of conventional enrichment techniques have been very successful. These techniques and the results obtained in various applications are discussed in the following sections.

A new technique, the patented Linde "A" Burner System was specifically designed to overcome the characteristic limitations of oxygen combustion. As a result, the "A" Burner System, utilizing 100% oxygen for combustion, has been successfully installed and operated in a number of process furnaces including ferrous and non-ferrous applications. The principles involved in the "A" Burner and recent non-ferrous installations are discussed in the following sections.

INDUSTRIAL OXYGEN COMBUSTION TECHNIQUES

In practice, oxygen is applied to industrial furnace applications primarily in two ways; the first way is to enrich the combustion air by oxygen, the second way is to use oxygen/fuel burners which can be used in addition to the existing burners or to replace some of the existing burners. The choice of the technique depends on different parameters; degree of energy savings or production increase desired, the construction of the furnace, the material to be treated, etc.

Enrichment of the combustion air

Oxygen is added to combustion air to obtain an oxygen-air mixture which contains more than 21% oxygen. Oxygen enrichment is applied to reduce the volume of combustion air and/or to obtain a production increase in furnaces. Figure 1 indicates that the most significant increases in flame temperature are obtained in the range of 21 to 30% oxygen. Therefore, the first additions of oxygen will give the greatest effects towards the increased production. In practice the oxygen enrichment is limited to 30% oxygen in the air-oxygen mixture. The enrichment process can be utilized with only minor changes to the existing equipment; an oxygen flow control panel and the installation of an oxygen sparger in the combustion air pipe. The process can hence be installed with a minor investment cost. From an operator's viewpoint, oxygen enrichment is easy to apply; only the air flow has to be adapted to the process. This also makes it easy to go back to the original setpoints of the operation as when production increase is no longer needed.

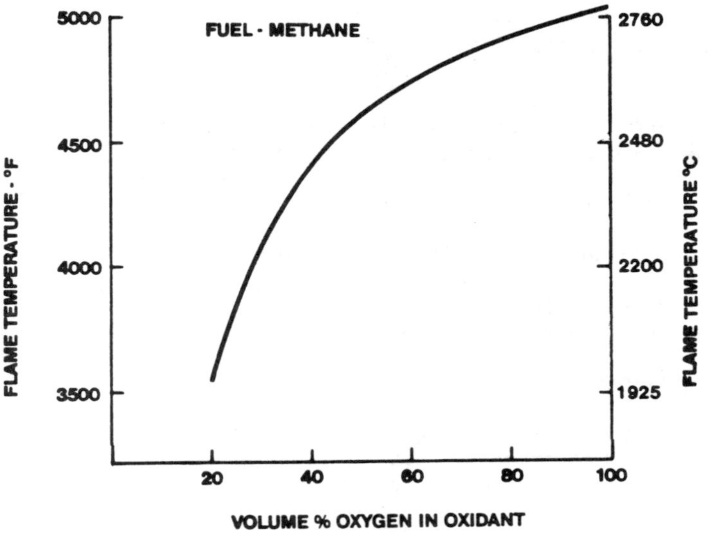

Fig. 1. Flame temperature as a function of oxygen concentration in the oxidant

Oxygen Burners

Union Carbide Corporation has developed oxygen/fuel burners for different types of fuels including gas, light oil and heavy oil. The characteristics of the burners (flame pattern and length, heating capacity) are defined for each application. The flame temperatures of these burners are characteristically high to release heat to specific locations for optimized productivity increases. The burners can be applied in two different ways:

To save energy: When oxygen can replace the combustion air on an economical basis, the maximum benefit can be reached when replacing the air completely. Usually the furnace construction (creation of hot spots) or the process (excessive oxidation of the charge) puts a limit on the amount of air replacement. Generally, conventional oxygen/fuel burners are not practical for high levels of furnace conversion due to high flame temperatures and low levels of gas circulations in the furnace. These effects tend to create furnace hot spots as well as charge oxidation.

To obtain a production increase: The increased flame temperature obtained with the oxygen burner is used to create an improved heat transfer in a local area of the furnace best suited to increase production. The application of auxiliary burners is the preferred method of oxygen enrichment for installations where there is a well defined heating zone.

PRACTICAL APPLICATIONS OF INDUSTRIAL TECHNIQUES

Table 1 gives an overview of the applications in industry. Nearly each high temperature process and each type of furnace can be considered for oxygen applications. Practical results are summarized in Table 2.

TABLE 1: REVIEW OF OXYGEN PROCESSES

Iron and Steel:

Oxygen burners in electric arc furnaces
Enrichment and direct injection of oxygen in cupolas.

Non-Ferrous Metals:

Enrichment in secondary melting of aluminum
Oxygen burners in lead industry
Oxygen burners in copper industry
Oxygen burners in antimony industry

Glass and Silicates:

Auxiliary oxygen burners in glass furnaces
Oxygen burners in enamel furnaces
Undershot oxygen lancing in glass furnaces.

Others:

Enrichment of rockwool cupolas
Enrichment of calcining furnaces

TABLE 2: PRACTICAL RESULTS OBTAINED WITH OXYGEN

PROCESS	OXYGEN APPLICATION	ENERGY SAVING, %	PRODUCTION INCREASE, %
AL Melting	3.5% Enrichment	38.1	47.4
	2% Enrichment	16.6	not wanted
Glass Melting	Auxiliary Burner	9.3	26.2
Enamel Furnace	Oxygen Burner	43.5	50
Lead Recuperation	Oxygen Burner	49.4	not wanted

Secondary Aluminum Melting

Aluminum ingots, used to produce rolled or extruded products, are produced by melting of primary aluminum or scrap. This melting is mostly done in a batch furnace in which the flame is in direct contact with the charge. The casting temperature is about 730°C (1350°F). The temperature of combustion products leaving the furnace vary between 730°C (1350°F) at the start of melting and 1180°C (2160°F) at the end of the cycle.

In order to obtain an increase in the melting rate and a decrease in the specific energy consumption, oxygen enrichment of the combustion air is applied.

In a 13 mT reverberatory furnace, the aluminum was melted using 4900 MJ/mT (4.2 MMBTU/T). By enriching the air stream by 3.5% (24.45% oxygen) the melting rate was increased by 47.4% which represented a total production rate increase of 28.7%. In the same furnace oxygen enrichment was also applied when no production increase was needed. The enrichment level was reduced to 2% in this case. The lower enrichment level and the fact that no production increase was wanted reduced the energy savings to only 16.6%, whereas 38.1% savings were obtained in the first case with the production increase.

Recuperation of Lead

Used car batteries contain lead as oxide and sulphate. This lead can be recuperated in a rotating furnace heated to 980 to 1130°C (1800 to 2070°F).

The air burner in a 10 mT furnace was replaced by an oil/oxygen burner resulting in an energy savings of nearly 50%. The cycle time was kept constant. Reduction of the off gases by more than 80% other advantages in the performance and maintenance of the filter installation.

RECENT ADVANCE IN COMBUSTION TECHNOLOGY: THE LINDE "A" BURNER SYSTEM

The Linde "A" Burner System has been designed to overcome the following undesirable consequences that are characteristic of high level oxygen enrichment in industrial furnaces:

1. The flame temperature increases markedly as the oxygen concentration in the oxidant increases. The high flame temperature can be undesirable in two respects:

    a) The heat transfer rate by radiation and conduction can be unusually high in the localized region around the flame. This can lead to hot spots causing damage to the furnace refractory and to uneven temperature distribution in the charge being heated.

    b) Nitrogen oxide emissions can increase. Reaction kinetics and equilibria for the formation of nitrogen oxides are favored by high temperatures. This is a problem even when using pure oxygen since there is usually sufficient nitrogen in the fuel and through air leakage into the furnace to form the nitrogen oxides.

2. The reduced volume of the oxidant per unit of fuel lowers the momentum of the flame. This, in turn, reduces the amount of mixing and recirculation of the gases within the furnace. Good gas circulation in the furnace is usually desired to obtain uniform heating of the charge and to prevent localized hot spots.

In order to overcome these disadvantages, the Linde "A" Burner was developed. The burner is described below using the operation of a conventional burner as a reference point.

In the case of a conventional air burner in a furnace, the fuel and air mix and burn close to the burner face or within a burner block. By varying the mixing pattern, a long or short flame can be achieved. The flame temperature is slightly below the theoretical adiabatic flame temperature of 1925°C (3500°F). The combustion products circulate through the furnace and leave through the flue opening. As a result of the gas recirculation, the temperature and heat transfer distributions within the furnace are quite uniform. If this type of burner were used with 100% oxygen, the flame temperature would be very high [about 2760°C (5000°F)] and the gas recirculation would be greatly reduced because of the low flow rate of oxygen as compared to air. The temperature and heat transfer distributions within the furnace would not be uniform. A key feature of the "A" Burner is that the furnace gases are aspirated into the oxidant jets <u>prior</u> to mixing with the fuel. A sketch of the "A" Burner is given in Fig. 2. In the following discussion, the operation of the "A" Burner is examined with respect to flame temperature, furnace gas recirculation, flame stability, and operating flexibility. For all of the examples and test results, the oxidant is 100% oxygen. However, it should be kept in mind that oxygen enriched air could also be used.

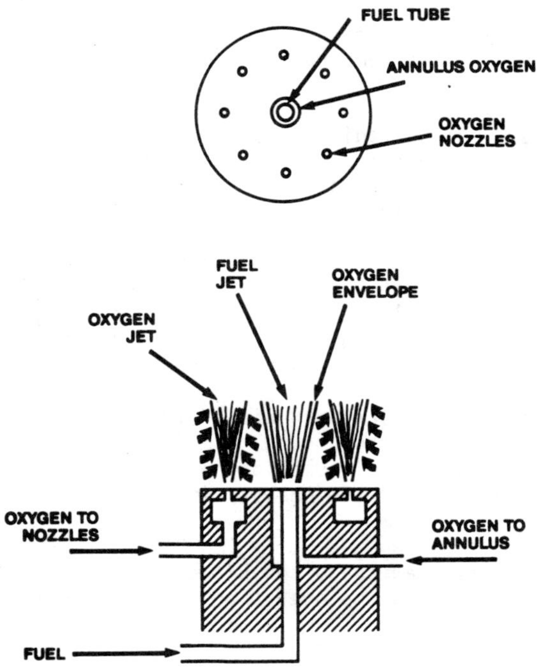

Fig. 2. Schematic sketch of the "A" Burner

Referring to Fig. 2, fuel gas is supplied at the burner axis as a relatively low velocity jet. Most of the oxygen (90-95%) is supplied as a ring of high velocity jets surrounding the center fuel stream. The purpose of the oxygen annulus (containing the remaining 5-10% oxygen) around the fuel stream is to stabilize the flame by initiating a low degree of combustion directly at the burner face. The aspiration of furnace gases into the oxygen jets prior to mixing with the fuel is indicated in Fig. 2 by the small arrows. By maintaining sufficient distance between the oxygen nozzles and fuel supply tube, enough of the furnace gases can be aspirated into the oxygen jet prior to mixing with the fuel so that the resulting flame temperature can be reduced to a value substantially below the theoretical flame temperature. At the point of combustion, both conventional air burners and the "A" Burner using 100% oxygen have an oxidant with a low concentration of oxygen. For the air burner, the diluent is nitrogen while for the "A" Burner using oxygen, the furnaces gases make up the diluent. A simplified schematic representation of the process steps for a conventional burner and the "A" Burner is given in Fig. 3.

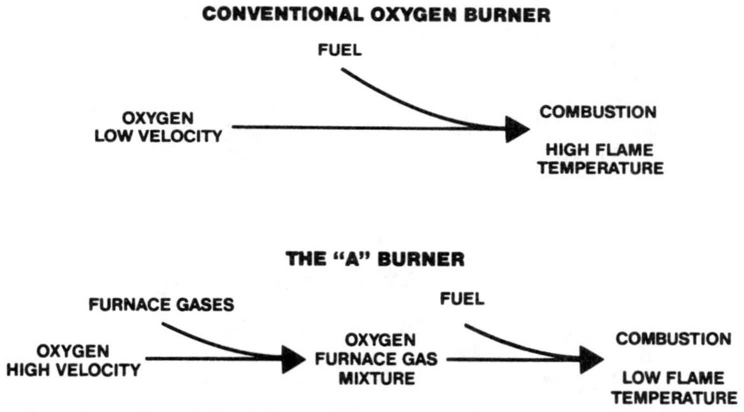

Fig. 3. Simplified burner process diagrams

An estimate of the flame temperature can be made using a simplified model of a jet. It is assumed that a given amount of furnace gas is mixed uniformly with the oxidant jet. Then the oxidant-furnace gas mixture is blended with the fuel and burned. Obviously, the actual operation is much more complex in practice considering the concentration gradients and fluctuations within a jet as well as the interaction between adjacent jets. However, for the purpose of explaining the essential features of the "A" Burner, the simplified model is useful.

For this simple case, the flame temperature would depend upon 1) the amount of furnace gas aspirated into the oxidant prior to mixing with the fuel jet, 2) the furnace gas temperature, and 3) the oxygen concentration in the oxidant. The amount of furnace gas aspirated into the oxidant jet can be designated by the recirculation ratio, R. This recirculation ratio is defined as the moles of furnace gas aspirated into the oxidant jet divided by the moles of oxidant. In Fig. 4, the calculated flame temperature with methane as the fuel is plotted versus the oxygen concentration in the oxidant for an average furnace gas temperature of 1315°C (2400°F) and for recirculation ratios of 0 to 6. When the recirculation ratio is zero (no aspiration of furnace gas into the oxidant jet prior to mixing with the fuel), the flame temperature is the theoretical adiabatic flame temperature. As furnace gas is aspirated into the jet (increasing values of R) the flame temperature decreases. As an example, when using 100% oxygen as the oxidant with an average furnace gas temperature of 1315°C (2400°F), the flame temperature decreases from 2780°C (5040°F) to 1945°C (3530°F) as the recirculation ratio increases from 0 to 6.

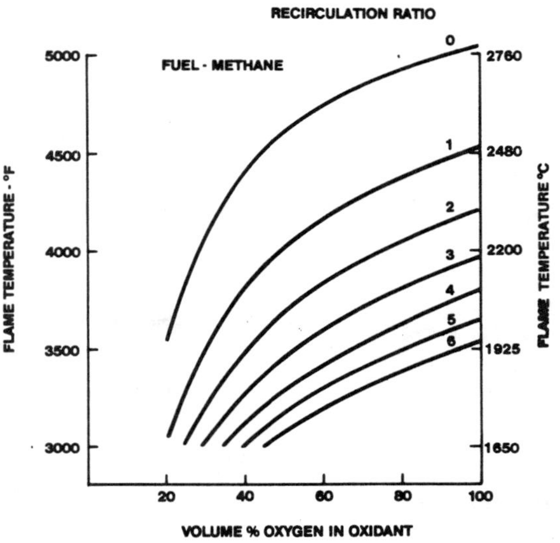

Fig. 4. Flame temperature as a function of oxygen concentration in the oxidant and recirculation ratio

Gas mixing and recirculation within the furnace are accomplished with the "A" Burner by using very high velocity oxygen jets. The velocity of the oxygen jets of the "A" Burner are generally 5 to 20 times the typical velocity of the air in a conventional air burner. The high momentum for an air burner is due to the high mass rate and moderate gas velocity. About the same momentum is achieved using the "A" Burner with a low mass rate and high gas velocity. As a result, the gas mixing and recirculation required to obtain uniform furnace temperature are achieved with the "A" Burner using 100% oxygen.

The "A" Burner has been designed with removable oxygen nozzles in the burner face. Each nozzle can be changed easily to alter the hole diameter and angle with corresponding changes in the jet velocity and angle. This feature of the "A" Burner has provided flexibility that has proven to be very useful for industrial applications. The angle and direction of the oxygen jets determine where the heat release by combustion will occur in the furnace. If a portion of the furnace is cool due to unusual heat losses (such as at a charging door), some of the oxygen jets can be directed to that area to compensate accordingly. In this way the "A" Burner can be tailored on site to fit the specific needs of a furnace. Within a few days of operation after installation of the burners, the desired temperature distribution in an industrial furnace is achieved by making adjustments in the nozzles.

Reaction kinetics and equilibria for forming nitrogen oxides are favored by high temperatures. The replacement of air with oxygen normally results in higher flame temperature with a corresponding increase in the concentration of NO_x providing nitrogen is present. With the "A" Burner, measured NO_x levels have been very low, indicating that the flame temperatures were low. Using natural gas as a fuel, emissions of thermal NO_x as low as 0.9 mg/MJ (0.002 lbs/10^6 BTU) were obtained with the "A" Burner in an experimental furnace at a flue gas temperature of 1540°C (2800°F).

In summary, the "A" Burner has been developed to use up to 100% oxygen in industrial furnaces. The key features of the burner are:

1) Low flame temperature
2) Low NO_x
3) Uniform temperature distribution in the furnace
4) Flexibility to change the heating patterns within the furnace as desired

The "A" Burner System usually consists of patented burners, oxygen and fuel flow control piping skids and a control console. The control console utilizes a programmable controller to optimally integrate all system components into a flexible combustion package while providing for easy operator interfacing and safety interlocking.

RESULTS FROM "A" BURNER APPLICATIONS

Overview

The Linde "A" Burner System has been demonstrated in various ferrous as well as non-ferrous heating and melting furnace applications. Ferrous applications have included soaking pits, batch reheat furnaces and continuous furnaces. Fuel savings in these applications have ranged between 44% for a batch reheat furnace originally with 540°C (1000°F) air preheat temperature to 60% for a batch reheat furnace originally fired with cold air. Temperature distributions were at least as uniform as the air fired practice, often with fewer "A" Burners installed than original air/fuel burners. Many of these results have been reported elsewhere (Browning, 1985).

The following paragraphs describe the recent applications of the "A" Burner System in non-ferrous applications.

"A" Burner Application in the Aluminum Industry

A demonstration of the Linde "A" Burner System was started in August 1986 in an aluminum remelt furnace of VAW Reinstmetall (pure metal), Grevenbroich, Germany. Aluminum was melted in a 30 mT reverberatory furnace by an air/heavy fuel oil burner installed at the furnace end opposite the charging door. The off gases leave the furnace through the stack near the charging door. In the air practice, a typical cycle consisted of charging, melting, overheating, treating, holding and casting. Because of the furnace and manpower planning, the holding period could be more than 16 hours, which made the total energy consumption larger than the energy needed for melting and heating the charge, as is indicated in Table 3 which represents the average data of a series of melts.

TABLE 3: ALUMINUM REVERB FURNACE: AIR PRACTICE PERFORMANCE		
Charge Weight	25992 kg	
Melting Rate	0.952 mT/h	
Oil Consumption	liter	liter/mT
Melting	2807	108.8
Total	4443	170.9

For the evaluation period, a 4" "A" Burner was used with a heating capacity of 200 Nm^3/hour of natural gas at a supply pressure of 500 m bar. The control panel installed allowed the manual or automatic regulation of the gas flow. The oxygen flow was automatically regulated using the gas flow signal and a ratio control station. The installation of the control panels, the piping and wiring was done while the furnace was still operating with the air/fuel burner. Switching of the burners occurred during the normal week-end shutdown so that normal operations could start again on Monday without any production loss.

Applying the "A" Burner, the first three heats were done at the same rate as was applicable with the air burner. Afterwards an increased melting rate was applied so that the total cycle took only 24 hours. In those cases the firing rate of the burner was regulated by means of the off gas temperature. The typical flow pattern obtained was as follows:

o With solid material in the furnace the burner was fired at full rate.

- o Once about 3/4 of the charge was molten, the firing rate was dropped to 80% of maximum capacity.
- o When the aluminum bath was flat, the natural gas flow was decreased slowly to 40%.
- o During holding and during casting periods, only 10% of the maximum flow was needed.

The burner maintained a stable flame throughout this flow range.

During the evaluation period which consisted of 25 heats, data on energy consumption and cycle times were monitored. The average values are summarized in Table 4.

TABLE 4: ALUMINUM REVERB FURNACE: "A" BURNER PRACTICE PERFORMANCE

Charge Weight	24092 kg	
Melting Rate	2.184 mT/h	
Gas Consumption	Nm^3	Nm^3/mT
Melting	1529.8	63.30
Total	1793.6	74.45

Energy savings: The expectations on energy savings were largely fulfilled. Referring to the air data, an overall savings of 61% was obtained, as can be seen in Table 5. The energy savings for just the charging, melting and heating periods averaged 48%. The increase to 61% for the total cycle is mainly due to a reduction in the holding time. One should notice, however, that the calculated energy savings are not a straightforward comparison; we compared oil performances with natural gas. Experience has shown that such a switch in fuels normally involves an increase in energy consumption of about 5%, when all other parameters remain constant. The obtained energy savings of 48% and 61% would have been higher if the same type of fuel was used.

Melting rate: A considerable increase in the melting rate was obtained by using the "A" Burner System; from 0.952 mT/h, the melting rate went up to 2.184 mT/h which represents on increase of 129.4%. Three factors influenced this increase. First, there is the improved heat exchange caused by the increased amount of CO_2 and H_2O in the offgas. Secondly, the reduction of the offgas volume per unit of available heat to the furnace allowed more fuel to be used with oxygen. In the air practice, the heating rate was limited because of the offgas volume and the existing stack. The last reason for the production increase is due to the flexibility of the system which made it possible to obtain a better man power utilization.

TABLE 5: ALUMINUM REVERB FURNACE: ENERGY CONSUMPTIONS AND SAVINGS		
	AIR	OXYGEN
Fuel Consumption (Total)	170.9 1/mT	74.45 Nm^3/mT
Energy Cons. (MJ/mT)		
Melting	3980	2070
Total	6250	2430
Energy Savings (%)		
Melting	47.9	
Total	61.1	

Dross formation: The measurement of dross formation remains a difficult item. Too many factors can influence the data, for example, the way the refining of the melt is done or the manner in which the furnace is charged. Data taken during the air practice as well those taken during the evaluation period vary a lot among themselves but no difference can be seen between the air and oxygen data.

Offgas composition: Spot analysis of dust and NO_x emissions were done on the off gases during both practices. Although the concentrations are increased by using the "A" Burner System, the hourly and the specific emission decreased tremendously, because of the increased production rate and the decrease in offgas volume.

Furnace construction and aluminum quality: During all heats the aluminum quality was checked by analysis and the furnace refractory was inspected visually. No negative effects could be noticed during the evaluation periods on either parameter.

Conclusion: This evaluation period has shown that the "A" Burner System is a valid solution for heating and melting aluminum in reverberatory furnaces. Technically, the "A" Burner System, because of its use of oxygen, allows for energy savings and production increases while avoiding quality and furnace deterioration and excess dross formation because of its low flame temperature. The energy savings and the production increase will lead to direct economic savings. The reduction of dust and NO_x emissions will reduce or eliminate the investment costs for new filter installations or reduce the operating and maintenance costs for existing filter installations.

Kennecott, Copper Anode Furnace

A demonstration of the patented Linde "A" Burner System was conducted during March of 1985 in the No. 2 Anode Furnace of the Utah Copper Division of Kennecott Minerals Company. Prior to this, a measurement program was conducted to record the base operating parameters of the furnace. The combustion system in this furnace was used only to hold temperature - all heat input to melt charged scrap and all refining was accomplished with the patented Linde "SMART" System that had been installed in September 1984 as part of a separate test program. As a result, the combustion system only makes up for furnace heat losses such as conduction losses through the walls and radiation losses through the furnace opening.

The installation of the "A" Burner System for the demonstration was accomplished very simply with all control piping and an electrical control console integrally mounted on a single skid. The furnace was operated normally with the air/fuel system during most of the preparations. The furnace was shut down only during the burner changeover of 1-2 hours immediately preceding the "A" Burner System start-up.

The "A" Burner System operated for the next 48 hours without any unplanned shutdowns. Initially, the furnace was empty of any blister copper charge, the previous charge having been cast prior to the changeover of the burners. The "A" Burner was operated in this empty furnace for the next 22 hours. During this time optical pyrometer readings were taken of six points on the inside furnace wall. Based on these measurements, the oxygen and fuel flows were adjusted. Steady state firing rate for the empty anode furnace was about 5490 MJ/hr (5.2 MMBTU/hr). The furnace temperature distribution as measured by the optical pyrometer was very uniform. End to end temperature difference in the furnace averaged only 5.5°C (10°F).

After about 22 hours, the furnace was charged with copper. The total charge consisted of 75% blister copper poured into the anode furnace charging mouth from ladles and 25% scrap copper. The scrap copper was melted and the total charge refined by the Linde SMART System. Normal anode furnace operation does not call for the use of the burners during the refining step so the "A" Burner System was not utilized for several hours.

After refining, the charge was held at its normal pour temperature by the "A" Burner System. The temperature of the bath was read by consumable probes over a period of 12 hours. Also monitored in this manner was the oxygen concentration of the bath. Kennecott was satisfied with both the temperature and the oxygen concentration of the bath during the time when the "A" Burner System was operating.

The steady state "A" Burner firing rate was about 4960 MJ/hr (4.7 MMBTU/hr) with the charge in the furnace, slightly lower than the case with no charge. This difference is due to the higher conductive heat losses from the furnace when it is empty.

This short demonstration of the "A" Burner in the anode furnace yielded successful results. Furnace temperatures were maintained and evenly distributed. The copper bath temperature and oxygen content were held at normal levels. A fuel savings of 64% was achieved. Also, the "A" Burner System operated in a safe and predictable manner throughout the demonstration period.

In summary, the results of this demonstration have shown the "A" Burner to be well-suited for application in an anode furnace. Table 6 summarizes these results.

TABLE 6: COPPER ANODE FURNACE: RESULTS WITH OXYGEN

	WITHOUT CHARGE	WITH CHARGE	(PREDICTION) WITH CHARGE & MOUTH COVER
Steady State Fuel Consumption, MJ/hr	5490	4960	2850
Savings Relative to Air/Fuel Case With Charge, %	-	64	80

Although not demonstrated during the test program, the significant reduction of flue gas volume, unique with combustion using 100% oxygen, allows for placement of a cover over the furnace mouth. The size of the furnace mouth is a key variable since the radiation heat loss through the mouth is a large fraction (60%) of the total furnace heat loss. Normally, a cover cannot be used as the large quantity of flue gases produced by combustion with air requires the large opening to exit the furnace. By reducing this mouth area with a cover during operation of the "A" Burner System, the steady state firing rate with a full furnace would be reduced to 2850 MJ/hr (2.7 MMBTU/hr). This corresponds to a fuel savings of 80% relative to the base case with air combustion. These results are also indicated in Table 6.

SUMMARY

The use of oxygen for combustion in process furnaces provides for process improvements such as higher productivity, fuel savings, reduced flue gas emissions and improved process control. A recently developed combustion technique, the Linde "A" Burner System, can be applied to overcome the traditional limitations associated with the use of oxygen in process furnaces. The "A" Burner System has been successfully applied in applications using 100% oxygen in non-ferrous industries where it achieved substantial fuel savings, productivity increases and other benefits.

ACKNOWLEDGEMENTS

The authors would like to express appreciation to the personnel of VAW and Kennecott Minerals Company for their valuable assistance and cooperation to make the demonstration programs successful.

This paper is based in part on an April 1986 Technical Paper authored by J. E. Anderson of Union Carbide Corporation, Linde Division, which was presented in Chicago, Illinois at the 1986 Symposium on Industrial Combustion Technologies.

REFERENCES

Anderson, J. E., U.S. Patent 4,378,205 (March 29, 1983), Oxygen Aspirator Burner Process for Firing a Furnace.

Anderson, J. E., U.S. Patent 4,541,796 (September 17, 1985), Oxygen Aspirator Burner for Firing a Furnace.

Anderson, J. E., A Low NO_x, Low Temperature Oxygen Fuel Burner, 1986 Symposium on Industrial Combustion Technologies, Chicago, IL (April 29-30, 1986).

Browning, R. A., The Linde "A" Burner System: Demonstrated Fuel Savings and Even Heating With 100% Oxygen, AISE Annual Convention, Pittsburgh, PA (September 23-26, 1986).

Masterson, I. F., U.S. Patent 4,657,586 (April 14, 1987), Submerged Combustion in Molten Materials.

OXYGEN USAGE IN THE KIDD CREEK SMELTER

C. J. Newman, Manager, Copper Operations
G. Macfarlane, Smelter Superintendent
K. Molnar, Senior Process Engineer

Kidd Creek Mines Ltd.
P.O. Bag 2002, Timmins, Ontario, Canada P4N 7K1

ABSTRACT

The Kidd Creek smelter is currently being expanded from its original design of 59,000 tonnes/year of copper to 100,000 tonnes/year. This is being achieved by the use of additional oxygen in the existing furnaces, allowing an increase in smelter capacity while maintaining the same gas flowrate through the original waste-heat boilers. This, in turn, has allowed the acid plant capacity to be increased by making relatively minor modifications. At current rates, excess heat is generated in both the smelting and converting furnaces. Present emphasis is on making use of this excess heat to treat low grade inert materials in the smelting furnace; use the maximum quantity of low silica gold ores that would otherwise give poor milling recoveries; feed a coarser flux; feed undried materials; recycle tankhouse scrap and other coarse materials to the converting furnace. The conversion has been made to the use of oxy-gas burners to maintain furnace temperatures when not feeding concentrates. This makes use of available oxygen and maintains a better bath melt condition than the original roof mounted oil burners. No fuel is burnt in the furnaces during normal operation. Present limitations and future potential for smelter capacity are discussed.

KEYWORDS

Oxygen, Mitsubishi Continuous Smelting Process, Copper, Smelting, Converting.

INTRODUCTION

The Kidd Creek Mines Ltd. metallurgical complex is located 24 km east of Timmins in Northern Ontario. Operations at this site commenced in 1966 with the construction of a 9,000 tonnes/day concentrator. Over the intervening years, the metallurgical facilities have been expanded and developed such that the concentrator now treats 13,000 tonnes/day of ore providing feed for a 134,000 tonnes/year zinc plant and a copper smelter and refinery.

Copper operations commenced in mid 1981 with the commissioning of the copper smelter and electrolytic refinery. The smelter, which utilizes the Mitsubishi Continuous Smelting Process to effect the continuous production of blister copper, has been comprehensively described in previous papers (1-7). From the initial design capacity of 59,000 tonnes/year of blister copper, the smelter has been expanded to a capability of 90,000 tonnes/year and is presently undergoing modifications which will

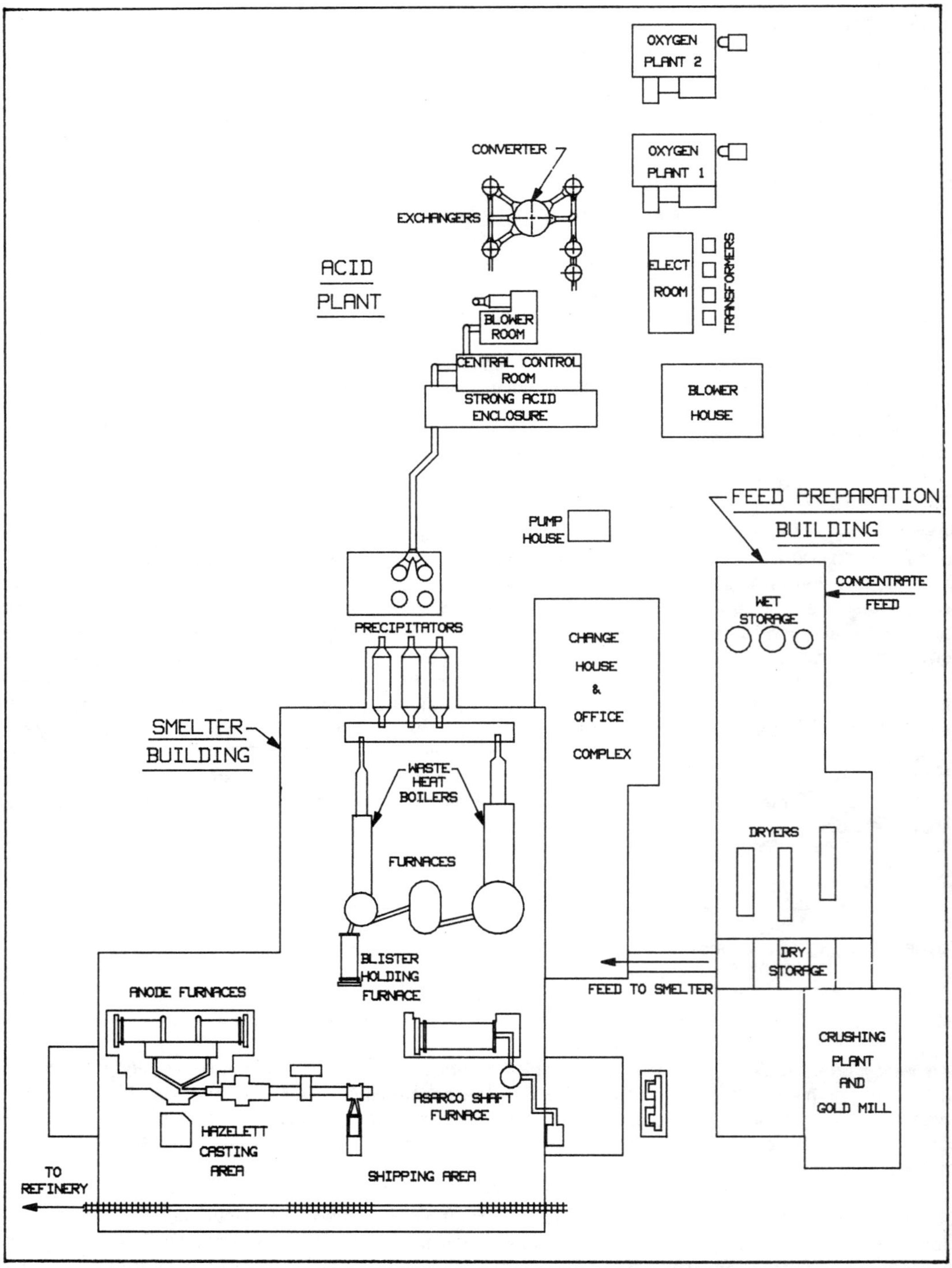

Fig. 1 Plan View of the Kidd Copper Smelter

allow a production rate of 100,000 tonnes/year.

Commencing with a brief general plant description, this paper will show how the increasing use of oxygen has played a fundamental role in the expansion of the smelter, the advantages of oxygen enrichment in this process and in ancilliary areas and latterly, discuss potential future developments.

PLANT DESCRIPTION

A schematic outline of the copper smelter is shown in Figure 1.

Furnace Feed

Smelting furnace feed consists of a mixture of dried copper concentrate, recycled converting furnace slag and silica flux. Copper concentrate is dried to less than 0.5% moisture in a rotary/flash dryer and slag is dried in a rotary dryer. The initial source of flux for the smelting furnace was tailings from the concentrator. In 1986, an existing 10 tonnes/hour crushing plant was expanded to service a newly built gold mill (8) and allow the treatment of gold bearing siliceous ores in the smelting furnace as flux. Treating these ores directly as a flux gives a greater gold recovery at a lower cost than alternative processing methods. These feed materials are each stored separately, individually weighed onto a common drag conveyor, and then transported by bucket elevators and drag conveyors to the smelting furnace charging system. Other materials treated at times are gold concentrates, smelter reverts, and a number of zinc plant residues. Typical analyses of the principal feed materials are shown in Table 1.

TABLE 1 Feed Materials to the Smelting Furnace (Wt %)

	Cu	Fe	S	Pb	Zn	CaO
Copper concentrate	26	28	31	0.6	4	-
Recycle slag	15	40	0.4	0.8	1.5	18
	SiO2	Al2O3	CaO	MgO	S	Au (g/Mg)
Flux from tailings	70	6	1	1.5	0.5	-
Gold ore flux	80	7	1	1	0.3	6

The Continuous Smelting Process

The 10.3 metre diameter smelting furnace is charged pneumatically through ten 38 mm diameter feed pipes which are each contained within 76 mm diameter steel process lances. Oxygen enriched air is introduced to the furnace via these consumable lances which are located approximately 500 mm above the molten bath. The melt, consisting of matte and slag, flows down a refractory lined launder to a 3 MVA slag cleaning furnace. Slag continuously overflows and is water granulated before being conveyed to a storage bin. Matte is syphoned from the furnace and allowed to flow, again by gravity, into the 8.2 metre diameter converting furnace. Oxygen enriched air is blown into the converting furnace through 6 process lances together with limestone as a flux. The slag formed is water granulated and conveyed to the feed preparation area for drying. It is then fed either to the smelting furnace or as a coolant to the converting furnace. Blister copper produced is continuously syphoned from the furnace and flows to a 250 tonne holding furnace. From here it is transferred in 18 tonne ladles by overhead crane to one of two 400 tonne anode furnaces. After fire-refining, the metal is cast into a continuous strip, 17.8 mm thick. Anodes are blanked from this strip and transported to the Kidd copper refinery.

An analysis of each of the furnace products is shown in Table 2 and a plan of the furnaces in the continuous smelting section in Figure 2.

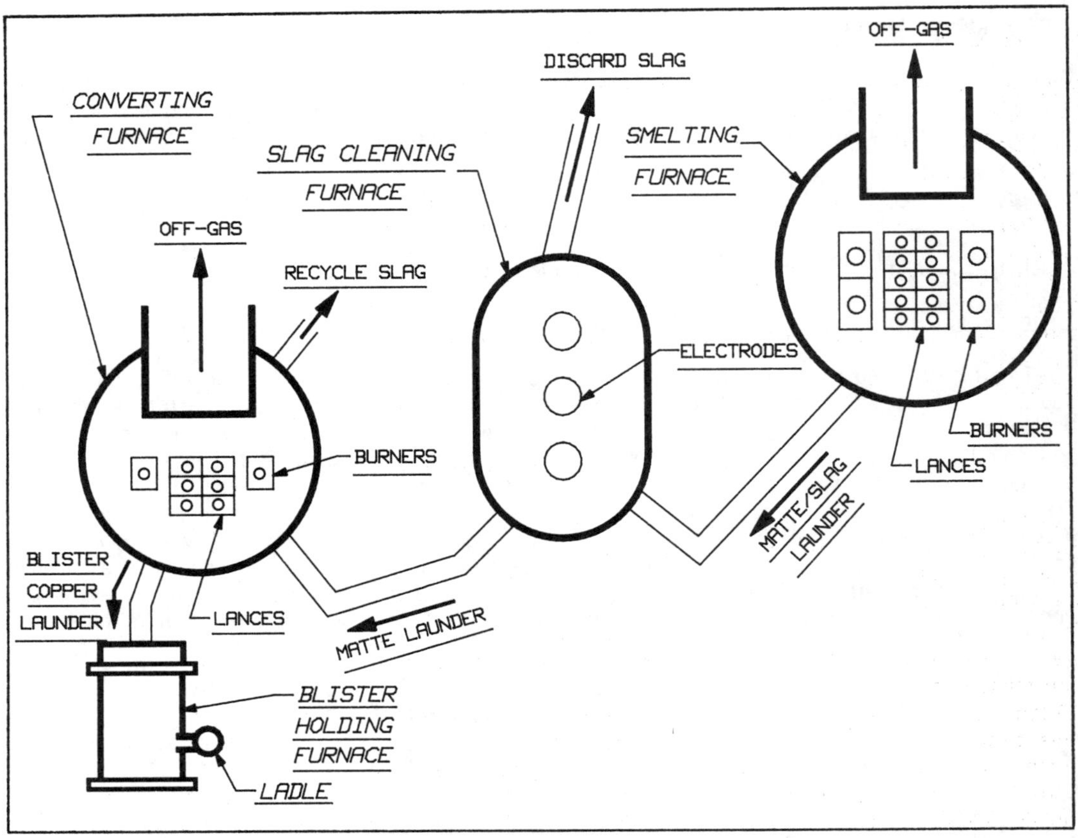

Fig. 2 Plan of the Continuous Smelting Section

Treatment of Furnace Off-Gas

The off-gases from the smelting and converting furnaces are cooled in two waste heat boilers before entering a common balloon flue. Gas cleaning is effected in three parallel hot electrostatic precipitators, a venturi scrubber and three parallel banks of mist precipitators. Dusts from the boilers and flues are conveyed to the feed preparation area for recycle to the smelting furnace, while the hot precipitator dust is removed from the system, slurried with jarosite water and transported to the leach section of the zinc plant. Here zinc and copper are recovered and the lead and silver may be recovered as a residue (9). The copper cement produced is returned to the copper smelter. The cleaned gas is dried and treated in a Monsanto double contact acid plant where overall sulphur fixation in excess of 99% is achieved.

TABLE 2 Typical Analysis of Furnace Products (Wt %)

	Cu	Fe	S	Pb	Zn
Matte	67	8	22	0.4	1.5
Discard slag	0.7	40	0.8	0.2	5
Blister copper	98.5	-	0.8	0.2	-

Oxygen Plants

Both oxygen plants were designed and built by Air Products. The original unit had a rated capacity of 190 tonnes/day of contained oxygen at 95% purity. Typically, production rates of between 210 and 230 tonnes/day are obtained depending on ambient conditions, with a further 20 tonnes/day achievable if the purity is dropped to 85%. The second plant was commissioned at the end of 1985 and produces 240 tonnes/day. The two plants are connected to maximize the use of available compressor capacity. Up to 160 tonnes of liquid oxygen is kept in four 40 tonne storage tanks which can vaporize oxygen for the smelter at up to 5 tonnes/hour. Delivery pressure of oxygen to the smelter is controlled at 350 kPa for the continuous process with up to 12 tonnes/day being vaporized to 700 kPa for other uses within the smelter. With the addition of the second plant, a reciprocating oxygen compressor was also commissioned. This boosts the pressure from 350 to 2,100 kPa and is capable of handling 50 tonnes/day. This oxygen is used in the zinc plant's pressure leach vessel with a small amount (approximately 1 tonne/day) being utilized in the copper refinery to accelerate slimes leaching.

THE ROLE OF OXYGEN IN THE SMELTER EXPANSION

The Kidd smelter was originally intended to include two continuous furnace lines having a total production capacity of 118,000 tonnes per year of copper. For economic reasons, only one line was built with an initial design capacity of 59,000 tonnes/year. Other sections of the smelter, including the feed preparation, melting and casting areas were built with a capacity of 118,000 tonnes/year to allow for subsequent expansion of the smelting area. Excess copper concentrates produced in the concentrator were to be toll smelted. Early in the commissioning phase of the first line, both the smelting and converting furnaces were operated separately at rates well in excess of the design concentrate feedrate of 35 tonnes/hour. These higher rates could not be achieved with both furnaces being blown at the same time because of an insufficient oxygen supply and acid plant capacity limitations. However, the demonstrated capacity of each furnace, supported by tests conducted at Mitsubishi Metal Corporation's Naoshima smelter at higher oxygen enrichment levels, made it clear at an early stage that the capacity of the single line Kidd smelter could be maintained at levels significantly above original design.

Table 3 shows the increase in copper production of the Kidd smelter between 1982 and 1986. The first three years were essentially a de-bottlenecking period with most of the increase resulting from an improvement in on-line time from 63% to 81%. In late 1983, a 160 tonne storage and vaporization facility for liquid oxygen was built and by early 1984, bulk oxygen was being purchased, increasing the feedrate to 40 tonnes/hour of concentrate. Liquid oxygen was trucked in from southern Ontario and despite the 2000 km round trip, proved to be a cost-effective way to increase smelting capacity and therefore reduce toll smelting charges. In August of 1984, the oxygen plant capacity increased from 190 to 220 tonnes/day by supplementing the plant's main blower with air from an available spare compressor. Despite this, oxygen still had to be purchased at a rate averaging 22 tonnes/day. This year proved that the main limitations on the capacity of the smelter were the oxygen supply as well as the acid plant capacity.

TABLE 3 Smelter Production History

	1982	1983	1984	1985	1986
Copper production, tonnes	28,700	43,900	64,500	82,900	86,736
Concentrate smelted, tonnes	118,700	169,300	244,400	308,400	336,289
On-line time, %	63.3	71.2	80.8	84.0	84.7
Concentrate feedrate, tonnes/hour	30.7	34.9	38.8	46.0	51.4
Oxygen used, tonnes/day	100	135	190	245	311
Bulk oxygen purchased, tonnes/day	0	6	22	54	0

Increased addition of oxygen was the only practical method of increasing smelter throughput without increasing furnace off-gas volumes. Any modifications to the waste heat boilers were considered to be too time consuming and costly. In May of 1985, the first phase acid plant expansion was completed along with upgrading the feed system to the furnaces. Following this change, bulk oxygen purchase averaged 54 tonnes/day until November 1985 when the new 240 tonnes/day oxygen plant was commissioned. By the end of 1985, copper was being produced at a rate in excess of 80,000 tonnes/year. In May 1986, the second phase of the acid plant expansion was completed and for the remainder of the year, production was at a rate of more than 90,000 tonnes/year.

The basis of the expansion in smelter capacity was not to exceed the design gas flowrates through the waste heat boilers in order to avoid both dust build-up problems and costly boiler modifications. The expansion in capacity was to be achieved through the use of additional oxygen. Table 4 compares original furnace design parameters and the expanded condition. The most notable point is that by increasing both the concentrate feedrate and the oxygen enrichment to both furnaces, the need to burn fuel oil no longer exists and the process becomes autogenous. Without the combustion products from burning fuel oil, lance blowing can be increased by an equivalent volumetric amount. This additional air through the lances allows an increase in concentrate feedrate of 10 tonnes/hour while increasing oxygen addition to the lance air gives a further 14 tonnes/hour.

TABLE 4 Furnace Design Conditions

	Original Design (59,000 tonnes/year)	Expanded Design (100,000 tonnes/year)
Smelting Furnace		
Feedrate, tonnes/hour	35	58.5
Lance air, Nm3/hr	20,300	23,600
Lance oxygen, Nm3/hr	4,000	8,900
Bunker C oil, kg/hr	600	0
Total off-gas flow, Nm3/hr	30,500	30,500
Oxygen enrichment, %	33	41
Oxygen/concentrate ratio, Nm3/t	228	228
Oxygen efficiency, %	99	99
Converting Furnace		
Lance air, Nm3/hr	10,400	13,400
Lance oxygen, Nm3/hr	800	2,200
Bunker C oil, kg/hr	250	0
Total off-gas volume, Nm3/hr	13,900	15,300
Oxygen enrichment, %	26.3	31.5
Oxygen/concentrate ratio, Nm3/t	84	84
Oxygen efficiency, %	93	93
Acid Plant Converter		
Gas flowrate, Nm3/hr	79,900	123,000
Composition - % SO2	10	13.1
- % O2	10.3	14.8
Acid production rate, tonnes/day	833 (normal)	1692 (maximum)
Equivalent concentrate feedrate	35	74

Higher production rates have been demonstrated during a recent period in which the feedrate averaged 58 tonnes/hour, almost equal to that required for the projected capacity of 100,000 tonnes/year. The average blister production of 331 tonnes/day was actually above the design figure but this was due to a higher on-line time then planned. Maximum concentrate feedrates exceeded 65 tonnes/hour with a total solid charge of more than 90 tonnes/hour to the smelting furnace. No smelting problems were encountered at these rates.

TABLE 5 Recent Actual Production Rates

	Recent Production Rates	100,000 tpy Design
Blister production, tonnes/day	331	310
Concentrate smelted, tonnes/hour	57.8	58.5
On-line time, %	92.9	83
Smelting furnace		
-lance blast, Nm3/hr	31,140	32,500
-oxygen enrichment, %	42.5	41
-recycle slag, tonnes/hour	4.6	5.3
Converting furnace		
-lance blast, Nm3/hr	16,800	15,600
-oxygen enrichment, %	30.3	31.5
-recycle slag , tonnes/hour	2	5.7

Total oxygen requirements for the expanded plant are shown in Table 6 . This was the basis for building the second plant of 240 tonnes/day capacity bringing the combined capacity of the plants to 460 tonnes/day. The bulk storage facility, which can vaporize up to 120 tonnes/day of oxygen, is kept full with excess liquid from the two producing plants as an emergency back-up.

TABLE 6 Oxygen Requirements at the Expanded Design (tonnes/day)

Smelting furnace	290
Converting furnace	70
Miscellaneous	20
Zinc plant	50
Contingency	30
Total	460

EFFECT OF INCREASED OXYGEN UTILIZATION

Operating Consumables

The expansion in smelting capacity was completed without adversely affecting major operating costs of the furnaces despite the higher oxygen enrichments. As shown in table 7, refractory consumption dropped from 1983 to 1985 as part of an ongoing refractory improvement program. It rose in 1986 when the smelting furnace was completely relined for the first time and the converting furnace received a new fused-cast hearth. Refractory consumption during a campaign, however, continues to drop. The addition of extra copper cooling blocks to the furnace sidewalls and most recently to the roofs has considerably improved refractory life in these areas. Furnace roof refractory now has an apparently unlimited life. The campaign length continues to increase with the current one planned to last 16 months, which is almost twice that achieved only two years ago.

Process lance consumption has tended to drop gradually on the smelting furnace. In 1985, mild steel lances were used instead of type 304 SS. Despite the slight rise in consumption, total costs dropped since the mild steel lances cost only one-third that of the stainless steel. Temperature control in the converting furnace has not been as effective as evidenced by the gradual increase in consumption as production capacity increased. It is expected that this will decrease with time as experience is gained with temperature control and the capacity of the coolant addition system is increased.

TABLE 7 Refractory and Lance Consumption

	1983	1984	1985	1986*
Refractory consumption kg/tonne conc	4.4	1.8	1.3	2.3
Process lance consumption kg/tonne conc				
- Smelting furnace	.51	.34	.34	.42
- Converting furnace	.45	.31	.54	.60

Treatment of Additional Materials

The expansion in capacity by increasing oxygen enrichment eliminated the need to burn additional fuel, except when the line is on stand-by status, and has in fact created a need for coolant addition to both the smelting and converting furnaces. This is shown in the heat balance for the furnaces at the 100,000 tonnes/year design figure in Table 8.

The smelting furnace presently treats dried converting furnace slag for copper recovery but this only provides half of the coolant requirements. The converting furnace can also be fed with this slag for temperature control. This recycle of slag to the converting furnace can in principal provide a limitless coolant supply but practically speaking is limited by slag drying and handling capacity. For short periods and depending on slag inventories, slag addition has averaged up to 10 tonnes/hour to the smelting furnace and 5 tonnes/hour to the converting furnace. This demonstrates the capacity to melt additional copper-bearing materials. Other miscellaneous materials routinely fed to the furnaces include cement copper at up to 20 tonnes/day, gold flotation concentrates at up to 50 tonnes/day, lead silver residue at up to 20 tonnes/day, low silica gold ores at up to 200 tonnes/day, and miscellaneous reverts (matte and boiler dusts) at up to 50 tonnes/day. The size of furnace feed material to the smelting furnace is no longer limited by the size of the charging lances. Oversize material from a vibrating screen prior to the charge tanks is directed to the smelting furnace through the roof and melts quickly in the bath due to the available excess heat.

In May 1987, 50 tonnes of tankhouse scrap was shredded to 150 x 150mm pieces and fed through a temporary chute to the converting furnace at a rate of 9 tonnes/hour or roughly double the average rate of refinery recycle scrap production. This testwork demonstrated that the converting furnace has sufficient capacity to melt all of the tankhouse scrap without requiring extra fuel. This would allow the present shaft melting furnace and adjacent holding furnace to be shut down, realizing a large fuel and maintenance cost saving. The feasibility of installing this equipment is currently being investigated. The shredder would also be used for feeding other miscellaneous copper-bearing scrap materials.

Oxygen/Natural Gas Burners

The smelting and converting furnaces were originally fitted with roof mounted air/oil burners; four on the smelting furnace and two on the converting furnace. Typical firing rates to maintain furnace temperatures when not feeding concentrates were 1,200 kg/hour on the smelting furnace and 600 kg/hour on the converting furnace. Once the need for these burners for normal operation disappeared, with the addition of more oxygen to the lance air, it was decided to convert to the use of small oxygen/natural gas burners to keep the furnaces hot when required. This avoided the problems involved in trying to keep the burner openings clean of buildup when they were not being used. The burners, which have a small diameter, can be inserted through either existing lance holes or burner holes. The major advantage of using these burners is that they are only needed when feed is not being smelted and so there is excess oxygen available. Reduced fuel requirements and lower natural gas prices compared to oil results in significant cost savings. Another significant side effect is that the more intense, localized flame from these burners allows far better control of the furnace bath melt condition and allows an easy restart of smelting even after a holding period lasting several days.

TABLE 8 Furnace Heat Balance

	Smelting furnace (1000 kcal/hr)	Converting furnace (1000 kcal/hr)
Heat of reaction	44,803	11,760
Heat of molten charge	-	5,500
Heat of combustion	-	-
INPUT	44,803	17,260
Heat content of off gases	15,221	7,160
Dust	1,192	308
Slag	13,503	1,626
Matte	5,500	-
Blister	-	2,826
Heat losses	6,053	3,503
OUTPUT	41,469	15,423
Heat Surplus	3,334	1,837
Capacity for solid feed		
- recycle slag, tonnes/hour	10.3	5.7
- copper, tonnes/hour	18.4	10
- solid matte, tonnes/hour	15.2	8.4

USE OF NITROGEN IN THE SMELTER

An interesting corollary to the increased utilization of oxygen throughout the smelter has been the use of the waste nitrogen from the air separation plants.

Kidd's copper concentrate is slightly pyrophoric, becoming increasingly more so with decreasing particle size. This was particularly evident in the baghouse servicing the furnace feed system. Since this dust collector was located immediately above the smelting furnace, water could not be used to control the initial bag fires. A line was run to the collector's inlet plenum to allow the introduction of nitrogen. Its use has been effective in minimizing the spread of fire to adjacent compartments. In a similar manner, a header containing nozzles was installed in the dry concentrate storage bin. Though the use of nitrogen does not necessarily extinguish a fire in this situation, its use arrests the roasting action until remedial action, such as lowering the bin level, can be completed.

The use of nitrogen has also been tested recently in two metallurgical applications. Testing is in progress to add concentrate as a magnetite reductant to the slag cleaning furnace, with the objective of minimizing copper losses to the discard slag. Nitrogen is used to stir the concentrate into the bath and has proven to be effective. A second use has been in the anode production area where ammonia is used as a deoxidant in the anode furnaces. Towards the end of the refining period, as the oxygen level within the charge approaches the end point, utilization efficiency of the ammonia decreases. Excess hydrogen from the ammonia decomposition burns in the furnace hood and off-gas ductwork, producing high temperatures (approximately 1,200°C) with subsequent steelwork damage. Simply reducing the ammonia flowrate is not an acceptable alternative since a minimum pressure of 180 kPa must be maintained at the two injecting tuyeres. It has been found that by using a 1:1 mixture of ammonia and nitrogen, this off-gas temperature is reduced to 400°C with no significant effect on reduction efficiency.

CONCLUSIONS

The increased use of oxygen in the Kidd smelter has resulted in the following improvements:
1. Allowed a 65% increase in smelter capacity for a small capital investment by maintaining design furnace off-gas flow-rates.
2. Reduced fuel consumption to zero except while the furnaces are on stand-by status.
3. Provided sufficient heat to melt various metal-bearing materials such as cement copper, silver/lead residue and gold ores without fuel addition.
4. Allowed conversion to oxy-gas burners resulting in significant cost savings.
The availability of nitrogen as a waste from the oxygen plant has lead to its use in several applications such as fire suppression and reduced heat generation in the anode furnace off-gas system.

FUTURE PLANS

Recent smelter operation at high throughputs has demonstrated no smelting problems and has indicated further potential for increasing production capacity above the 100,000 tonnes/year rate. This will be achieved by further equipment de-bottlenecking, increases in instantaneous feed-rates, improvements to process control and increases in furnace on-line time and campaign length. The adaptability of the process to treat miscellaneous metal-bearing materials has been well demonstrated. Efforts will continue in this direction to handle more of these materials to make use of the excess heat available. Since the increase in concentrate smelting rates and the increase in feed-rates of miscellaneous materials will be achieved with no additional manpower and no additional fuel, there will be a significant improvement in smelter productivity.

ACKNOWLEDGEMENTS

The authors wish to express their appreciation of the dedicated efforts of the technical, operating and maintenance personnel in the copper smelter in their contribution to maximizing the productivity of the plant. Thanks are given to the management of Kidd Creek Mines for permission to publish this paper.

REFERENCES

1. M. P. Amsden, R. M. Sweetin and D. G. Treilhard, "Selection and Design of Texasgulf Canada's Copper Smelter and Refinery", Journal of Metals, 30 (7) 1978 pp 16-26.
2. R. M. Sweetin, C. J. Newman and A. G. Storey, "The Kidd Smelter Start-Up and Early Operation", TMS/AIME International Sulphide Smelting Symposium, San Francisco, November, 1983.
3. L. E. Laforest, U. Nakano and M. deLaplante, "Kidd Smelter, Commitment to a Clean Environment", TMS/AIME, International Sulphide Smelting Symposium, San Francisco, November, 1983.
4. R. M. Sweetin, C. J. Newman and D. J. Kemp, "The Kidd Copper Smelter and Refinery - Description and Early Operation", CIM Bulletin, May, 1984.
5. C. J. Newman, A. G. Storey and R. J. St. Eloi, "Anode Casting Practice at the Kidd Smelter", CIM 23rd Annual Conference, Quebec City, August, 1984.
6. G. Macfarlane, A. G. Storey, V. Kelly and G. G. Marrs, "The Control of Lead and Selenium in Copper Produced at Kidd Creek Mines", CIM 24th Annual Conference of Metallurgists, Vancouver, August, 1985.
7. C. J. Newman, A. G. Storey, and K. Molnar, " Expansion of the Kidd Creek Copper Smelter", TMS/AIME, New Orleans, March 1986, A86-41.
8. G. Macfarlane, "Conversion of a Crushing Plant to a Gold Mill in a Copper Smelter", CIM Annual General Meeting, Toronto, May, 1987.
9. K. Donyina, "Impurity Control Through Recycling at Kidd Creek Mines Ltd", CIM 15th Annual Hydrometallurgical Meeting, Vancouver, August, 1985.

INCREASING PRODUCTIVITY BY THE USE OF TONNAGE OXYGEN AT THE HUELVA SMELTER

D. de la Villa, J. Contreras and P. Barrios

Río Tinto Minera, S.A., Huelva, Spain

ABSTRACT

Since 1975 the Huelva Smelter is operating with the Outokumpu Flash Smelting process. In 1981 oxygen enriched air operation was adopted with the purpose of increasing productivity. Other modifications including the installation of a new central jet type concentrate burner, changes in the reaction shaft and cooling system, off-gas recirculation and various energy savings have been carried out. As a result of these improvements, the Huelva Smelter has achieved a higher productivity. This paper presents the significant results gained with the oxygen enriched air operation.

KEYWORDS

Oxygen in Flash Furnace, oxygen in converters, smelter productivity, energy savings, cost reduction, operating improvements.

INTRODUCTION

Rio Tinto Minera's Huelva Smelter is situated on the South West coast of Spain, a few kilometers from the town of Huelva. The Huelva Complex assumes a millenarian tradition in the mining and smelting of copper, gold and silver ores from the Rio Tinto Mines (1).

The Huelva Complex started up in 1970, with two Momoda type blast furnaces (2). In 1975, the Outokumpu Flash Smelting process was adopted and the smelting capacity was increased (3).

During the last years, Huelva Smelter adopted the oxygen enriched air operation in order to increase the treatment capacity. Recently, Rio Tinto Minera carried out a series of programs to improve the productivity and to reduce costs.

Detailled descriptions of recent improvements and operating results are given below.

ADOPTION OF OXYGEN ENRICHED AIR OPERATION

The oil crisis in 1975 caused an extraordinary increase in oil prices and the smelters were forced to the substitution of oil. The smelters had to choose between oxygen enrichment, coal or other energy resources.

At that time fuel oil and electricity prices were under Government control in Spain and the ratio between electricity and oil cost permitted us to continue with oil consumption. Meanwhile, a great part of the smelters had changed to oxygen enrichment operation.

In 1981 the Huelva Smelter started testing the use of oxygen enriched air in the Flash smelting furnace. The pure liquid oxygen was supplied by an oxygen producer. Liquid oxygen was evaporated and introduced into the process air stream before the inlet to the air preheater by a tuyere. The oxygen amount was raised gradually to get a maximum rate of 2,000 Nm^3/h.

A large series of tests were carried out at the smelter to provide useful information and permitted, after detailled studies, to take up the introduction of oxygen consumption on a large scale. Since the start up the matte grade produced at the Flash smelting furnace was generally in the range of 45-50% Cu. During the test period the oxygen enrichment levels were raised to 25% approximately and the matte grade achieved 50-55% Cu.

Table 1 shows the typical operating data and heat balance in the reaction shaft under different conditions during these tests.

TABLE I Mass and Heat Balances Before Use of Tonnage Oxygen

MASS BALANCE

		Atmospheric air	Enriched air
INPUT			
Concentrate	tpd	900	1,000
Flux	tpd	117	138
OUTPUT			
Matte	tpd	426	454
Slag	tpd	457	532
Matte grade	%	50	52
Process air	Nm^3/h	48,500	42,100
Oxygen enrichment	%	21.0	24.8
Gas flow	Nm^3/h	56,600	49,700
Anode production	tpd	300	330

HEAT BALANCE

		Atmospheric air	Enriched air
INPUT			
Chemical Reactions	Mcal/h	23,800	27,500
Fuel-oil	Mcal/h	18,500	13,000
Preheated air	Mcal/h	5,500	4,600
TOTAL	Mcal/h	47,800	45,100
OUTPUT			
Heat in matte	Mcal/h	3,800	4,000
Heat in slag	Mcal/h	7,500	8,800
Heat in dust	Mcal/h	1,300	1,500
Heat in gases	Mcal/h	29,900	25,500
Heat losses	Mcal/h	5,300	5,300
TOTAL	Mcal/h	47,800	45,100
Steam production	t/h	49	41
Fuel-oil Consumption (air preheater included)	Kg/h	2,700	2,000

The results were satisfactory as expected and the use of oxygen enriched air showed the following effects,

- Increased smelting capacity.

- Reduction in fuel-oil consumption.

- Reduction in exhaust gas volume from the Flash Furnace resulting in a better temperature control at the inlet of the electro-precipitators.

- Reduction in dust troubles in the waste heat boilers.

USE OF TONNAGE OXYGEN

Because of the above advantages the tonnage oxygen operation was adopted in 1983. During 1984 a 300 tonnes per day oxygen plant was erected and started to operate in January 1985, supplying oxygen over the fence. From January 1985 to June 1985, when the second furnace campaign ended, new trials were carried out and the oxygen enrichment was raised step by step to 40% O_2. During these operations all the advantages found in the earlier tests were improved. Also it was decided to increase the matte grade and 55-60% Cu was achieved. This operation had no adverse effect and the stability of the process allowed to make these changes to be made in a short period of time.

Further Modifications

During the general shut down major modifications were carried out in order to improve the installation and to achieve a more efficient operation. All these modifications have already been reported and the reasons explained in great detail in recent papers (4,5), thus only a brief description of the modifications is made,

Concentrate Burner

The four venturi type concentrate burners were substituted by a single Outokumpu central jet type burner.

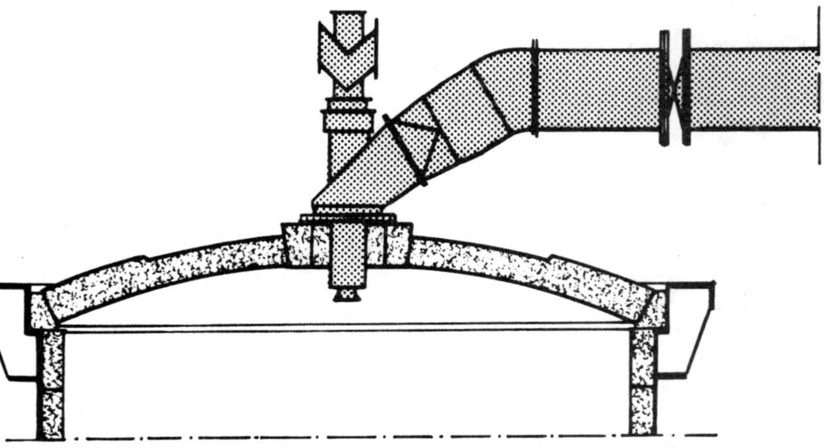

Fig. 1. New arrangement of a central jet distributor type burner

Changes in the Shaft and Cooling System

Reaction shaft has been shortened by 3 meters.

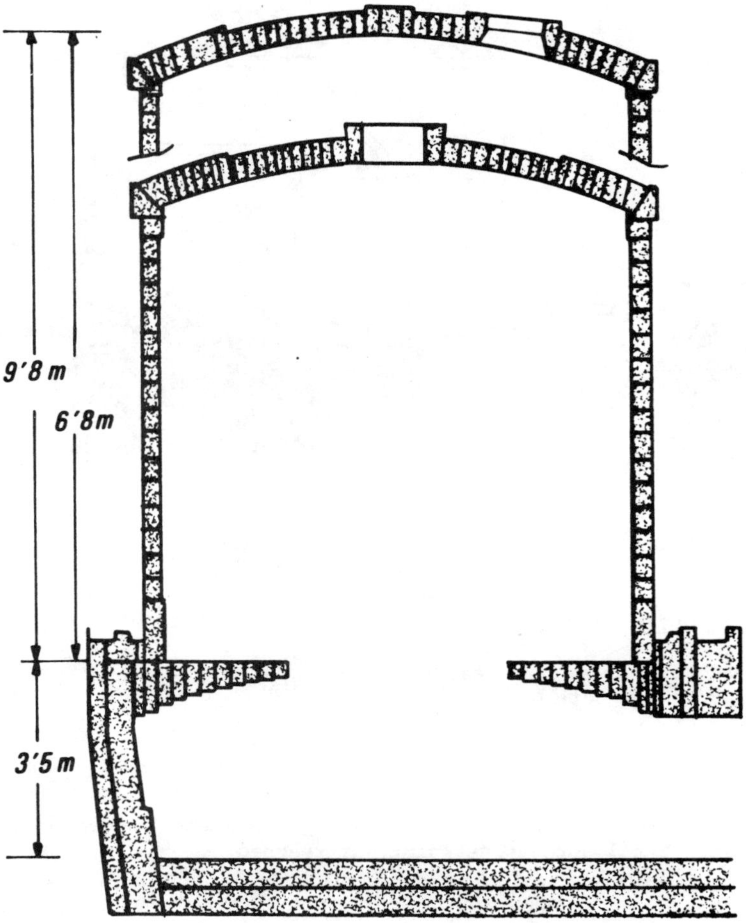

Fig. 2. Shortening of the reaction shaft

The shaft cooling spray water system was improved and the reaction shaft-settler arch junction was redesigned.

Off-gas Recirculation

The gas handling system has been modified and the gases can be recirculated from the outlet of the electroprecipitators back to the radiation section of the waste heat boiler.

Figure 3 shows the new arrangement.

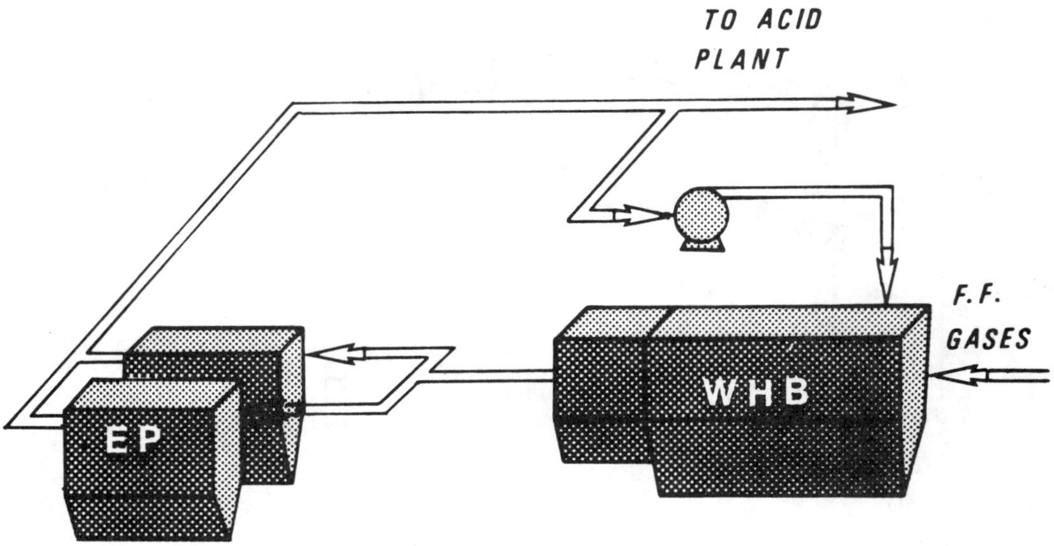

Fig. 3. Off-gas recirculation system

Exhaust Gases Heat Recovery

The exhaust gases of the steam superheater are sent to the dryer and the waste heat is recovered for concentrate drying.

RECENT OPERATION

The 300 tpd oxygen plant was commisioned in January 1985 and oxygen enrichment was raised step by step until autogenous point was reached.

As a result of the use of tonnage oxygen in the Flash Furnace some of the operating parameters have been modified.

- Smelting capacity increased as shown in fig. 4.

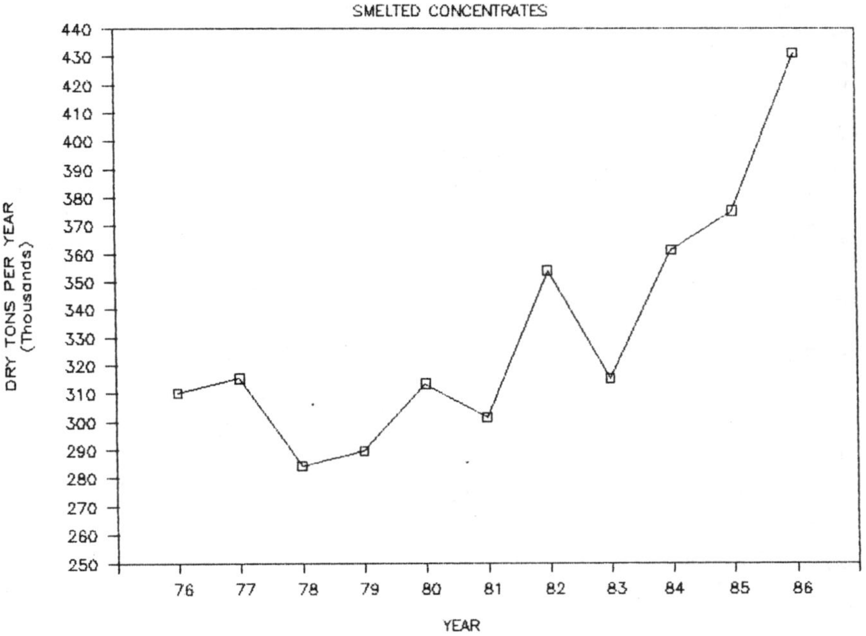

Fig. 4. Smelting capacity from 1976

- The range of enrichment is 35-50% depending on composition of feed blend.

- Fuel consumption. Although the flash smelting process principles did not change when oxygen enrichment was adopted, the heat balance of reaction shaft was modified. The lower amount of nitrogen flowing through the process as an inert gas reduces proportionally the fuel-oil necessary to heat this gas.

Figure 5 shows unit fuel consumption from 1976.

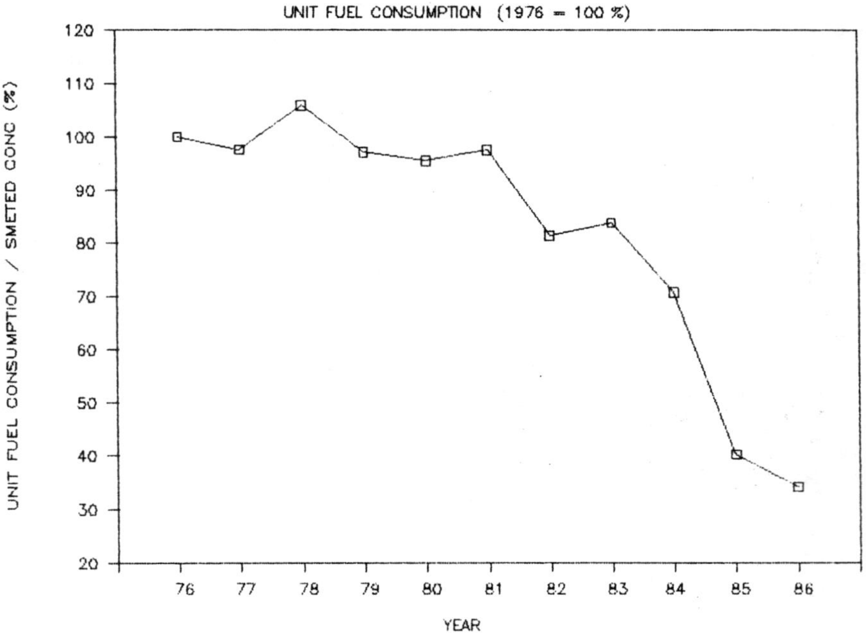

Fig. 5. Oil consumption from 1976

- Process air temperature has been reduced from 345 to 150 °C, by shutting down the oil-fired stage of the air preheater, keeping the steam exchangers stage in operation.

- Lower dust production has been observed in 1986, about 20% less than that produced in recent years.

- Reduction of electric energy production, as a result of the lower off-gas flow in Flash Furnace.

- Copper and magnetite in Flash Furnace slags increased when using a higher oxygen enrichment. In order to reduce the effect in the Electric Furnace, coke and silica additions have been modified.

TABLE II Typical Mass and Energy Balances for the Recent Operation

MASS BALANCE

		Enriched air (after shortening Reaction shaft)
INPUT		
Concentrate	tpd	1,275
Flux	tpd	196
OUTPUT		
Matte	tpd	524
Slag	tpd	744
Matte grade	%	57
Process air	Nm^3/h	31,100
Oxygen enrichment	%	36.2
Gas flow	Nm^3/h	37,200
Anode production	tpd	430
Acid production	tpd	1,160

HEAT BALANCE

INPUT		
Chemical Reactions	Mcal/h	37,400
Fuel-oil combustion	Mcal/h	4,000
Preheated air	Mcal/h	1,500
TOTAL	Mcal/h	42,900
OUTPUT		
Heat in matte	Mcal/h	4,600
Heat in slag	Mcal/h	12,200
Heat in dust	Mcal/h	1,900
Heat in gases	Mcal/h	19,900
Heat losses	Mcal/h	4,300
TOTAL	Mcal/h	42,900
Steam production	t/h	31
Fuel-oil Consumption (air preheater included)	Kg/h	450

Use of Oxygen in Converters

The new oxygen plant supplies not only the low-pressure oxygen used in the Flash Furnace, but a medium-pressure amount of oxygen to be used in Converters, introducing it into the blowing air flow by a tuyere.

Due to the high temperatures produced during the copper blow stage oxygen has only been used during the slag blow stage.

The oxygen enrichment in Converters is in the range of 23-25% oxygen.

The effects of the use of enriched air have been the following,

Shortening of the Blowing Time

In order to achieve an optimal standard operation at the plant, some modifications were done. Besides the use of oxygen in converters to reduce the blowing time in slag blow stage, the height of one of the slag taping holes in Flash Furnace was raised 200 mm. Both changes allow the furnace to hold matte during a complete converter cycle. This allows the use of only one hot converter.

Table III shows some operating data of converters.

TABLE III

		1984	1986
Oxygen enrichment	%	21.0	25.0
Matte treated	tpd	460	477
Hot converters		2	1

Increase in the Reverts Smelting Capacity

In contrast to other plants, RTM has no flotation units to treat the produced reverts. All of them must be treated in converters and the Electric Furnace. The use of enriched air in converter allows the smelting of a larger amount of reverts during the slag blow stage. At the same time, a model to achieve a more reasonable ladles operation was built, reducing the production of reverts to values below 10% of material transported.

The larger usage of reverts at the Converters and the lower production allow to reduce the energy required in Electric Furnace for smelting reverts.

As a result of these changes in operating conditions, the stock of reverts have been reduced drastically.

- A lower off-gas flow to acid plants brings an energy saving in blowers consumption and a better temperature control.

- A higher SO_2 content during slag blow stage brings a more stable operation in acid plants.

Figure 6 shows the increase of the Converter productivity.

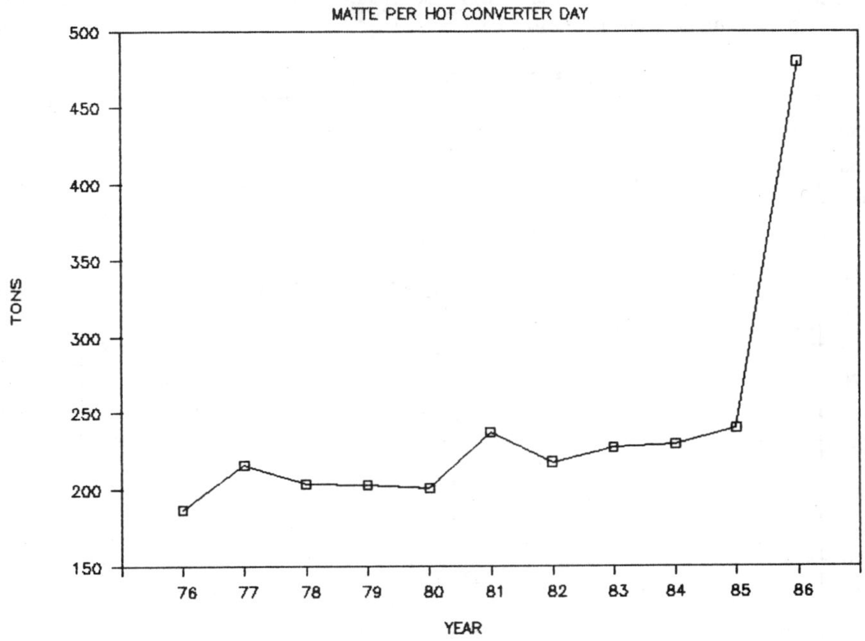

Fig. 6. Converter productivity

Summary of Oxygen Effects in Recent Operation

During the oxygen enriched air operation the expected benefits have been fully achieved.

- Smelting capacity has been increased by 30%
- Fuel consumption decreased from 100% in 1976 to 34% in 1986.
- Converters capacity for reverts smelting has been increased, reducing the amount of reverts smelted in Electric Furnace. Therefore the energy requirements have been reduced.
- Converters slag blowing time has been reduced in about 20%.
- Energy requirement for anode production has been decreased to 2.5×10^6 Kcal/t.

<u>Productivity Increase</u>

The whole effect of these innovations can be measured by the reduction of the unit operating costs, as shown in fig. 7.

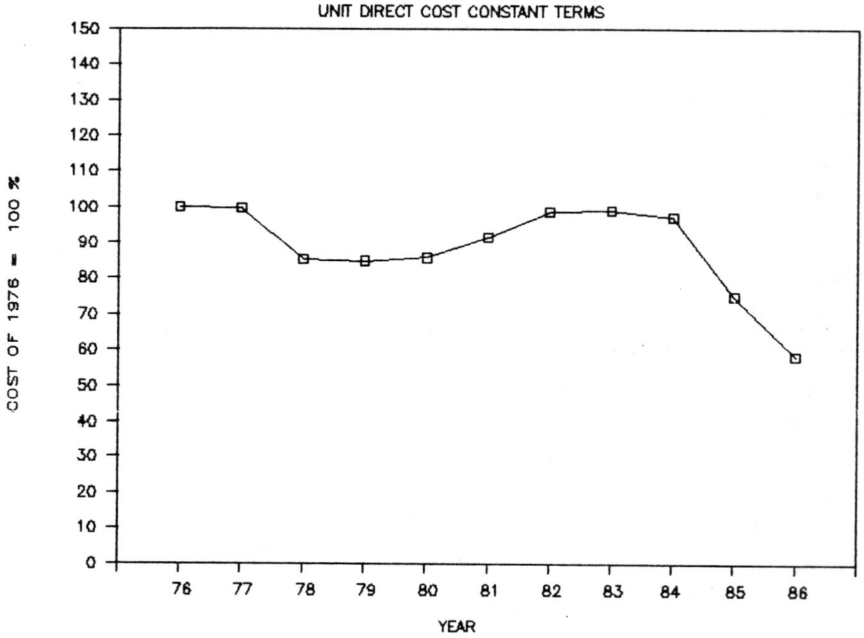

Fig. 7. Direct costs from 1976

Productivity can also be used as a measure of the recent improvements at the Huelva Smelter. Values are shown in fig. 8.

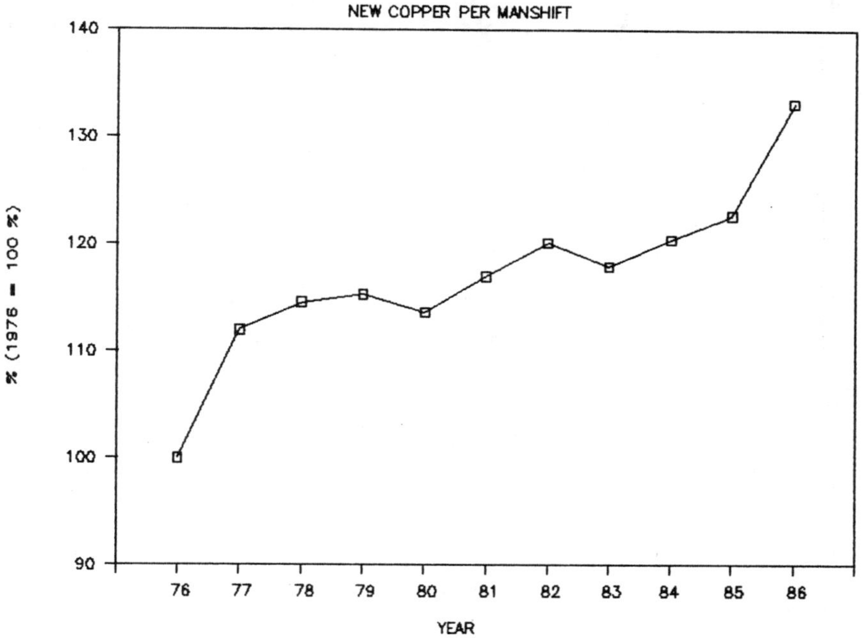

Fig. 8. Productivity from 1976

CONCLUSIONS

The introduction of the oxygen enriched air operation at the Huelva Smelter was carried out together with other modifications in the process, in order to increase the smelting capacity, reducimg fugitive emissions and operating costs.

The use of the tonnage oxygen and the modifications and improvements have resulted in a more efficient operation by the increase of the productivity.

ACKNOWLEDGEMENT

The authors thank Río Tinto Minera for having permitted the publication of this paper. Likewise, they thank the whole staff in Huelva Smelter for their valuable contribution.

REFERENCES

1. Avery, D. (1974). Not on Queen Victoria's Birthday (The Story of the Rio Tinto Mines). Collins, 21-23.

2. Luzón, F. (1969). Una nueva planta integrada de cobre electrolítico y de ácido sulfúrico. Revista de Metalurgia, Vol. 5, n° 6, pags. 697-708.

3. Palacios, C. (1977). Huelva Flash Smelter. The Third International Flash Smelting Congress, Helsinki-Hamburg-Huelva, May 16-20.

4. de la Villa, D., Cevallos, F. and Barrios, P. (1986). Recent innovations at Huelva smelter. AIME Conference 1986. New Orleans. March 2-6.

5. de la Villa, D., Contreras, J. and Barios, P. (1986). Use of oxygen enrichment at the Huelva smelter. Fifth International Flash Smelting Congress. Helsinki-Wroclaw. May 18-24.

AUTHOR INDEX